Aquatic and Standing Water Plants of the Central Midwest

Books in the Aquatic and Standing Water Plants of the Central Midwest Series by Robert H. Mohlenbrock

Cyperaceae: Sedges

Filicineae, Gymnospermae, and Other Monocots, Excluding Cyperaceae: Ferns, Conifers, and Other Monocots, Excluding Sedges

Acanthaceae to Myricaceae: Wild Petunias to Myrtles

Nelumbonaceae to Violaceae: Water Lotuses to Violets

Other Southern Illinois University Press Books by Robert H. Mohlenbrock

Guide to the Vascular Flora of Illinois, revised and enlarged edition

Distribution of Illinois Vascular Plants, with Douglas M. Ladd

A Flora of Southern Illinois, with John W. Voigt

In the Illustrated Flora of Illinois Series

Ferns, 2nd edition

Flowering Plants: Basswoods to Spurges

Flowering Plants: Flowering Rush to Rushes

Flowering Plants: Hollies to Loasas

Flowering Plants: Lilies to Orchids

Flowering Plants: Magnolias to Pitcher Plants

Flowering Plants: Nightshades to Mistletoe

Flowering Plants: Pokeweeds, Four-o'clocks, Carpetweeds, Cacti, Purslanes, Goosefoots, Pigweeds, and Pinks

Flowering Plants: Smartweeds to Hazelnuts

Flowering Plants: Willows to Mustards

Grasses: Bromus to Paspalum, 2nd edition

Grasses: Panicum to Danthonia, 2nd edition

Sedges: Carex

Sedges: Cyperus to Scleria, 2nd edition

Aquatic and Standing Water Plants of the Central Midwest

Filicineae, Gymnospermae, and Other Monocots, Excluding Cyperaceae

Ferns, Conifers, and Other Monocots, Excluding Sedges

Robert H. Mohlenbrock

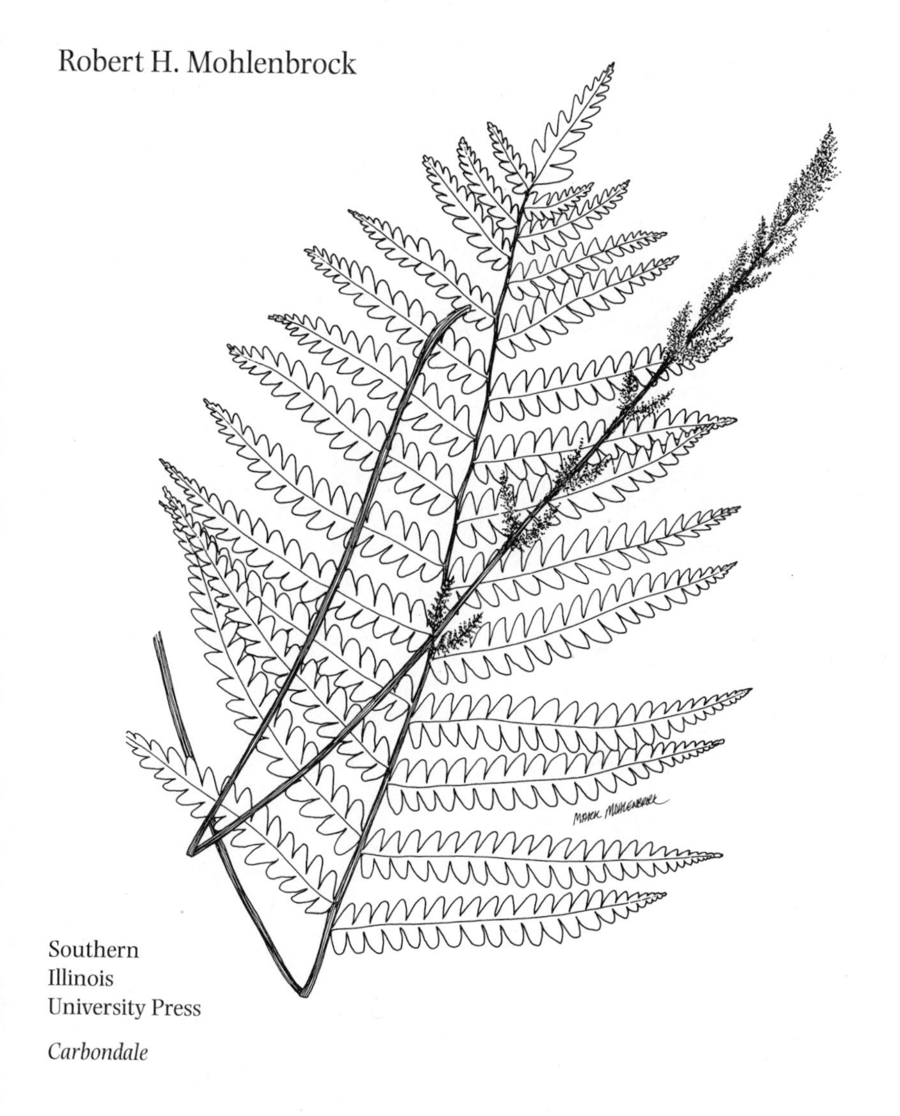

Southern
Illinois
University Press

Carbondale

Printed in the United States of America
09 08 07 06 4 3 2 1

Library of Congress Cataloging-in-Publication Data
Mohlenbrock, Robert H., 1931–
Filicineae, Gymnospermae, and other monocots, excluding Cyperaceae : ferns, conifers, and other monocots, excluding sedges / Robert H. Mohlenbrock.
p. cm. — (Aquatic and standing water plants of the central Midwest ; v.2)
Includes index.
1. Ferns—Middle West—Identification. I. Title. II. Series: Mohlenbrock, Robert H., 1931– .
Aquatic and standing water plants of the Central Midwest ; v.2.

QK525.5.M54M64 2005

587'.3'0978—dc22

0-8093-2670-1 (cloth : alk. paper) 2005023854

Printed on recycled paper. ♻

The paper used in this publication
meets the minimum requirements
of American National Standard
for Information Sciences—
Permanence of Paper for
Printed Library Materials,
ANSI Z39.48-1992. ∞

This book is dedicated to Mr. John "Rick" Stetter,

who believes in and encourages my work, and to

Kathleen Kageff, Barbara Martin, and Kristine Priddy,

tireless professionals at Southern Illinois University Press

whose meticulous work has made my life easier.

Contents

Illustrations

Series Preface

The purpose of the four books in the Aquatic and Standing Water Plants of the Central Midwest series is to provide illustrated guides to plants of the central Midwest that may live at least three months a year in water, though a particular plant may not necessarily live in standing water during a given year. The states covered by these guides include Iowa, Illinois, Indiana, Ohio, Kansas, Kentucky, Missouri, and Nebraska, except for the Cumberland Mountain region of eastern Kentucky, which is in a different biological province. Since 1990, I have taught week-long wetland plant identification courses in all of these states on several occasions.

The most difficult task has been to decide what plants to include and what plants to exclude from these books. Three groups of plants are within the guidelines of these manuals. One group includes those aquatic plants that spend their entire life with their vegetative parts either completely submerged or at least floating on the water's surface.

This group includes obvious submerged aquatics such as *Ceratophyllum*, the Najadaceae, the Potamogetonaceae, *Elodea, Cabomba, Brasenia, Nymphaea,* some species of *Ranunculus, Utricularia,* and a few others.

Plants in a second group are called emergents. These plants typically are rooted under water, with their vegetative parts standing above the water surface. Many of these plants can live for a long period of time, even their entire life, out of the water. Included in this group are *Sagittaria, Alisma, Peltandra, Pontederia, Saururus, Justicia,* and several others.

The most difficult group of plants that I had to consider is made up of those wetland plants that live most or all of their lives out of the water, but which on occasion can live at least three months in water. I concluded that I would include within these books only those species that I personally have observed in standing water for at least three months during a year, or which have been reported in the literature as living in standing water.

In this last group, for example, I have included *Poa annua,* since Yatskievich, in his *Steyermark's Flora of Missouri* (1999), indicates that this species may occur in standing water, even though I have not observed this myself. I have included most plants of bogs, fens, and marshes if I have observed these plants actually to be in water.

It is likely that I failed to include some plants that should have been included, but that I had not observed myself.

The nomenclature I have used in these books reflects my own opinion as to what I believe the scientific names should be. If these names differ from those used by the U.S. Fish and Wildlife Service, I have indicated this. A partial list of synonymy is included for each species, particularly accounting for synonyms that have been in use for several decades.

After the description of each plant, I have indicated the habitats in which the plant may be found, followed by the states in which the plant occurs. I have indicated the U.S. Fish and Wildlife wetland designation for each species for the states that each occurs in. In 1988, the National Wetlands Inventory Section of the U.S. Fish and Wildlife Service attempted to give a wetland designation for every plant occurring in the wild in the United States. The states covered by these aquatic manuals occur in three regions of the Fish and Wildlife Service. Kentucky and Ohio are in region 1; Illinois, Indiana, Iowa, and Missouri are in region 3; and Kansas and Nebraska are in region 5. Definitions of the Fish and Wildlife Service wetland categories are:

OBL (Obligate Wetland). Occur almost always under natural conditions in wetlands, at least 99% of the time.

FACW (Facultative Wetland). Usually occur in wetlands 67%–99% of the time), but occasionally found in non-wetlands.

FAC (Facultative). Equally likely to occur in wetlands or non-wetlands 34%–66% of the time.

FACU (Facultative Upland). Usually occur in non-wetlands 67%–99% of the time, but occasionally found in wetlands.

UPL (Upland). Occur in uplands at least 99% of the time, but under natural conditions not found in wetlands.

NI (Not Indicated). Due to insufficient information.

A plus or minus sign (+ or -) may appear after FACW, FAC, and FACU. The plus means leaning toward a wetter condition; the minus means leaning toward a drier condition.

Although the Fish and Wildlife Service made changes to the wetland status of several species in an updated version in 1997, this later list has never been approved by Congress.

Following this is one or more common names currently employed in the central Midwest. A brief discussion of distinguishing characteristics and nomenclatural notes is often included. Illustrations accompany each species, showing the diagnostic characteristics. In some of the illustrations, a gap in the stem signifies that a portion of the stem has been omitted due to space limitations.

The sequence of families in these aquatic manuals is as follows:

1. Azollaceae
2. Blechnaceae
3. Equisetaceae
4. Isoetaceae
5. Lycopodiaceae
6. Marsileaceae
7. Onocleaceae
8. Osmundaceae
9. Thelypteridaceae
10. Pinaceae
11. Taxodiaceae
12. Acoraceae
13. Alismataceae
14. Araceae
15. Butomaceae
16. Cyperaceae
17. Eriocaulaceae
18. Hydrocharitaceae
19. Iridaceae
20. Juncaceae
21. Juncaginaceae
22. Lemnaceae
23. Maranthaceae
24. Najadaceae
25. Orchidaceae
26. Poaceae
27. Pontederiaceae
28. Potamogetonaceae
29. Ruppiaceae
30. Scheuchzeriaceae
31. Sparganiaceae
32. Typhaceae
33. Xyridaceae
34. Zannichelliaceae
35. Acanthaceae
36. Aceraceae
37. Amaranthaceae
38. Anacardiaceae
39. Apiaceae
40. Apocynaceae
41. Aquifoliaceae
42. Asclepiadaceae
43. Asteraceae
44. Balsaminaceae
45. Betulaceae

46. Boraginaceae
47. Brassicaceae
48. Cabombaceae
49. Caesalpiniaceae
50. Callitrichaceae
51. Campanulaceae
52. Caprifoliaceae
53. Ceratophyllaceae
54. Cornaceae
55. Cuscutaceae
56. Elatinaceae
57. Ericaceae
58. Escalloniaceae
59. Fabaceae
60. Gentianaceae
61. Grossulariaceae
62. Haloragidaceae
63. Hippuridaceae
64. Hypericaceae
65. Juglandaceae
66. Lamiaceae
67. Lauraceae
68. Leitneriaceae
69. Lentibulariaceae
70. Lythraceae
71. Malvaceae
72. Menyanthaceae
73. Myricaceae
74. Nelumbonaceae
75. Nymphaeaceae
76. Nyssaceae
77. Oleaceae
78. Onagraceae
79. Parnassiaceae
80. Plantaginaceae
81. Podostemaceae
82. Polemoniaceae
83. Polygonaceae
84. Primulaceae
85. Ranunculaceae
86. Rhamnaceae
87. Rosaceae
88. Rubiaceae
89. Salicaceae
90. Sarraceniaceae
91. Saururaceae
92. Saxifragaceae
93. Scrophulariaceae
94. Solanaceae
95. Styracaceae
96. Ulmaceae
97. Urticaceae
98. Valerianaceae
99. Verbenaceae
100. Violaceae

This volume consists of families 1–15 and 17–34. A previous volume has covered family 16, the Cyperaceae. The dicots are also in two volumes, the first covering families 35–73, the second covering families 74–100.

The Central Midwest

Descriptions and Illustrations

1. AZOLLACEAE—MOSQUITO FERN FAMILY

Azolla is the only genus worldwide.

1. **Azolla Lam.**—Mosquito Fern

Plants aquatic, floating, with slender roots; stems usually prostrate, glabrous; leaves with minute hairs on upper surface of upper lobe; sporocarps paired; microspores in masses, covered by arrowlike barbs (glochidia); megaspores globose, containing 3 floats and a blue-green algal colony.

Azolla consists of seven species in the tropics and temperate parts of the world. The majority of plants in a colony at a given time do not bear sporocarps. It has been traditional since 1944 to use the number of septae in the glochidia of the microspores to identify the species, but this character is not reliable.

Within cavities of the upper leaf lobe are colonies of the blue-green alga *Anabaena azollae*, which is a nitrogen-fixing organism. As a result, *Azolla* is often used as an agricultural fertilizer or mixed with livestock food as a nutritional supplement.

1. Plants dark green to reddish; stems 0.5–1.0 cm long; megaspores densely covered with tangled hairs .... 1. *A. caroliniana*
1. Plants green or blue-green to reddish; stems 1.0–1.5 cm long; megaspores sparsely covered with tangled hairs .... 2. *A. mexicana*

1. **Azolla caroliniana** Willd. Sp. Pl. 5(1):541. 1810. Fig. 1.

Plants dark green to reddish, free-floating or occasionally forming mats; stems 0.5–1.0 cm long; upper lobe of leaf about 0.5 mm long; megaspores densely covered with tangled hairs.

Stagnant or slow-moving water in ponds, lakes, and streams.

IA, IL, IN, KS, KY, MO, NE, OH (OBL).

Mosquito fern.

2. **Azolla mexicana** C. Presl, Abh. Konigl. Bohm. Ges. Wiss., ser. 5, 3:150. 1845. Fig. 1.

Plants green, blue-green, to reddish, free-floating or occasionally forming mats; stems 1.0–1.5 cm long; upper lobe of leaf about 0.7 mm long or longer; megaspores sparsely covered by a few short filaments.

Stagnant or slow-moving water in ponds, lakes, and streams.

IA, IL, KS, MO, NE (OBL).

Mosquito fern.

2. BLECHNACEAE—CHAIN-FERN FAMILY

Leaves monomorphic or dimorphic; blades pinnatifid to bipinnate; veins of leaves usually united; sori elongated along vein; indusia opening from the side; spores reniform.

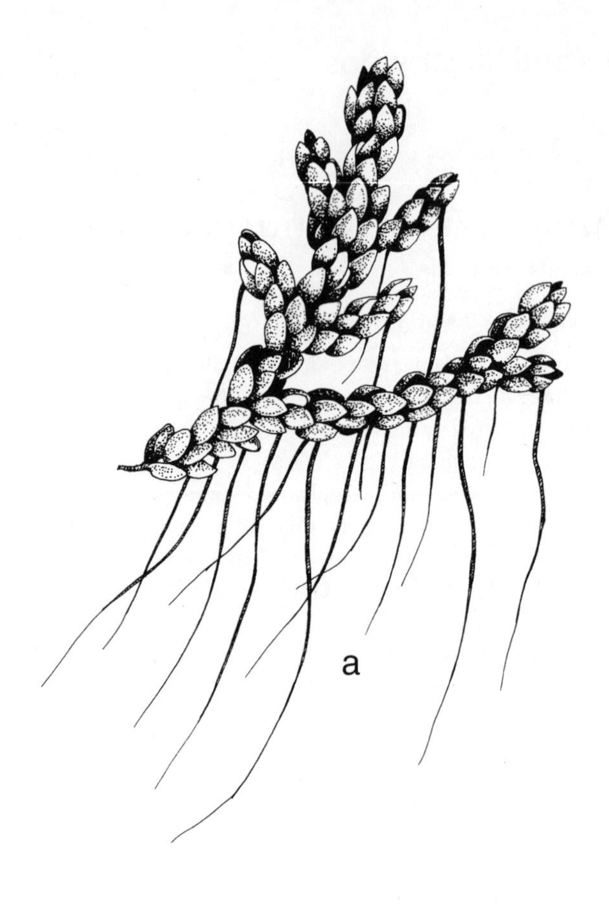

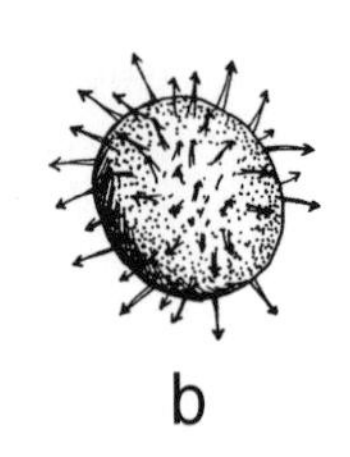

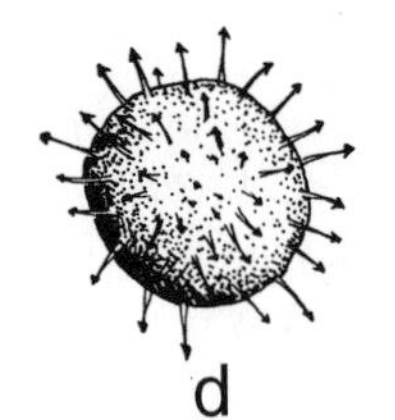

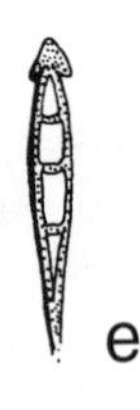

1. *Azolla caroliniana*
(Mosquito fern).
a. Habit.
b. Microsporocarp.
c. Glochidium.
Azolla mexicana.
d. Microsporocarp.
e. Glochidium.

This family consists of ten genera, with only *Woodwardia* occurring in the central Midwest. There are approximately 250 species in the family worldwide.

1. **Woodwardia** Sm.—Chain-fern

Stems long-creeping, with brown scales; leaves monomorphic or dimorphic, the sterile ones pinnatifid or pinnate; rachis scaly; sori in chainlike rows along the veins.

Of the fourteen species that comprise this genus, two are found in wetlands in the central Midwest. Despite the fact that both species are placed in the same genus, they are quite different in that one species is monomorphic and the other is dimorphic.

1. Leaves dimorphic; sterile blades pinnatifid .... 1. *W. areolata*
1. Leaves all of one type, pinnate .... 2. *W. virginica*

1. **Woodwardia areolata** (L.) Moore, Index Fil. 45. 1857. Fig. 2.

Leaves dimorphic, deciduous; sterile leaves deeply pinnatifid, to 60 cm long, often nearly as broad, bright green, nearly glabrous when mature, the segments entire; petiole pale, becoming red-brown at base; fertile leaves pinnate, up to 30 cm long, the pinnae linear, up to 5 mm wide; sori linear-oblong, sunken into the blades.

Bogs, seeps, wet woods, wet sandstone ledges.

IL, IN, KY, MO, OH (OBL).

Netted chain-fern.

This species is more similar in appearance to the sensitive fern *(Onoclea sensibilis)* than to the other species of *Woodwardia.* The sterile leaves are most readily distinguished from the sterile leaves of *Onoclea sensibilis* by its segments being entire rather than finely toothed.

2. **Woodwardia virginica** (L.) Sm. Mem. Acad. Roy. Sci. (Turin) 5:412. 1793. Fig. 3.

Leaves monomorphic, deciduous, up to 1 m long, pinnate, the blade with some glands; petioles pale, dark purple near the base; pinnae linear to narrowly lanceolate, up to 3.5 cm wide; sori elongated, linear, usually appearing confluent.

Bogs, swamps, marshes.

IL, IN, OH (OBL).

Virginia chain-fern.

This species is somewhat similar to young fronds of cinnamon fern *(Osmunda cinnamomea)*, but the base of the petiole in the Virginia chain-fern is dark purple, while in cinnamon fern it is covered by a cinnamon-colored wool. Virginia chain-fern is also somewhat similar in appearance to marsh fern *(Thelypteris palustris)*, but the marsh fern has open leaf venation and lacks brown scales on the petioles.

3. EQUISETACEAE—HORSETAIL FAMILY

Equisetum is the only genus worldwide.

1. **Equisetum** L.—Horsetail; Scouring Rush

Rhizome stout, branched; aerial stems erect, hollow, ridged, conspicuously jointed at the nodes; branches and reduced scale leaves borne in whorls at the nodes, the leaves comprising the sheath; sporangia borne in terminal cones; spores alike.

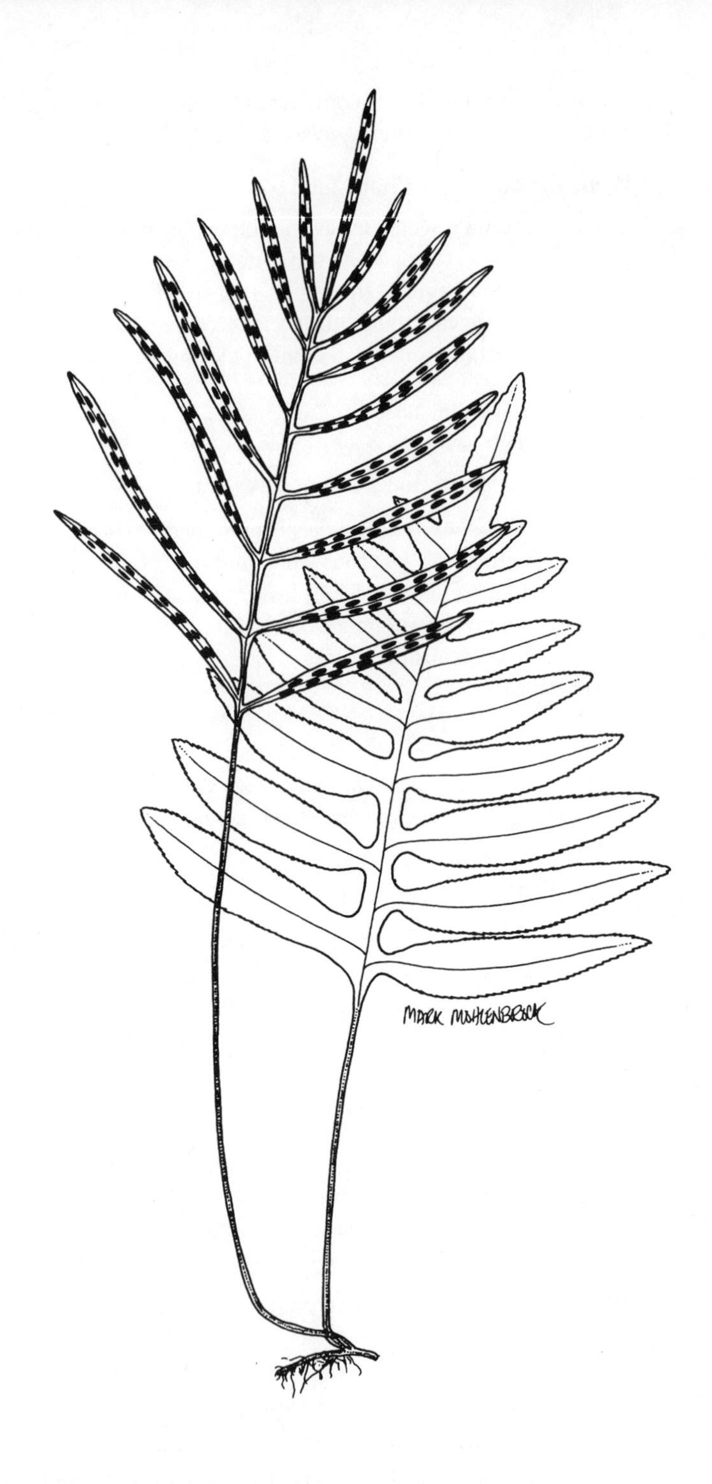

2. ***Woodwardia areolata*** (Netted chain fern). Sterile frond (right), fertile frond (left).

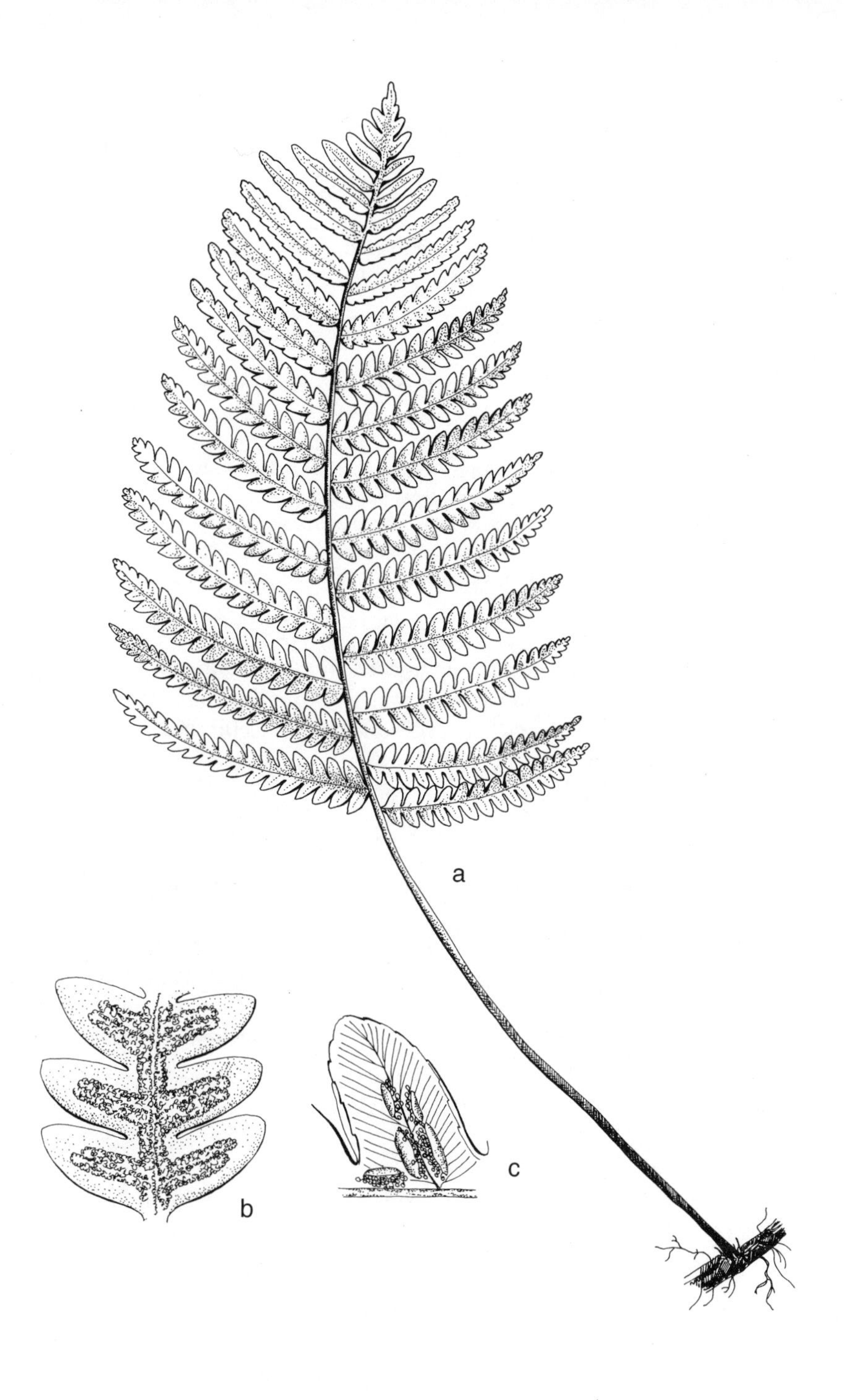

3. *Woodwardia virginica* (Virginia chain fern).

a. Leaf.
b. Portion of pinna.
c. Pinna with sori.

There are about twenty-five species of *Equisetum* in the world. The hollow, jointed stems may be branched or unbranched, and both conditions may occur in the same species. A sheath is present at each joint on the stem and usually bears slender, scalelike teeth. Between the nodes, the stem is ridged and commonly contains deposits of silica in 1–2 rows of tubercles.

Although there are thirteen species and named hybrids of *Equisetum* in the central Midwest, only a few are sometimes found in standing water.

1. Aerial stems persisting 1 year or less; stomates on upper surface, or absent; cone apex rounded.
 2. Aerial stems unbranched.
  3. Teeth of sheaths more than 11, often black or with narrow white margins .... 1. *E. fluviatile*
  3. Teeth of sheaths fewer than 11, with prominent white margins .... 2. *E. palustre*
 2. Aerial stems branched with regular whorls of branches .... 2. *E. palustre*
1. Aerial stems persisting for more than 1 year; stomates sunken; cone apex pointed.
 4. Teeth of sheaths 3; aerial stems crooked .... 3. *E. scirpoides*
 4. Teeth of sheaths 5–10; aerial stems straight .... 4. *E. variegatum*

1. **Equisetum fluviatile** L. Sp. Pl. 1062. 1753. Fig. 4.

Stems annual, erect, branched or unbranched, to 1 m tall, smooth, with up to 25 rounded ridges, with a large central cavity about five-sixths the diameter of the stem; sheaths brown or green, with 15–20 (–22) free, persistent teeth; cone 10–25 mm long, long-pedunculate, not apiculate.

In shallow water of lakes, streams, and swamps.

IA, IL, IN, NE, OH (OBL).

Water horsetail.

The sheaths of the unbranched stems have more than 11 teeth, these often black.

2. **Equisetum palustre** L. Sp. Pl. 1061. 1753. Fig. 5.

Stems annual, regularly branched above the base, to 35 cm tall, with 10–18 more or less rounded, roughened ridges, with a central cavity less than one-half the diameter of the stem; sheaths green, with 5–10 persistent, white-margined teeth with a dark central stripe; cone 10–20 mm long.

Banks of rivers and streams, marshes, swamps.

IL (FACW).

Marsh horsetail.

The distinguishing features of this horsetail are the small central cavity of the stem, the 5–10 persistent sheath teeth, and the branched stems.

3. **Equisetum scirpoides** Michx. Fl. Bor. Am. 2:281. 1803. Fig. 6.

Stems evergreen, very slender, curled, usually unbranched, to 15 cm tall, with 3–6 smooth ridges, without a central cavity but with 3 cavities between center of stem and outer wall; sheaths green below, dark above, with three teeth, the tip of each tooth usually deciduous; cones 2–5 mm long, sessile or nearly so, apiculate.

Moist, shaded woods, wet woods, bogs.

IA, IL (FAC+).

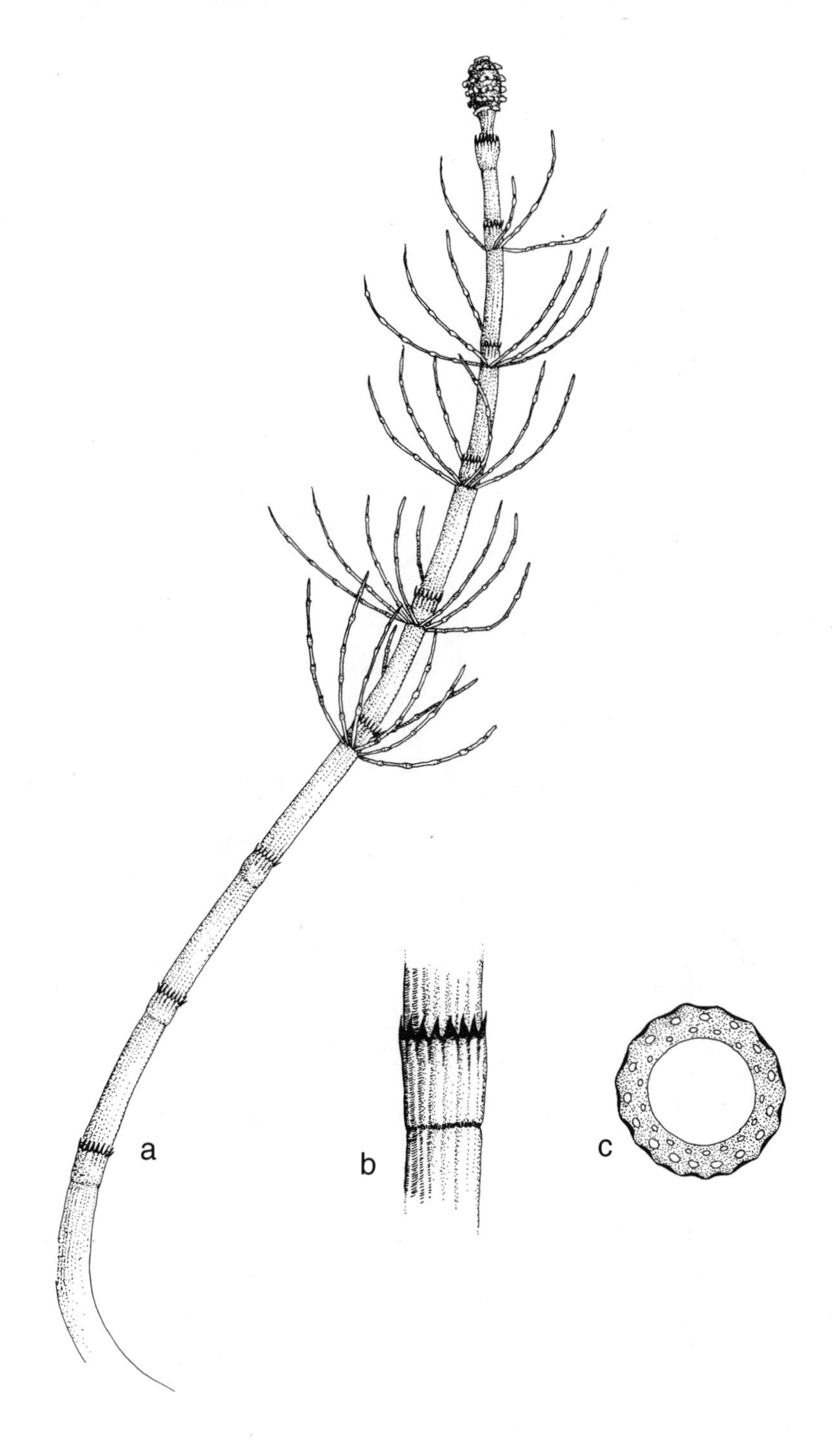

4. *Equisetum fluviatile* (Water horsetail).

a. Habit.
b. Node and stem.
c. Cross-section of stem.

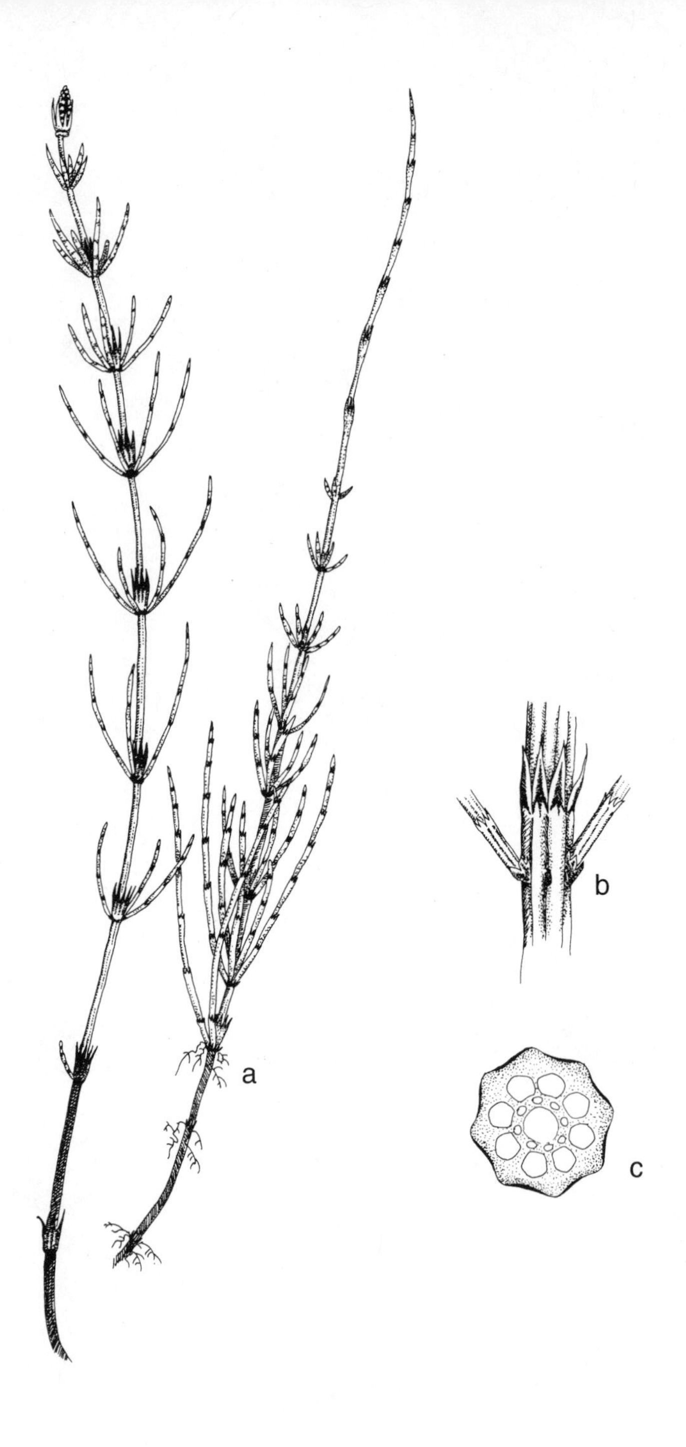

5. *Equisetum palustre* (Marsh horsetail).

a. Habit.
b. Node and stem.
c. Cross-section of stem.

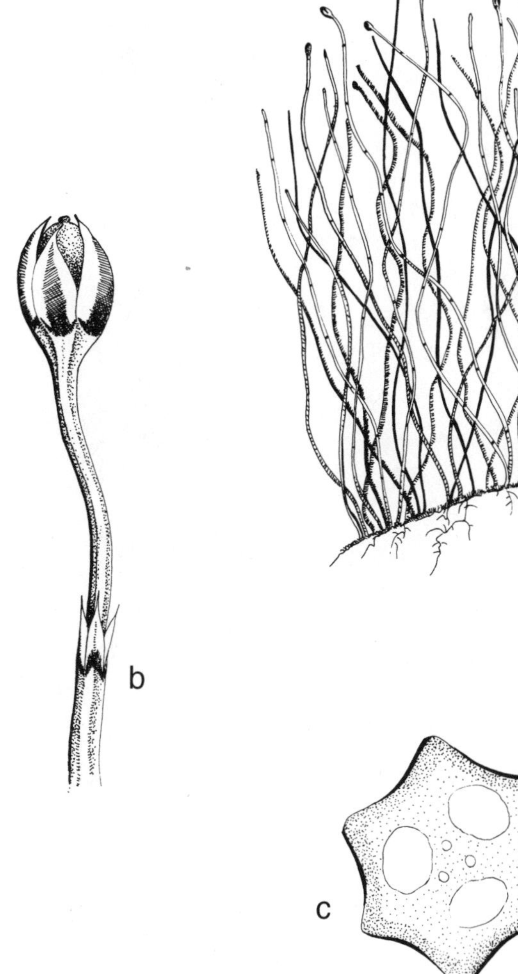

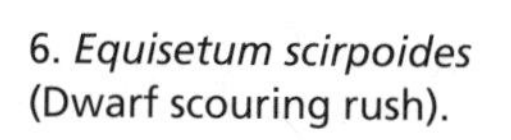

6. *Equisetum scirpoides* (Dwarf scouring rush).

a. Habit.
b. Cone, stem, and node.
c. Cross-section of stem.

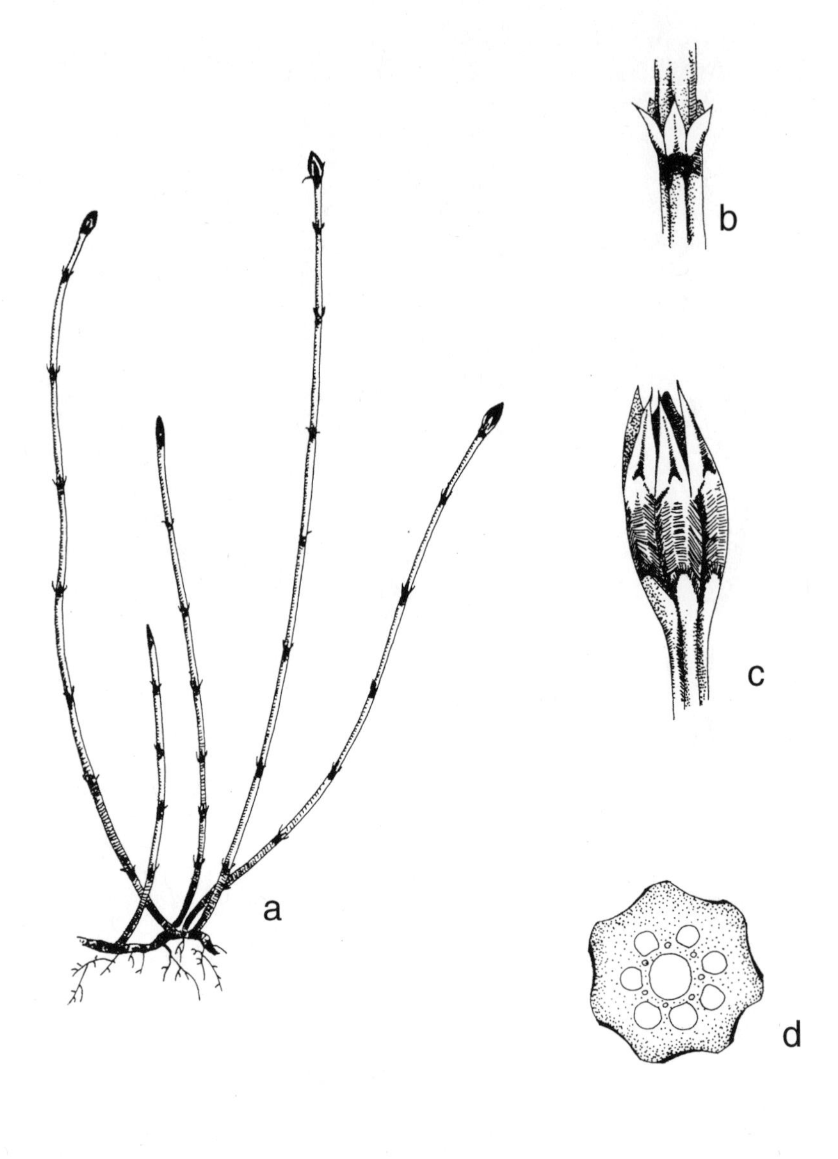

7. *Equisetum variegatum* (Variegated scouring rush). a. Habit. b. Node and stem. c. Cone. d. Cross-section of stem.

Dwarf scouring rush.

This tiny species is the only *Equisetum* with curly, crooked stems. It also produces very small cones.

4. **Equisetum variegatum** Schleich. Usteri, Neue Ann. Bot. 21:120–135. 1797. Fig. 7.

Stems evergreen, erect, firm, unbranched above the base, slender, strongly grooved, to 25 cm tall, with 4–10 angled ridges, with each ridge bearing tubercles in two rows, with a central cavity less than one-half the diameter of the stem; sheaths green below, dark above, with 5–10 teeth, the tip of each tooth usually deciduous and with a central groove throughout; cones 5–10 mm long, apiculate.

Shores, banks, swamps, ditches.

IL, IN (FACW).

Variegated scouring rush.

The teeth of the sheaths are 5–10 in number.

4. ISOETACEAE—QUILLWORT FAMILY

Isoetes is the only genus, and it is nearly worldwide in distribution.

1. **Isoetes** L.—Quillwort

Stem a fleshy, lobed corm; leaves grasslike, spirally arranged, broadened at base; ligule present just above sporangium; sporangia embedded in the bases of the leaves, brown-punctate or -striate, partially covered above; spores of 2 kinds.

Isoetes is comprised of about 150 species. In addition to the three described below from the central Midwest, *I. butleri* is a terrestrial species that is known from Illinois, Kentucky, and Missouri.

1. Megaspores echinate .... 1. *I. echinospora*
1. Megaspores reticulate or tuberculate.
    2. Sporangium punctate or striate; megaspores tuberculate; microspores spinulose .... 3. *I. melanopoda*
    2. Sporangium neither punctate nor striate; megaspores reticulate, with narrow ridges; microspores smooth or minutely roughened .... 2. *I. engelmannii*

1. **Isoetes echinospora** Durieu, Bull. Soc. Bot. France 8:164. 1861. Fig. 8.

Corm 2-lobed; leaves to 25 (–40) cm long, bright green to reddish green, pale toward base; sporangium brown-streaked; megaspores 400–550 mμ in diameter, echinate; microspores 20–30 mμ in diameter, smooth to spinulose.

Emergent or in shallow water of lakes, ponds, and streams.

OH (OBL).

Spiny-spored quillwort.

Water in which this quillwort is found is usually slightly acidic.

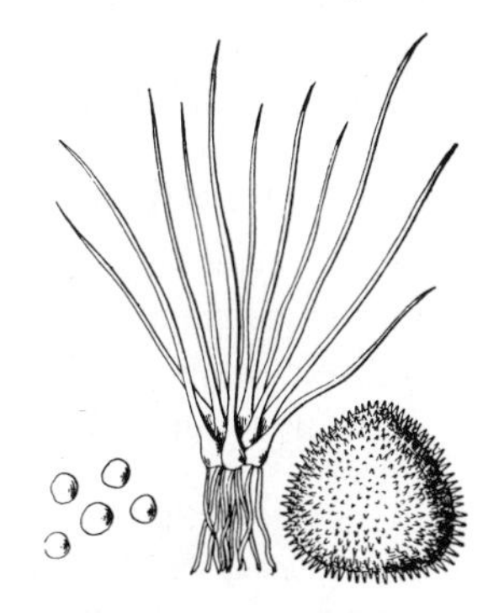

8. *Isoetes echinospora* (Spiny-spored quillwort). Habit. Megaspore (lower right).

2. **Isoetes engelmannii** A. Br. Flora 29:178. 1846. Fig. 9.

Corm 2-lobed; leaves 15–60 (–100), 13–50 cm long, slender; sporangia oblongoid, 6–13 mm long, neither punctate nor striate; megaspores 400–615 mμ in diameter, reticulate with narrow ridges; microspores 21–33 mμ in diameter, smooth to minutely roughened.

Shallow water of ponds and ditches.

IL, KY, MO, OH (OBL).

Engelmann's quillwort.

3. **Isoetes melanopoda** Gay & Dur. Bull. Soc. Bot. France 11:102. 1864. Fig. 10.
Isoetes melanopoda Gay & Dur. var. *pallida* Engelm. Trans. St. Louis Acad. Sci. 4:387. 1882.
Isoetes melanopoda Gay & Dur. f. *pallida* (Engelm.) Fern. Rhodora 51:103. 1949.

Corm 2-lobed; leaves 15–85, 15–40 cm long, black and shining at base, or occasionally pale (f. *pallida*); sporangia oblongoid, 0.5–3.0 cm long, brown-punctate; megaspores 280–440 mμ in diameter, low-tuberculate; microspores 20–30 mμ in diameter, spinulose.

Shallow water of ponds and ditches.

IA, IL, IN, KS, KY, MO, NE (OBL).

Black quillwort.

Specimens with pale rather than dark leaf bases have been called f. *pallida*.

5. LYCOPODIACEAE—CLUBMOSS FAMILY

Horizontal stems present or absent, either on the surface of the ground or subterranean; upright shoots branched or unbranched, leafy at least at base; lateral shoots present or absent; leaves on subterranean stems flat, appressed, nongreen, scale-like; leaves on aerial stems appressed, ascending, or spreading, 1-veined, remote or densely arranged; cones sessile or pedunculate, erect or nodding; sporangia solitary, axillary; subtending leaves leaflike and sometimes aggregated into cones; spores of one kind.

There may be as many as fifteen genera in the family, distributed over much of the world. In addition to the genus below, there are three other genera of Lycopodiaceae in the central Midwest.

1. **Lycopodiella** Holub—Bog Clubmoss

Horizontal stems on surface of ground; upright shoots unbranched, forming leafy peduncles; gemmae absent; cone solitary on a leafy peduncle, linear-lanceolate, usually somewhat serrulate.

There are ten species in the genus, with most of them occurring in wet habitats.

1. Leaves entire; fertile stems up to 6 cm tall .... 3. *L. inundata*
1. Leaves sparsely serrulate; fertile stems usually more than 6 cm tall.
    2. Leaves appressed; cone at most only 2 mm thicker than peduncle .... 2. *L. appressa*
    2. Leaves spreading to ascending; cone at least 3 mm thicker than peduncle .... 1. *L. alopecuroides*

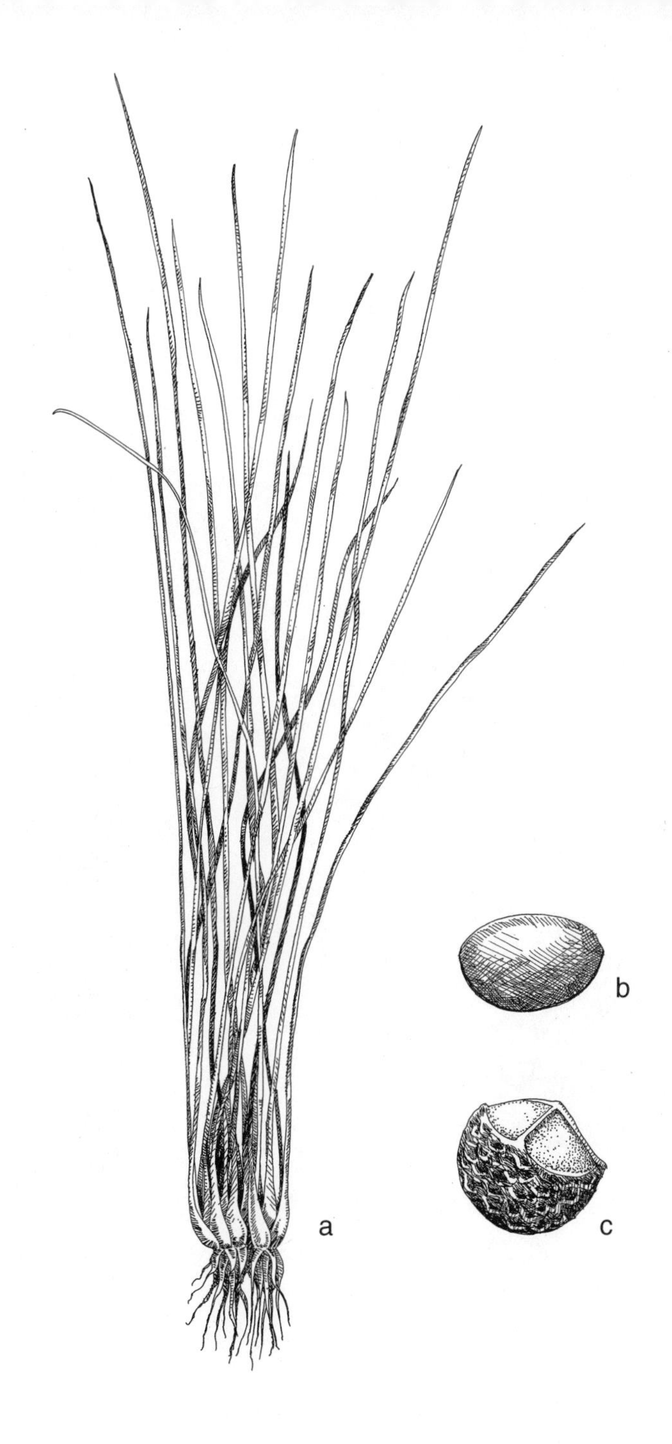

9. *Isoetes engelmannii* (Engelmann's quillwort).

a. Habit.
b. Microspore.
c. Megaspore.

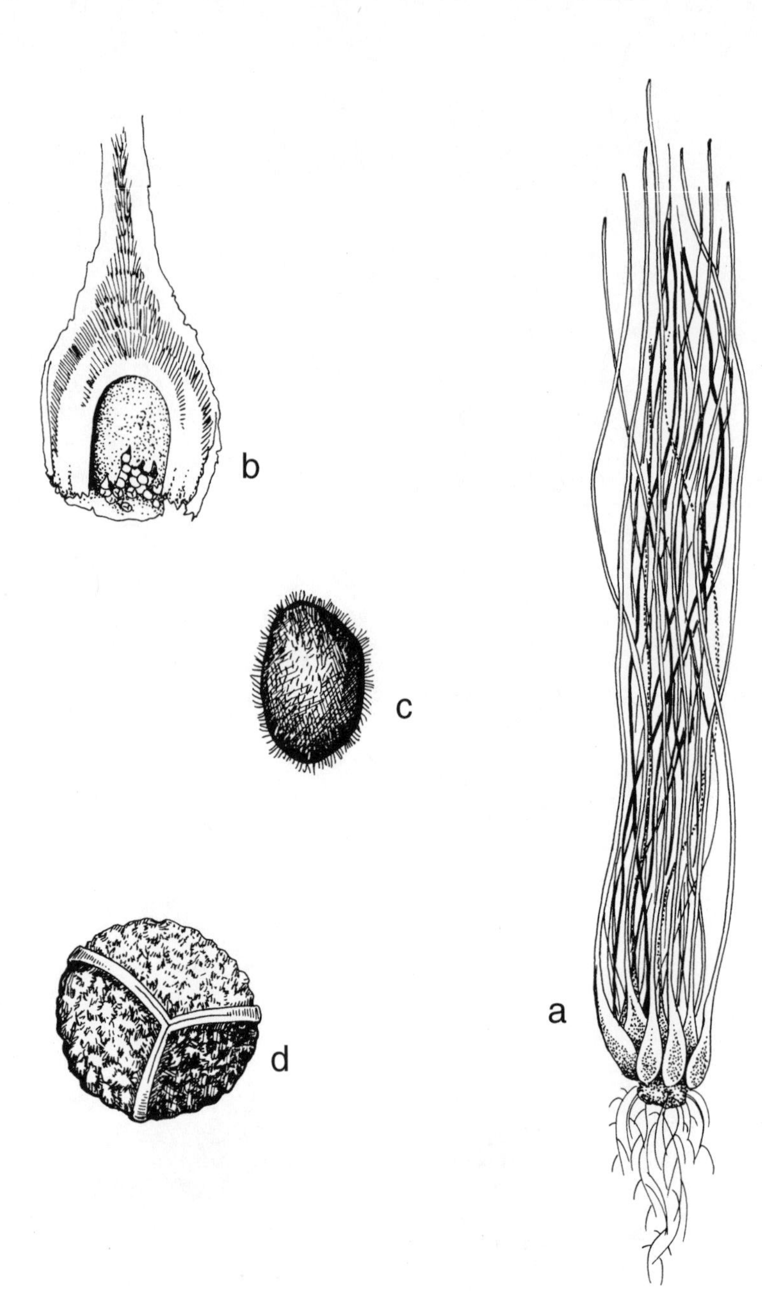

10. ***Isoetes melanopoda*** **(Black quillwort).** a. Habit. b. Sporangium. c. Microspore. d. Megaspore.

1. **Lycopodiella alopecuroides** (L.) Cranfill, Am. Fern Journ. 71:97. 1981. Fig. 11.
Lycopodium alopecuroides L. Sp. Pl. 2:1102. 1753.

Horizontal stems arching, to 40 cm long, 2–4 mm in diameter; leaves spreading to ascending, 5–7 mm long, less than 1 mm wide, sparsely serrulate; peduncle 1 (–3), to 30 cm long, to 3 mm broad; cone 2–6 cm long, 1.2–2.0 cm broad.

Bogs, marshes, ditches.

KY (FACW+).

Foxtail bog clubmoss.

This species has thicker cones than the other species of *Lycopodiella* in the central Midwest.

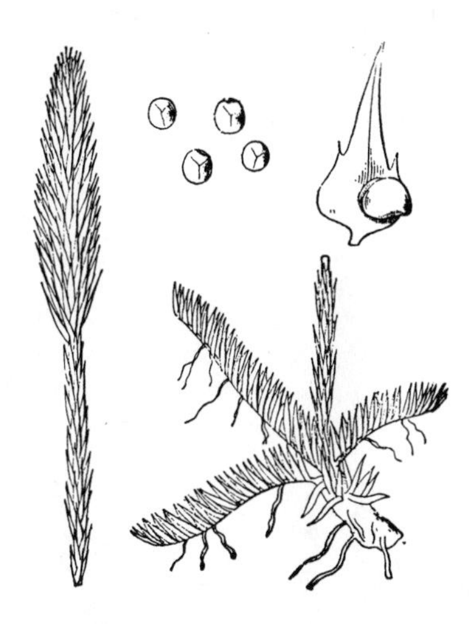

11. *Lycopodiella alopecuroides* (Foxtail bog clubmoss). Habit. Sporangia subtended by bracts (above).

2. **Lycopodiella appressa** (Chapm.) Cranfill, Am. Fern Journ. 71:97. 1981. Fig. 12.
Lycopodium inundatum L. var. *appressum* Chapm. Bot. Gaz. 3:20. 1878.
Lycopodium appressum (Chapm.) F. E. Lloyd & L. Underwood, Bull. Torrey Club 27:153. 1900.

Extensively creeping plant with forked, prostrate stems; leaves linear-lanceolate, attenuate at apex, usually with up to 7 teeth on each margin, appressed, 6- to 10-ranked, 5–7 mm long; gemmae absent; sporangia borne in a terminal cone; cone 2.5–6.0 cm long, 3.5–5.0 cm thick; sporophylls green, leafy, lanceolate, appressed, without teeth on each margin; spores 45–55 mμ in diameter.

Bogs, marshes, shores, ditches.

IL, MO (FACW), KS (NI).

Bog clubmoss.

The appressed leaves readily distinguish this species of *Lycopodiella.*

3. **Lycopodiella inundata** (L.) Holub, Preslia 36:21. 1964. Fig. 13.
Lycopodium inundatum L. Sp. Pl. 2:1102. 1753.

Extensively creeping plant with forked, prostrate stems; leaves linear-lanceolate, attenuate at apex, entire or spinulose, somewhat spreading, 8- to 10-ranked, 5–8 mm long; gemmae absent; sporangia borne in a terminal cone; cone 1–4 cm long, 0.5–1.0 cm thick; sporophylls green, leafy, lanceolate, entire or serrulate near base; spores papillate at apex, reticulate at base, 45–55 mμ in diameter.

Bogs, marshes, shores.

IL, IN, OH (OBL).

Bog clubmoss.

The spreading leaves and very short upright stems are the distinguishing features of this bog clubmoss.

12. ***Lycopodiella appressa*** (Bog clubmoss). Habit.

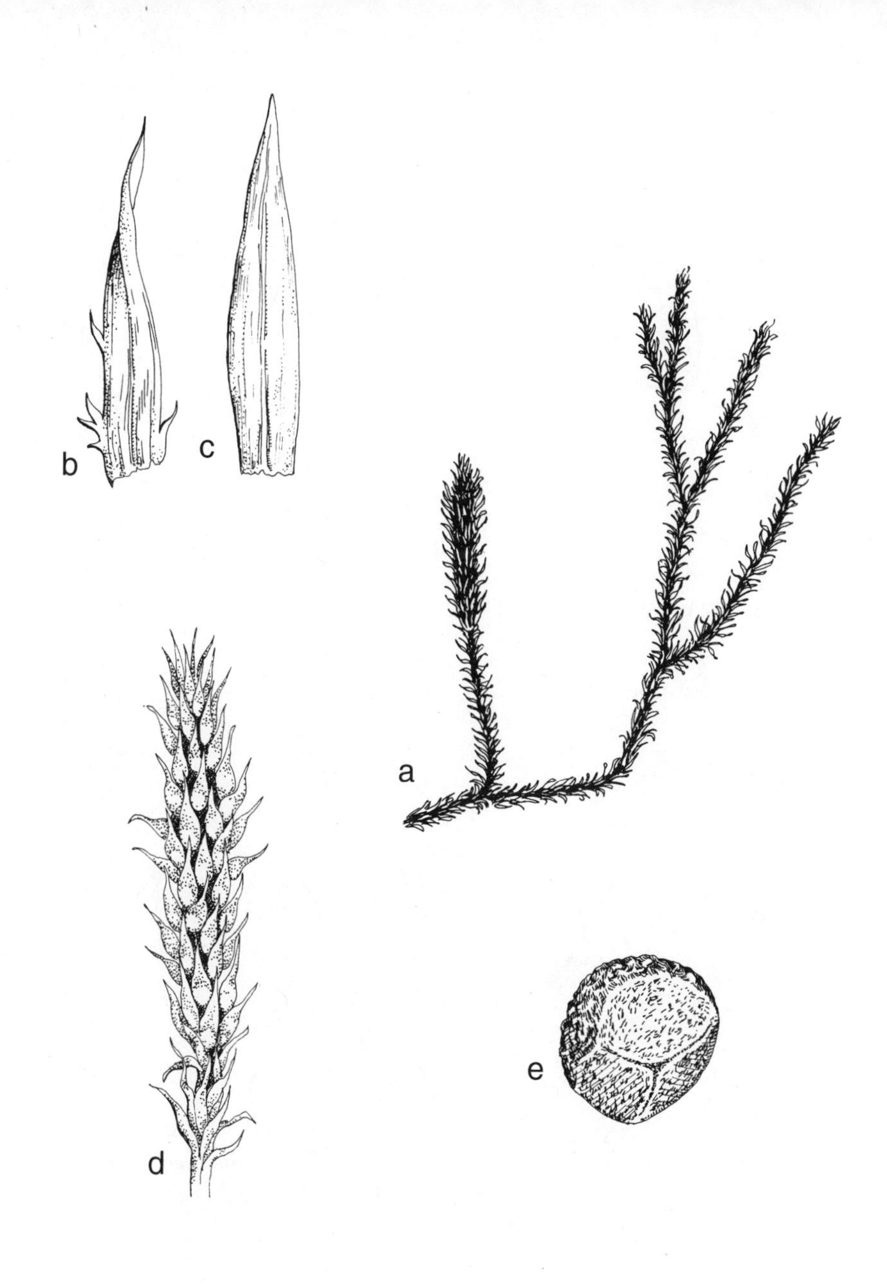

13. *Lycopodiella inundata* (Bog clubmoss).

a. Habit.
b., c. Leaves.
d. Cone.
e. Spore.

14. *Marsilea quadrifolia* (Water-clover). a. Habit. b. Leaf.

6. MARSILEACEAE—WATER-CLOVER FAMILY

Rhizomatous aquatics; leaves petiolate, 4-parted or filiform; sori borne in hard sporocarps, with two kinds of sporangia; megasporangia with a single megaspore; microsporangia with up to 64 microspores; gametophytes remaining within the spores.

Three genera comprises this family, consisting of about fifty species. Two aquatic genera occur in the central Midwest.

1. Leaves with 4 leaflets, appearing like a four-leaf clover .... 1. *Marsilea*
1. Leaves filiform .... 2. *Pilularia*

1. **Marsilea** L.—Water-clover

Aquatic plants, usually in colonies; leaves deciduous, divided into four leaflets; sporocarps borne on stalks near base of petioles, more or less hairy.

Forty-five species comprise this genus, two of them in the central Midwest.

1. Roots present along internodes as well as at nodes .... 1. *M. quadrifolia*
1. Roots present only at nodes .... 2. *M. vestita*

1. **Marsilea quadrifolia** L. Sp. Pl. 2:1099. 1753. Fig. 14.

Roots present on the nodes and the internodes; blades glabrous or nearly so, 4-parted, up to 20 mm across, the petioles 15–30 cm long; sporocarps ellipsoid, 4–5 mm long, 3–4 mm thick, glabrous, punctate, long-stalked, the stalk arising from near the base of the petiole. Sporocarps June–October.

Ponds, lakes.

IA, IL, IN, KY, MO, OH (OBL).

Water-clover.

The roots that form on both the nodes and internodes distinguish this water-clover from the other species in the central Midwest.

2. **Marsilea vestita** Hook. & Grev. Icon. Filic. 2:plate 159. 1830. Fig. 15.
Marsilea mucronata A. Braun, Am. Journ. Sci., ser. 2, 3:55. 1847.

Roots present only on the nodes; blades pubescent or glabrous, 4-parted, up to 16 mm across, the petioles up to 20 cm long; sporocarps ellipsoid, 3.5–7.5 mm long, 3.0–6.5 mm thick, pubescent at first, becoming glabrous, punctate, long-stalked, the stalk arising from near the base of the petiole. Sporocarps April–October.

Ponds, wet depressions.

IA, KS, NE (OBL).

Water-clover.

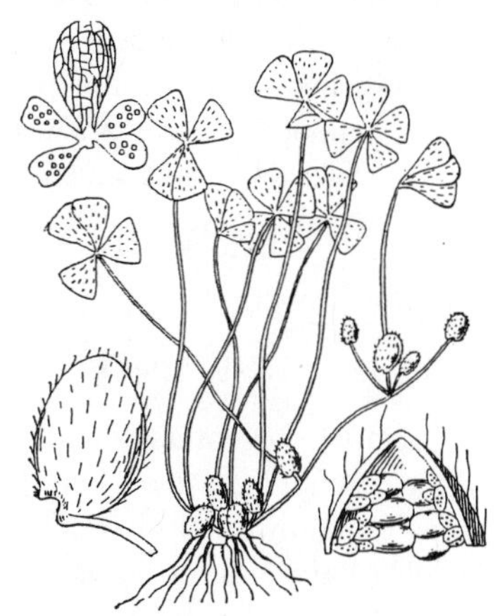

15. *Marsilea vestita* (Water-clover). Habit. Sporangium (right).

This more western species differs from *M. quadrifolia* by its roots forming only at the nodes.

2. **Pilularia** L.—Pillwort

Aquatic perennials; leaves elongated, grasslike; sporangia in hard cases attached to the tip of the stalk, globose, pubescent.

The species of this aquatic fern genus have leaves that resemble those of grasses. There are six species in the genus, only the following known from the central Midwest.

1. **Pilularia americana** A. Braun, Monatsber. Konigl. Preuss. Akad. Wiss. Berlin 1863:435. 1863. Fig. 16.

Aquatic, rooting at the nodes; leaves up to 10 cm long, filiform, bladeless, glabrous or nearly so; sporangium case 1.5–3.0 mm in diameter, globose, pubescent, at least at first, pendulous. Sporocarps produced May–October.

Lakes, ponds.

MO, NE (OBL).

Pillwort.

This fern, because of its obscurity, may be more common than the collections indicate. The young leaves are curled when first produced.

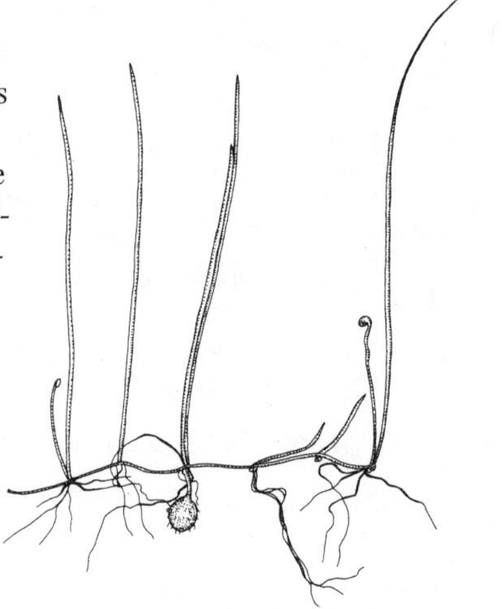

16. *Pilularia americana* (Pillwort). a. Habit.

7. ONOCLEACEAE—SENSITIVE FERN FAMILY

Stems branched or unbranched, scaly; leaves strongly dimorphic; sterile leaves pinnatifid to pinnate-pinnatifid, the pinnae entire to serrulate, the veins free or reticulate; sori enclosed by hardened pinnae, often beadlike; spores usually green.

Four genera, all sometimes placed in the Dryopteridaceae but differing by having strongly dimorphic leaves and usually having green spores, comprise this family. Only the following genus sometimes occurs in standing water in the central Midwest.

1. **Onoclea** L.—Sensitive Fern

Only the following species comprises the genus.

1. **Onoclea sensibilis** L. Sp. Pl. 1062. 1753. Fig. 17.

Deciduous perennial from creeping, branching rhizomes; leaves dimorphic, the sterile pinnatifid, to 50 cm long, with a winged rachis, the fertile bipinnate, the stiff pinnules tightly enrolled to enclose the sporangia; sori round, covered by a delicate, hoodlike indusium; fertile pinnules at maturity dry, hard, globose, eventually cracking to liberate the spores.

Low woods, marshes, swamps.

IA, IL, IN, KS, KY, MO, NE, OH (FACW).

Sensitive fern.

17. ***Onoclea sensibilis*** (Sensitive fern). Sterile leaves (left), fertile leaf (right).

This plant is sometimes confused with the netted chain fern *(Woodwardia areolata)* since the sterile fronds of both are deeply pinnatifid. The pinnules of *Onoclea* are serrulate while those of the *Woodwardia* are entire.

The common name "sensitive fern" is due to the blackening of the leaves at the very first frost of the season.

8. OSMUNDACEAE—ROYAL FERN FAMILY

Stems creeping; leaves dimorphic; fertile leaves erect, smaller than the sterile leaves; blades once- or twice-pinnate, the pinnae pinnatifid to pinnate.

There are three genera in the family, consisting of between sixteen and thirty-six species. Only the following genus occurs in the United States:

1. **Osmunda** L.—Royal Fern

Large, perennial ferns with fibrous roots; leaves dimorphic, the long, scaleless petiole winged at the very base; fertile leaves (or parts of them) not leaflike; sporangia large, stipitate, pear-shaped, bivalved; spores numerous, green. This genus consists of ten species, with two of them often occurring in standing water in the central Midwest.

1. Leaves bipinnate, the pinnules serrulate; sporangia borne on upper half of leaf .... 2. *O. regalis*
1. Leaves always pinnate-pinnatifid, the pinnules entire; sporangia borne on a separate fertile leaf .... 1. *O. cinnamomea*

1. **Osmunda cinnamomea** L. Sp. Pl. 1066. 1753. Fig. 18.

Erect perennial about 1 m tall from dark brown rootstocks; sterile blades lanceolate to elliptic, pinnate-pinnatifid, with up to 30 pairs of pinnae, each pinna with up to 35 pairs of lobes, without serrulations, glabrous; petiole, rachis, midvein, and leaf axils lanate, the wool cinnamon-colored; fertile leaves borne separately in the center of a ring of sterile leaves, much reduced, erect, the pinnules bearing many crowded brown sporangia.

Wet areas, swamps, seeps, sometimes on wet sandstone ledges.

IA, IL, IN, KY, MO, OH (FACW).

Cinnamon fern.

In the sterile condition, this fern is sometimes difficult to distinguish from *O. claytoniana,* which is never aquatic, but cinnamon fern has more wool on its stipes and has acuminate pinnules. The pinnules in *O. claytoniana* are more or less obtuse.

2. **Osmunda regalis** L. var. **spectabilis** (Willd.) Gray, Man. 600. 1856. Fig. 19.
Osmunda spectabilis Willd. Sp. Pl. 5:98. 1810.

Erect perennial to nearly 1 m tall; sterile blades narrowly ovate, bipinnate, with 2–8 pairs of pinnae, each pinna bearing up to 21 pinnules, the pinnules alternate, oblong to lanceolate, subacute or obtuse at apex, asymmetrical at the rounded base, serrulate, glabrous, to 5 cm long, to 2.5 cm broad; petiolules about 1 mm long; fertile portions on upper part of leaf strongly contracted, with densely crowded, yellow-brown sporangia.

18. *Osmunda cinnamomea* (Cinnamon fern). Sterile leaf with fertile frond superimposed.

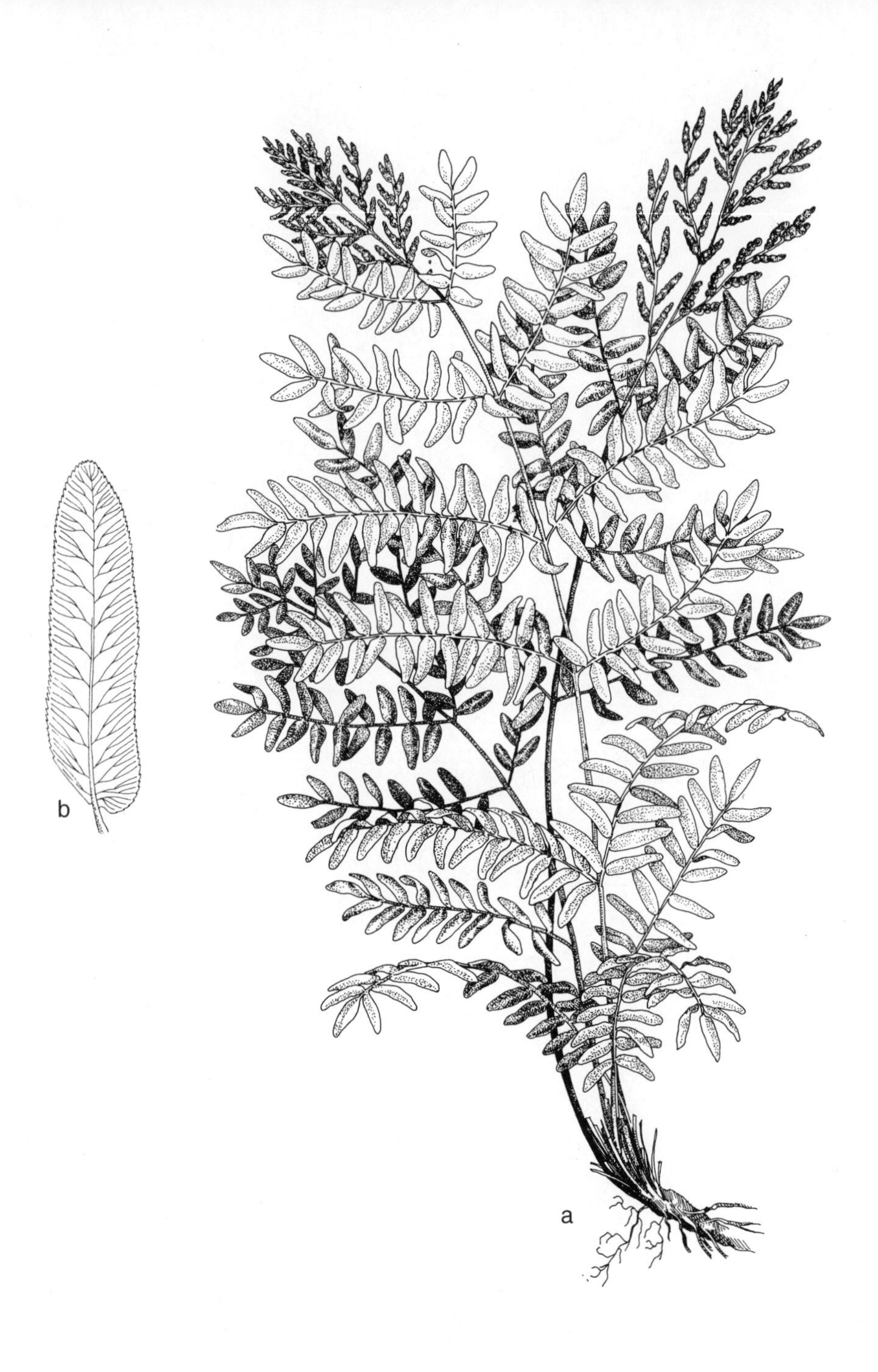

19. *Osmunda regalis* var. *spectabilis* (Royal fern). a. Habit. b. Pinna.

Swamps, rarely on moist, sandstone ledges.
IA, IL, IN, KY, MO, OH (OBL).
Royal fern.
Our plants belong to var. *spectabilis.* Typical var. *regalis* is native to Europe and Asia.

9. THELYPTERIDACEAE—THELYPTERIS FAMILY

Leaves usually monomorphic; blade usually pinnate to pinnate-pinnatifid; veins free or connecting, running to the margin; sori round to oblong, variously placed on the back of the frond.

Members of this family have usually been placed with genera now in the Dryopteridaceae. Fern experts have recently pointed out several technical differences between the two families. As a result, the exact number of genera to be assigned to the Thelypteridaceae is not currently known, although there may be as many as thirty genera and more than nine hundred species. Of these, only *Phegopteris* and *Thelypteris* are known from the central Midwest, with only *Thelypteris* containing a species that sometimes occurs in standing water.

1. **Thelypteris** Schmidel—Thelypteris

Rootstocks slender, creeping or erect; leaves pinnate-pinnatifid; rachis unwinged; sori round; indusium cordate. The genus consists of about 875 species. Only the following occurs in standing water in the central Midwest.

1. **Thelypteris palustris** Schott var. **pubescens** (Laws.) Fern. Rhodora 31:34. 1929. Fig. 20.
Lastrea thelypteris var. *pubescens* Laws. Edinb. New Philos. Journ. II 19:277. 1864.

Rather fragile, deciduous perennial from slender, nearly scaleless rhizomes; leaves somewhat dimorphic, the sterile pinnate-pinnatifid, membranaceous, to 50 cm long, puberulent on both surfaces, with up to 40 pairs of pinnae; fertile leaves pinnate-pinnatifid, more firm, to 40 cm long, puberulent on both surfaces, with up to 25 pairs of pinnae; petiole slender, glabrous at maturity; sori round, usually confluent, borne on the back of the leaf segments; indusium eglandular.
Marshes, swamps, bogs, banks, wet ditches.
IA, IL, IN, KS, KY, MO, OH (FACW+).
Marsh fern.
Our variety differs from the typical variety of Europe and Asia by having both the upper and lower leaf surfaces puberulent. Marsh fern differs from most other ferns with pinnate-pinnatifid leaves by having nearly glabrous petioles.

10. PINACEAE—PINE FAMILY

Evergreen or deciduous, monoecious trees; bark smooth or scaly; needles borne singly or in clusters; male cones solitary or clustered, with overlapping sporophylls; female cones woody, with flat, overlapping scales, each scale with 2 seeds; seeds winged.

There are ten genera and two hundred species in this family, with *Larix* being the only genus with a species occurring in standing water.

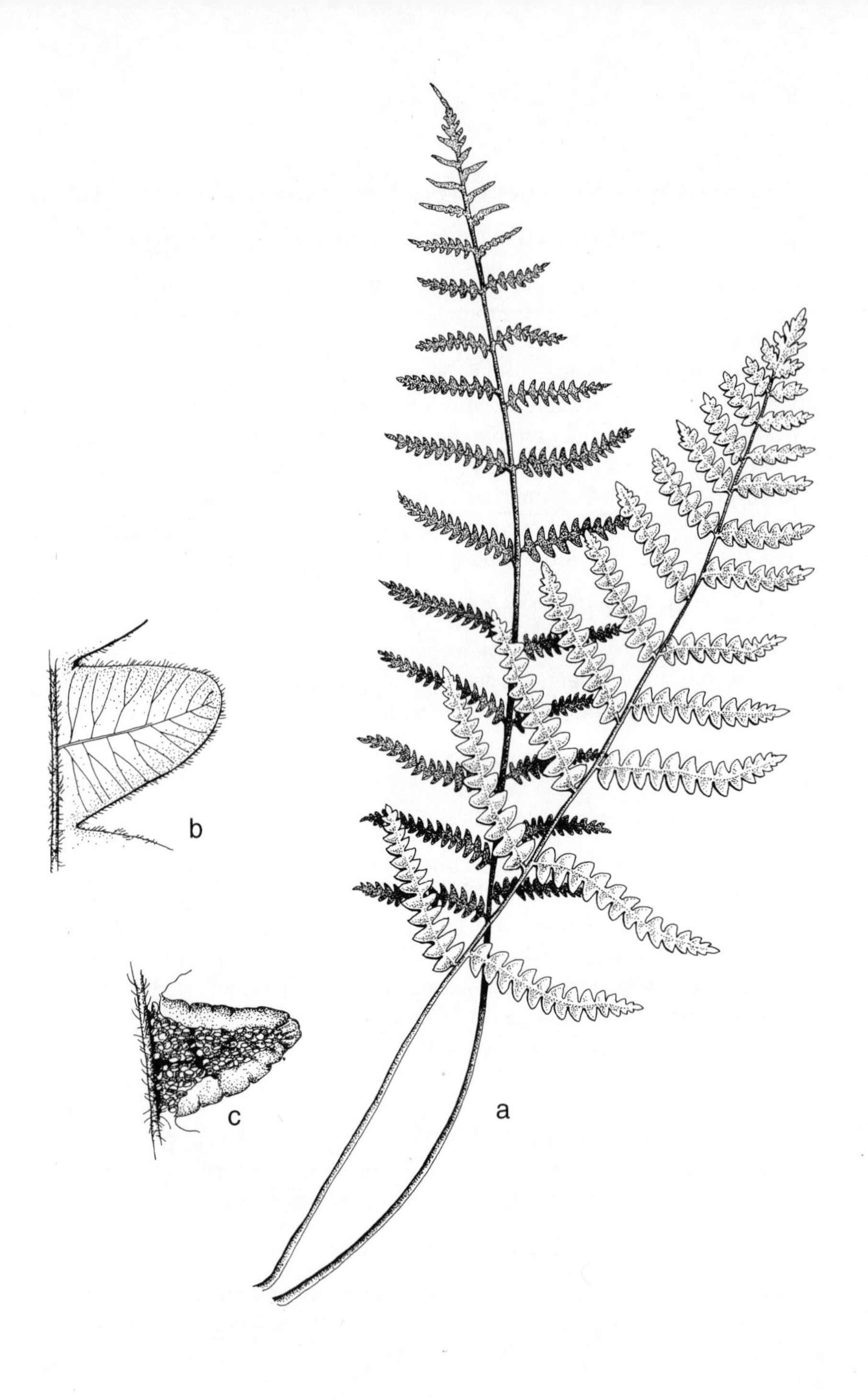

20. ***Thelypteris palustris*** **(Marsh fern).** a. Leaves. b. Sterile pinna. c. Fertile pinna.

1. **Larix** Mill.—Larch; Tamarack

Deciduous trees with open crown; branches whorled; needles borne in clusters of 20 or more; short shoots present; twigs smooth or somewhat roughened by elevated leaf scars; male cone solitary; female cones terminal, on short shoots, with persistent scales; seeds winged.

Ten species comprise this genus.

1. **Larix laricina** (DuRoi) K. Koch, Dendrologie 2 (2):263. 1873. Fig. 21.
Pinus laricina DuRoi, Diss. Observ. Bot. 49. 1771.

Tree to 50 feet; branches gray, smooth, becoming red-brown and scaly; needles in clusters of 20 or more, deciduous, 1–2 cm long, 0.3–0.5 mm broad, pale blue-green; female cones 1–2 cm long, 5–10 mm broad, on curved stalks, or sometimes sessile, with 10–30 scales; seeds winged.

Bogs.

IL, IN, OH (OBL).

American larch; tamarack.

The deciduous needles borne in clusters distinguish this tree from any other in our area.

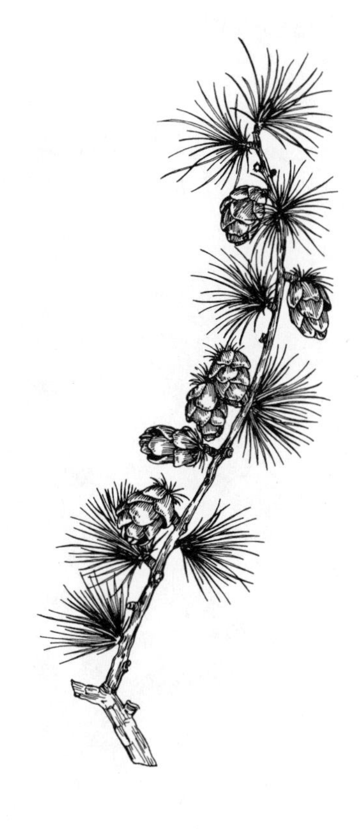

21. *Larix laricina* (American larch). Branch with leaves and cones.

11. TAXODIACEAE—BALD CYPRESS FAMILY

Deciduous (in the central Midwest) or evergreen trees; leaves needlelike or scalelike, spirally arranged or 2-ranked; male cones with spirally arranged sporophylls; female cones with overlapping scales; seeds winged.

Four genera and seven species comprise this family.

1. **Taxodium** Rich.—Bald Cypress

Deciduous trees; leaves 2-ranked, flattened, needlelike; female cones globose, maturing in one season.

There are four species in this genus, with only the following in the central Midwest.

1. **Taxodium distichum** (L.) Rich. Ann. Mus. Nat. Hist. Nat. 16:298. 1810. Fig. 22.
Cupressus disticha L. Sp. Pl. 2:1003. 1753.

Tree to 120 feet, often swollen to form a buttressed base; bark red-brown to pale brown, flaky; branches usually horizontally spreading; leaves needlelike, flat, linear, 2-ranked, up to 15 mm long, the base of the needle usually twisted; female cones at ends of twigs, pendulous, up to 3 cm thick.

Swamps; frequently grown as an ornamental.

IL, IN, KY, MO (OBL).

Bald cypress.

When this species grows in standing water, it often forms pneumatophores, or "knees", which are starch-storing organs and serve to anchor the trees in their watery habitat.

22. *Taxodium distichum* (Bald cypress). Branch with leaves and fruit.

12. ACORACEAE—SWEET FLAG FAMILY

Characters of the genus.

This only genus in the family is often placed in the Araceae but differs from the Araceae in lacking a spathe associated with the inflorescence and by its linear, grasslike leaves.

1. **Acorus** L.—Sweet Flag

Plants rhizomatous; leaves linear, grasslike, with 1–3 prominent veins; true spathe absent, but represented by the extension of the stem; spadix elongate, completely covered with perfect flowers; perianth parts 6; stamens 6; ovary superior, 2- to 3-celled; fruit dry, 1- to 3-seeded.

The three species of *Acorus* occur in Asia, Europe, and North America. Two species occur in the central Midwest.

1. Midvein and 2–several lateral nerves prominent, not excentric .... 1. *A. americanus*
1. Only the midvein prominent and excentric, the lateral nerves obscure .... 2. *A. calamus*

1. **Acorus americanus** (Raf.) Raf. New Fl. & Bot. N. Am. 1:57. 1836. Fig. 23.
Acorus calamus L. var. *americanus* Raf. Med. Fl. 1:25. 1828.

Rhizome stout, creeping, aromatic; leaves numerous, basal, linear, to 1 (–2) m long, to 2.0 (–2.5) cm broad, with 2–several prominent parallel veins; scape flattened, leaflike, to 1 m long, extended up to 60 cm beyond the spadix as a modified spathe; inflorescence crowded into a dense spadix to 10 cm long, 0.8–2.0 cm thick; flowers small, brownish yellow, the perianth segments concave; fruits obpyramidal, indehiscent, fertile. May–mid-August.

Marshes, ditches, fens, spring branches, sloughs, often in standing water.

IA, IL, IN, NE (OBL).

Sweet flag.

Although Rafinesque described this species as distinct from *A. calamus*, most botanists except Wilson (1960) and Thompson (1995) considered the two to be the same. *Acorus calamus*, however, is native to Asia and Europe but widely distributed in North America. *Acorus americanus* is readily distinguished by its two prominent, nonexcentric veins in each leaf and by its fertile seeds.

The leaves of *Acorus americanus* resemble those of some grasses, some sedges, irises, bur-reeds, and young cat-tails, but the strongly sweet-scented rhizomes are distinctive.

The rhizomes have a history of uses as a medicine and an additive to perfumes.

2. **Acorus calamus** L. Sp. Pl. 324. 1753. Fig. 24 and Fig. 23d.

Rhizome stout, creeping, aromatic; leaves numerous, basal, linear, to 1 (–2) m long, to 2.5 cm broad, with a single, excentric midvein and obscure lateral veins; scape flattened, leaflike, to 1 m long, extended to 60 cm beyond the spadix as a modified, open spathe; spadix to 10 cm long, 0.8–2.0 cm thick; flowers small, brownish yellow, the perianth segments concave; fruits obpyramidal, indehiscent, shriveled, infertile. May–mid-August.

23. Acorus americanus (Sweet flag).
a. Habit.
b. Spadix.
c. Leaf, showing vein in center of leaf.
d. Leaf of *Acorus calamus* showing vein excentric.

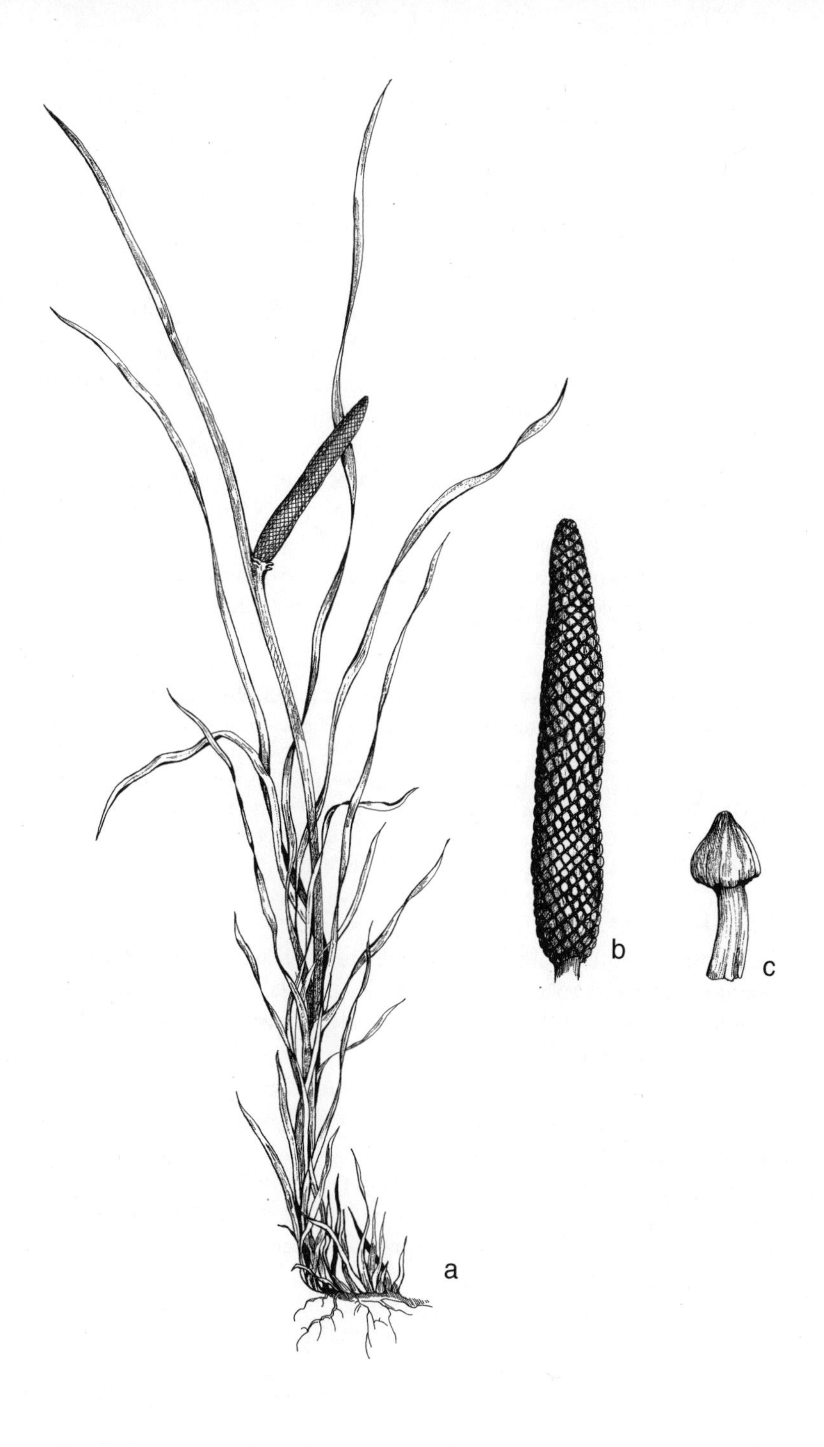

24. Acorus calamus (Sweet flag).

a. Habit.
b. Spadix.
c. Fruit.

Marshes, ditches, fens, spring branches, sloughs, often in standing water.

IA, IL, IN, KS, KY, MO, NE, OH (OBL).

Sweet flag.

The single, excentric vein of each leaf is distinctive for this species. In the United States, the seeds are apparently infertile.

The rhizomes have a history of use as a medicine and as an additive to perfumes.

13. ALISMATACEAE—WATER PLANTAIN FAMILY

Annuals or perennials; leaves basal with a well-developed petiole sheathing at the base, the blade sometimes reduced; inflorescence paniculate, racemose, or umbellate; flowers radially symmetrical, perfect or unisexual, subtended by bracts; sepals 3, green, separate; petals 3, white or rarely pinkish, separate, caducous; stamens 6–40; ovaries 10 or more, free, each 1-celled, superior; fruit an achene.

Care should be taken when collecting specimens of this family since both flowers and fruits are generally needed for positive identification.

This family consists of eleven genera and nearly one hundred species found throughout the world. Most of the species live in or near water. Three genera occur in the central Midwest.

1. Receptacle flat, bearing a single ring of pistils; stamens 6–9; flowers perfect .... 1. *Alisma*
1. Receptacle convex, bearing several rows of pistils; stamens 12–numerous (6–9 in *Echinodorus tenellus*); flowers unisexual or perfect.
    2. Achenes not winged; base of whorled inflorescence branches bearing 3 bracts and several bracteoles; flowers perfect; stamens never more than 21 .... 2. *Echinodorus*
    2. Achenes winged; base of whorled inflorescence branches bearing 3 bracts and no bracteoles; flowers unisexual; stamens usually more than 21 .... 3. *Sagittaria*

1. **Alisma** L.—Water Plantain

Perennials with rhizomes; leaves basal, usually petiolate, sometimes reduced or even absent in submerged plants; inflorescence paniculate with whorled branches; flowers perfect, bracteate; sepals 3, green, separate, persistent; petals 3, white, separate, caducous; stamens 6 or 9; ovaries 10–25, free, arranged in a single series on a flat receptacle, superior; achenes flattened, unwinged, grooved on the back.

This genus is distinguished readily from *Echinodorus* and *Sagittaria* by its ovaries arranged in a single series on the flat receptacle.

There are nine species of *Alisma,* all in the northern hemisphere. Three species occur in the central Midwest. The European *A. plantago-aquatica* is not known from the central Midwest, although it has been previously reported. All species are important food plants for waterfowl.

1. Blades of leaves above the water linear to narrowly elliptic, 3–5 times longer than broad .... 1. *A. gramineum*
1. Blades of leaves above the water broadly elliptic to ovate-lanceolate, up to 3 times longer than broad.
    2. Petals 1.0–2.5 mm long; flowers at most 3.5 mm broad; achenes 1.5–2.2 mm long .... 2. *A. subcordatum*
    2. Petals 3.5–6.0 mm long; flowers at least 7 mm broad; achenes 2.1–3.0 mm long .... 3. *A. triviale*

1. **Alisma gramineum** C. C. Gmel. Fl. Bedford. 4:256. 1826. Fig. 25.

Rooted perennial; submerged leaves up to 50 cm long, up to 12 mm broad, linear; emergent leaves (when present) up to 5 cm long, 3–5 times longer than wide, linear to narrowly elliptic, glabrous, petiolate; scape (primary stalk of inflorescence) erect, bearing a panicle with the secondary branches borne in whorls of 2–8; bracts of the inflorescence usually elliptic; flowers 2.5–4.0 mm broad; sepals obtuse, 1.3–2.3 mm long, green; petals white, often with a pinkish tinge, 2.5–3.5 mm long; stamens 6–9; ovaries 11–20, free from each other; fruiting heads 3.4–3.8 mm in diameter; achenes with 2 parallel grooves on the back, 1.8–2.6 mm long. June–August.

Cool streams, along spring branches, around ponds, usually in standing water.

IA, MO, NE (OBL).

Grass-leaved water plantain.

This species is distinguished easily from the other two species in the central Midwest by its linear to narrowly elliptic leaves and its achenes with a pair of parallel grooves on the back.

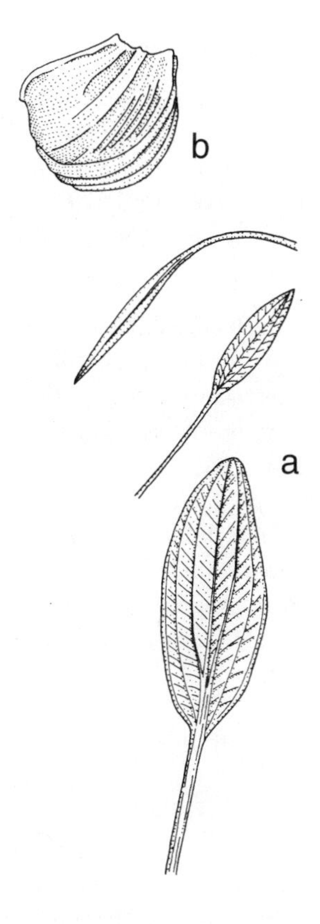

25. *Alisma gramineum* (Grass-leaved water plantain). a. Leaves. b. Achene.

2. **Alisma subcordatum** Raf. Med. Rep. N. Y. II 5:362. 1808. Fig. 26.
Alisma plantago Bigel, Fl. Bost. 87. 1814.
Alisma parviflorum Pursh, N. Am. Sept. 1:252. 1814.
Alisma plantago Bigel var. *parviflorum* (Pursh) Torr. Fl. N. & Mid. U.S. 1:382. 1824.
Alisma plantago-aquatica L. var. *parviflorum* (Pursh) Farwell, Ann. Rep. Comm. Parks & Boulev. Det. 11:44. 1900.

Rooted perennial; leaves basal, the blade elliptic to ovate, membranous, to 12 cm long, to 6.5 cm broad, subacute at apex, rounded to cordate at base, glabrous, the petioles much longer than the blades; scape erect, bearing a panicle with much-branched secondary branches borne in whorls of 3–10; bracts of inflorescence lanceolate; flowers at most 3.5 mm broad; sepals obtuse, 1–3 mm long; petals white, 1.0–2.5 mm long; stamens 6–9, slightly longer than the ovaries; styles at anthesis less than half as long as the 10–25 ovaries; fruiting heads 3–4 mm in diameter; achenes 1.5–2.5 mm long, with one groove on the back. June–August.

Marshes, sloughs, margins of ponds, often in standing water.

IA, IL, IN, KS, KY, MO, NE, OH (OBL).

Small-flowered water plantain.

The leaves of this species are very similar to those of *A. triviale*, but the flowers and fruits are smaller.

3. **Alisma triviale** Pursh, Fl. Am. Sept. 1:252. 1814. Fig. 27.
Alisma plantago-aquatica L. var. *americanum* Roem. & Schultes, Syst. 7:1598. 1830.
Alisma plantago Bigel var. *triviale* (Pursh) BSP. Prel. Cat. N. Y. Pl. 58. 1888.
Alisma plantago-aquatica L. var. *triviale* (Pursh) Farwell, Ann. Rep. Comm. Parks & Boulev. Det. 11:44. 1900.

Rooted perennial; leaves basal, the blade elliptic to ovate, membranous, to 15 cm long, to 8 cm broad, subacute at apex, rounded to cordate at base, glabrous, the petioles much longer than the blades; scape (primary stalk of inflorescence) erect, bearing a panicle with the secondary branches borne in whorls of 3–10; bracts of inflorescence ovate or narrowly ovate; flowers at least 7 mm broad; sepals green, obtuse, 3–4 mm long; petals white, 3.5–6.0 mm long; stamens 6–9, at least twice as long as the ovaries; styles at anthesis as long as the 10–25 ovaries; fruiting heads 5–7 mm in diameter; achenes broadly rounded on back, 2.5–3.0 mm long, with usually one groove on the back. June–August.

Marshes, ditches, margins of ponds, along creeks, often in standing water.

IA, IL, IN, KS, MO, NE, OH (OBL).

Large-flowered water plantain.

The U.S. Fish and Wildlife Service calls this species *Alisma plantago-aquatica.*

2. **Echinodorus** Rich.—Burhead

Annuals or perennials with rhizomes; leaves linear to ovate, palmately veined, the submerged ones usually reduced; inflorescence racemose or umbellate, with a few whorled branches, bracteate; flowers perfect, with both bracts and bracteoles; sepals 3, green, separate, persistent; petals 3, white, separate, caducous; stamens 6–

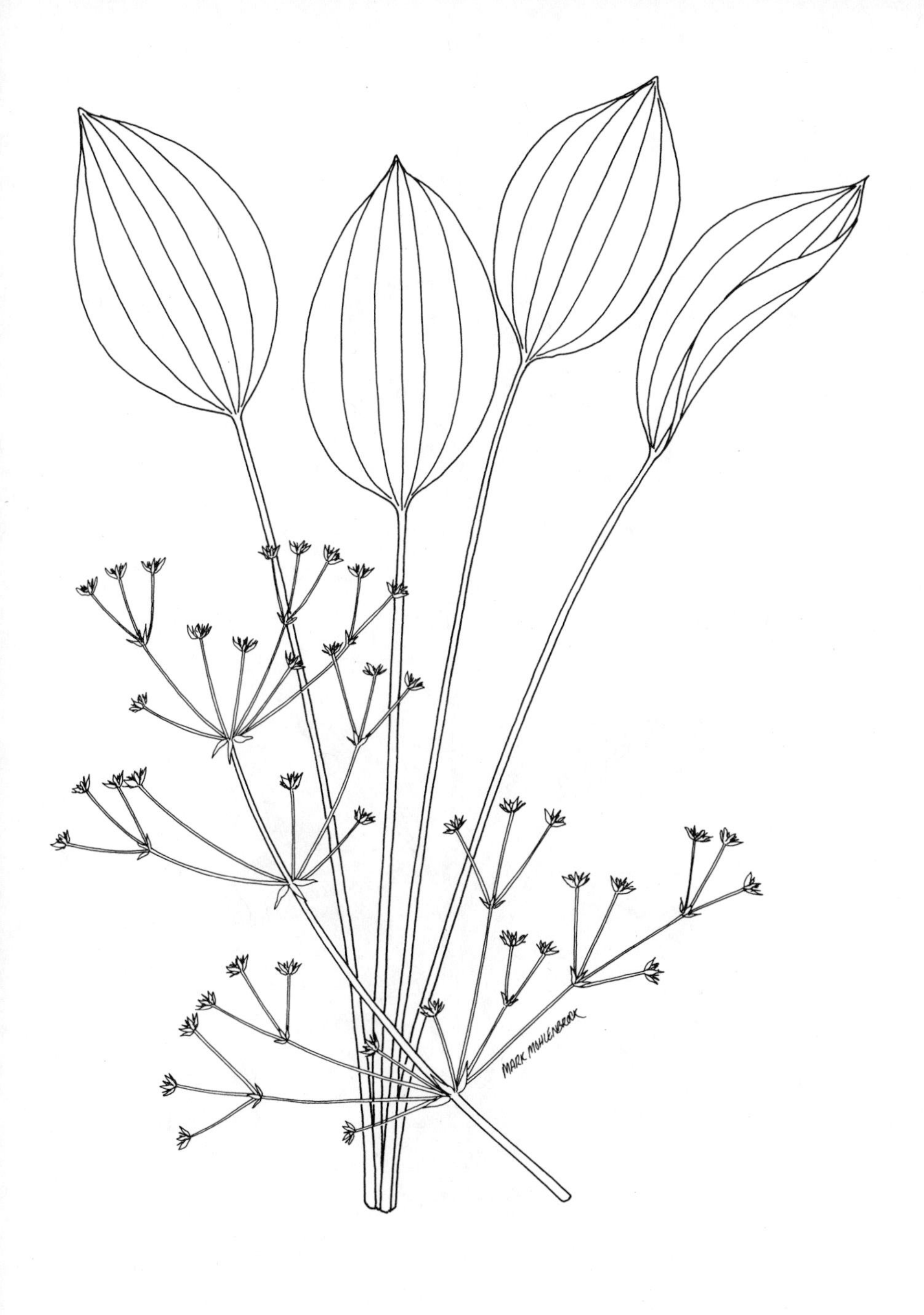

26. ***Alisma subcordatum*** (Small-flowered water plantain). Habit.

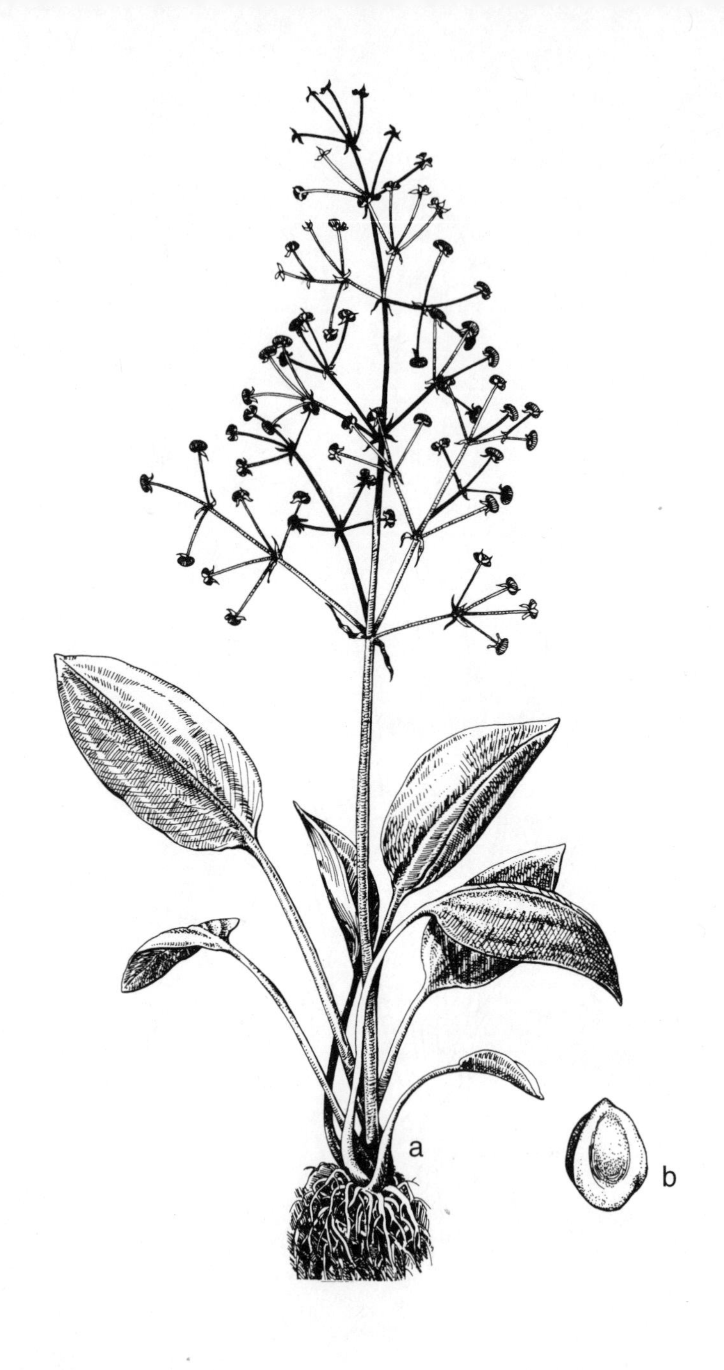

27. *Alisma triviale* (Large-flowered water plantain). a. Habit. b. Achene.

20; pistils 10–many, free from each other, in several series in a headlike club on a convex receptacle; achenes wingless, turgid, ribbed throughout.

Echinodorus consists of about fifty species distributed worldwide, with many in the tropics. Three species occur in the central Midwest. All provide food for waterfowl, and some of them are used as ornamental aquatics and in aquaria.

1. Plants erect, less than 10 cm tall; leaves linear to lanceolate; flowers at most only 6 mm broad; stamens 6–9; achenes 10–20, beakless or nearly so .... 3. *E. tenellus*
1. Plants erect and more than 10 cm tall, or plants creeping or arching; leaves broadly ovate, rarely lanceolate; flowers at least 8 mm broad; stamens 12 or 21; achenes more than 40, beaked.
   2. Scape creeping or arching; stamens 21; style shorter than the ovary; beak of achene incurved .... 2. *E. cordifolius*
   2. Scape erect; stamens 12; style longer than the ovary; beak of achene straight .... 1. *E. berteroi*

1. **Echinodorus berteroi** (Spreng.) Fassett var. **lanceolatus** (Wats. & Coult.) Fassett, Rhodora 57:144. 1955. Fig. 28.
Alisma rostratum Nutt. Trans. Am. Phil. Soc. 5:159. 1837.
Echinodorus rostratus (Nutt.) Engelm. Gray, Man. 460. 1848.
Echinodorus rostratus (Nutt.) Engelm. var. *lanceolatus* Engelm. ex Wats. & Coult. Gray, Man. 556. 1891.
Echinodorus cordifolius (L.) Griseb. var. *lanceolatus* (Wats. & Coult.) Mack. & Bush, Man. Fl. Jackson Co., Mo. 10. 1902.
Echinodorus cordifolius (L.) Griseb. f. *lanceolatus* (Wats. & Coult.) Fern. Rhodora 38:73. 1936.
Echinodorus rostratus (Nutt.) Engelm. f. *lanceolatus* (Wats. & Coult.) Fern. Rhodora 49:108. 1947.

Erect, rooted perennial; leaves basal, the blade broadly ovate, rarely lanceolate, obtuse, cordate or truncate at base, to 15 cm long, to 10 cm broad, glabrous, the petioles longer than the blade; scape usually more than 10 cm tall, with (1–) several clusters of 3–8 flowers; bracts 3–6 mm long; flowers 8–10 mm broad; sepals ovate, acute, green, persistent in fruit, 4–5 mm long; petals ovate to suborbiculate, acute, white, deciduous, 5–10 mm long; stamens 12; styles longer than the ovaries; achenes 2.5–3.5 mm long, with a straight beak about 1 mm long. July–September.

Ditches, sloughs, along rivers, around ponds and lakes, often in shallow, standing water.

IA, IL, IN, KS, KY, MO, NE, OH (OBL).

Tall burhead.

Superficially the leaves and flowers of this plant and of *E. cordifolius* look very much alike, but the two are easily distinguished because *E. berteroi* has an erect stem, 12 stamens, and a straight beak on each achene, while *E. cordifolius* has a creeping stem, 21 stamens, and an incurved beak on each achene.

2. **Echinodorus cordifolius** (L.) Griseb. Abh. Ges. Wiss. Gott. 7:257. 1857. Fig. 29.
Alisma cordifolia L. Sp. Pl. 1:343. 1753.
Sagittaria radicans Nutt. Trans. Am. Phil. Soc. 5:159. 1837.
Echinodorus radicans (Nutt.) Engelm. Gray, Man. 460. 1848.

28. *Echinodorus berteroi* var. *lanceolatus* (Tall burhead). a. Habit. b. Flower and buds. c. Achene.

Creeping or arching perennial; leaves upright, broadly ovate, obtuse, cordate, to 20 cm long, to 17 cm broad, glabrous, conspicuously cross-veined on the lower surface; scape prostrate, creeping, over 50 cm long, bearing many whorls of flowers; bracts 10–25 mm long; flowers more than 12 mm broad; sepals ovate to suborbicular, obtuse, green, persistent in fruit, 5–7 mm long; petals obovate, obtuse, white, deciduous, 6–12 mm long; stamens 21; styles shorter than the ovaries; achenes 1.8–2.2 mm long, ribbed, with an incurved beak about 0.5 mm long. July–September.

Marshes, swamps, ditches, sloughs, along rivers, around ponds and lakes, often in shallow, standing water.

IA, IL, IN, KS, KY, MO (OBL).

Creeping burhead.

This species is recognized by its creeping stems and its large, cordate, palmately veined leaves.

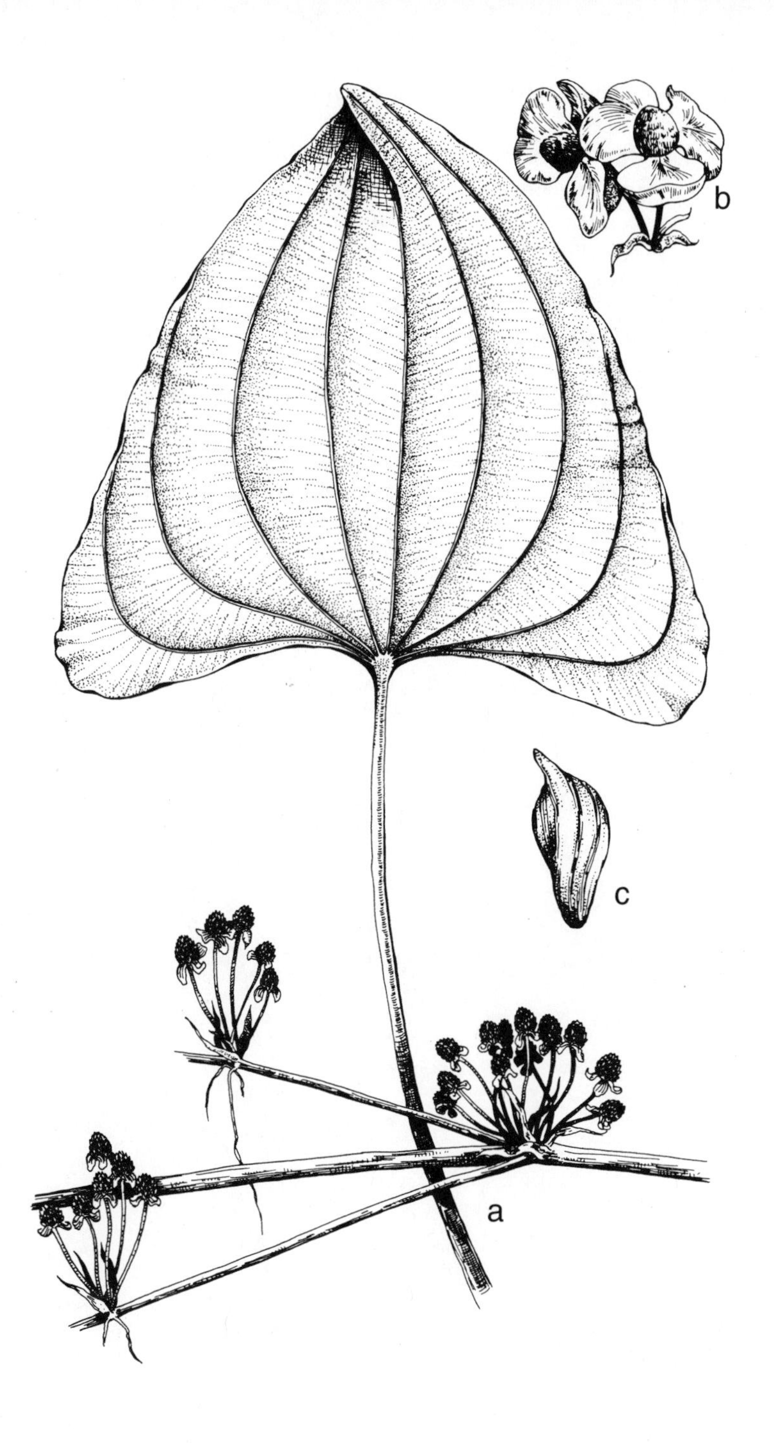

29. *Echinodorus cordifolius* (Creeping burhead).

a. Inflorescence and leaf.
b. Flowers.
c. Achene.

3. **Echinodorus tenellus** (Mart.) Buch. var. **parvulus** (Engelm.) Fassett, Rhodora 27:185. 1955. Fig. 30.
Alisma tenellum Mart. in Roem. & Schult. Syst. 7:1600. 1830.
Echinodorus parvulus Engelm. in Gray, Man. Bot. 438. 1856.
Echinodorus tenellus (Mart.) Buch. Abh. Naturw. Ver. Bremen 2:21. 1868.
Helianthium parvulus (Engelm.) Small, N. Am. Fl. 17:45. 1909.

Erect, rooted perennial, with creeping shoots frequently present; leaves basal, the blade linear to lanceolate, acute, tapering to base, to 3 cm long, glabrous, the petiole longer than the blade; scape to 10 cm tall, bearing a single whorl of 2–8 flowers; bracts 1–3 mm long; flowers at most only 6 mm across; sepals suborbicular, subacute, green, persistent in fruit, 1–2 mm long; petals suborbicular, subacute, white, deciduous, 1–3 mm long; stamens 6–9; styles shorter than the ovaries; achenes 1.0–1.5 mm long, ribbed, beakless or with a minute lateral beak 0.3 mm long, glabrous.

Around ponds, particularly sinkhole ponds.

IL, KS (OBL).

Dwarf burhead.

The U.S. Fish and Wildlife Service calls this plant *E. parvulus.* It is much less common than the preceding two species. The dwarf stature of this plant, the few number of stamens, and the narrow, grasslike leaves distinguish this plant from the others in the genus.

30. *Echinodorus tenellus* var. *parvulus* (Dwarf burhead). a. Habit. b. Flower. c. Achene.

3. **Sagittaria** L.—Arrowhead

Mostly aquatic, perennial herbs, with milky sap; leaves emersed, floating, or submersed, with sheathing petioles, the submersed leaves sometimes reduced to phyllodia; flowers whorled, bracteate, mostly unisexual, with the staminate flowers above the pistillate; sepals 3, greenish, separate, persistent, those of the staminate flowers reflexed, those of the pistillate flowers appressed, spreading, or reflexed; petals 3, white or rarely pinkish, separate, caducous; stamens in a tight spiral, mostly numerous; pistils numerous, distinct, 1-celled, spirally arranged on a dome-shaped receptacle, the ovaries superior; achenes flat, winged, beaked.

Lophotocarpus T. Dur. is included within *Sagittaria* since Bogin (1955) has pointed out that the usual distinguishing characteristics of *Lophotocarpus* of annuals with staminate flowers above and pistillate flowers below are not reliable.

Sagittaria is distinguished from *Echinodorus* by its flattened achenes and usual presence of separate staminate and pistillate flowers.

All species have achenes that are important food sources for waterfowl.

1. Pedicels of fruits recurved, rarely spreading; sepals of pistillate flowers mostly erect.
    2. Fruiting heads 1.2–2.0 cm in diameter; leaves hastate to sagittate .... 4. *S. calycina*
    2. Fruiting heads up to 1.2 cm in diameter; leaves linear-ovate to lance-elliptic .... 10. *S. platyphylla*
1. Pedicels of fruits spreading to ascending or absent; sepals of pistillate flowers mostly spreading to recurved.
    3. Filaments pubescent.
        4. Pistillate flowers sessile or nearly so .... 11. *S. rigida*
        4. Pistillate flowers pedicellate.
            5. Rhizomes present; stolons and corms absent .... 7. *S. graminea*
            5. Rhizomes absent; stolons and corms present .... 5. *S. cristata*
    3. Filaments glabrous.
        6. Emersed leaves linear to ovate, tapering or rounded at base, rarely sagittate.
            7. Leaves lanceolate to ovate; emersed plants with erect to ascending petioles .... 1. *S. ambigua*
            7. Leaves linear to occasionally sagittate; emersed plants with recurved petioles .... 6. *S. cuneata*
        6. Emersed leaves cordate, sagittate, or hastate.
            8. Bracts free or connate for less than 1/4 their length.
                9. Petioles winged in cross-section; beak of achene strongly recurved .... 2. *S. australis*
                9. Petioles ridged in cross-section; beak of achene ascending, not recurved .... 3. *S. brevirostra*
            8. Bracts connate for at least 1/4 their length.
                10. Basal lobes of leaves longer than the terminal lobe .... 9. *S. longiloba*
                10. Basal lobes of leaves shorter than the terminal lobe.
                    11. Beak of achene 1–2 mm long, horizontal .... 8. *S. latifolia*
                    11. Beak of achene less than 1 mm long, erect or curved .... 6. *S. cuneata*

1. **Sagittaria ambigua** J. G. Sm. N. Am. Sagittaria 22–23, t. 17. 1894. Fig. 31.

Plants monoecious or dioecious, erect, to 50 cm all, the flowering stem equaling or slightly shorter than the leaves; petiole slender, rounded, with expanded base; blade ovate to lanceolate, never sagittate, prominently nerved, acute to acuminate,

31. *Sagittaria ambigua* (Plains arrowhead). a. Habit. b. Achene.

tapering or slightly rounded at the base, to 22 cm long, to 7 cm broad; peduncle slender, erect, unbranched, the flowers in whorls of 3, the upper whorls staminate or pistillate, the lower 1–5 whorls pistillate; bracts 3, opposite the pedicel, usually free to the base, narrowly lanceolate, to 2 cm long; pedicel slender, terete, to 4 cm long; sepals 3, basally connate, short, green, separate, persistent, reflexed in fruit, minutely papillose, ovate-lanceolate, acute, 5–7 mm long, 1–3 mm broad; petals 3, free, white, a little longer than the sepals, caducous, narrowing to a slender claw at base, to 2 cm long, sometimes nearly as broad; stamens numerous, the filaments to 3 mm long, not dilated at base, glabrous, shorter than the anther; fruiting head globose, up to 8 mm in diameter when mature; achenes obovoid, 1.5–2.0 mm long, 1 mm broad, with thin narrow dorsal and ventral wings, the beak to 0.2 mm long, triangular and spreading at a right angle to the body of the fruit. June–August.

Ditches, borders of lakes and ponds, sloughs, often in standing water.

IL, IN, KS, MO (OBL).

Plains arrowhead.

This species is found in wetlands usually in areas of prairies. It is one of the species of *Sagittaria* that does not form sagittate leaves. It differs from *S. cuneata* specimens, which lack sagittate leaves, by the erect, rather than recurved, petioles.

2. **Sagittaria australis** (J. G. Sm.) Small, Fl. SE. U.S. 46. 1903. Fig. 32.
Sagittaria sagittifolia var. *longirostra* Micheli in DC. Monogr. Phan. 3:69. 1881, misapplied.
Sagittaria longirostra (Micheli) J. G. Sm. Mem. Torrey Club 5:26. 1894, misapplied.
Sagittaria longirostra (Micheli) J. G. Sm. var. *australis* J. G. Sm. Mohr, Bull. Torrey Club 24:20. 1897.
Sagittaria engelmanniana J. G. Sm. ssp. *longirostra* (Micheli) Bogin, Mem. N. Y. Bot. Gard. 9:223. 1955.

Plants monoecious or dioecious, erect, to 60 cm tall, the flower stalk usually a little longer than the leaves; petiole long, stout, sharply ridged; blade sagittate, prominently nerved, obtuse, to 13 cm long, to 10 cm broad, the basal lobes broad; peduncle stout, ridged, densely flowered, the flowers in whorls of 5–12, the upper flowers staminate or pistillate, the lower flowers usually all pistillate; bracts opposite the pedicels, free or basally connate, firm, arched-ascending, broadly lanceolate, acute, to 2.5 cm long; pedicels rather long, slender, terete, to 2.5 cm long; pedicels rather long, slender, terete, to 2.5 cm long; sepals free, short, persistent, and usually reflexed after flowering, to 1.5 cm long, nearly as broad; petals free, a little longer than the sepals, with a short claw; stamens numerous, the filaments to 5 mm long, slender, glabrous; fruiting head about 1.5 cm thick, echinate; achenes cuneate-obovoid, 2.3–3.2 mm long, 1.6–2.3 mm broad, with a ventral narrow wing and a broad dorsal wing, with 1–2 facial wings, the beak oblique, strongly recurved, 1.0–1.5 mm long. July–September.

Low, springy woods, occasionally in standing water.

IA, IL, IN, KY, MO, OH (OBL).

Appalachian arrowhead.

The very sharply winged or ridged petiole is very distinctive for this species.

32. *Sagittaria australis* (Appalachian arrowhead). a. Leaves and inflorescence. b. Achene.

3. **Sagittaria brevirostra** Mack. & Bush, Mo. Bot. Gard. Rep. 16:102. 1905. Fig. 33.
Sagittaria engelmanniana J. G. Sm. ssp. *brevirostra* (Mack. & Bush) Bogin, Mem. N. Y. Bot. Gard. 9 (2):244. 1955.

Plants monoecious or dioecious, erect, 40–70 cm tall, the flowering stem equaling or slightly exceeding the leaves, robust in appearance; petiole long, stout, ridged, with expanded base; blade sagittate, hastate, or very rarely unlobed, prominently nerved, the blade usually tapering to the acute apex, with median constrictions along each margin opposite the attachment of the petiole, 10–22 cm long, 3–12 cm broad, basal lobes lanceolate to ovate-lanceolate, tapering to the acute, bristle-tipped apex, equaling or slightly exceeding the blade in length, 9.5–22.0 cm long, 2.5–7.5 cm broad; peduncle stout, ridged, densely flowered, the flowers in whorls of 3 or 4, the lower whorls occasionally replaced by branches, the upper flowers either staminate or pistillate, the lower flowers entirely pistillate; bracts as many as and opposite the pedicels, free or slightly connate basally, firm, prominently nerved with a hyaline margin, equaling or exceeding the immature flower head, arched-ascending with reflexed tips, lanceolate to ovate-lanceolate, long-attenuate or acute, 1.5–4.0 cm long, 0.5–1.0 cm broad; pedicels long, slender, terete, those of the staminate flowers being slightly longer than those of the pistillate, the staminate 1.5–3.0 cm long, the pistillate 1–2 cm long; sepals free, short, with hyaline margins, persistent and usually reflexed after flowering, ovate-lanceolate with acute or obtuse apex, 0.6–1.5 cm long, 0.6–1.5 cm broad; petals free, exceeding the sepals, orbicular with a short claw, obtuse, 1–2 cm long, 1.0–1.5 cm broad; stamens numerous, the filaments to 5 mm long, slender, glabrous, equaling the anthers; fruiting head globose, 2–3 cm in diameter when mature; achenes cuneate-ovoid, 2.5–3.0 mm long, 1–2 mm broad, with a ventral narrow wing and a broad, crested dorsal wing, with 1–3 facial wings with one usually well developed, the broad-based beak usually oblique, rarely erect, recurved at apex, 0.5–1.5 mm long. July–September.

Ditches, sloughs, shorelines, usually in standing water.

IA, IL, IN, KS, KY, MO, NE, OH (OBL).

Midwestern arrowhead.

The nearly straight beak on the achene distinguishes this species from the similar *S. australis* and *S. latifolia*. *Sagittaria cuneata*, which also may have a nearly straight beak and sagittate leaves, has its bracts connate for more than one-fourth of their length. The similar *S. longiloba* of Kansas and Nebraska has the basal lobes of the leaves much longer than the terminal lobe.

4. **Sagittaria calycina** Engelm. in Torrey, Bot. Mex. Bound. Surv. 212. 1859. Fig. 34.
Sagittaria calycina Engelm. var. *maxima* Engelm. in Torrey, Bot. Mex. Bound. Surv. 212. 1859.
Lophotocarpus calycinus (Engelm) J. G. Sm. Mem. Torrey Club 5:25. 1894.
Lophotocarpus depauperatus J. G. Sm. Ann. Rep. Mo. Bot. Gard. 11:148. 1899.
Lophotocarpus calycinus (Engelm.) J. G. Sm. f. *maximus* (Engelm.) Fern. Rhodora 38:73. 1936.
Sagittaria montevidensis ssp. *calycina* (Engelm.) Bogin, Mem. N. Y. Bot. Gard. 9:197. 1955.

33. *Sagittaria brevirostra* **(Midwestern arrowhead).**
a. Leaves and inflorescence.
b. Inflorescence.
c. Staminate flower.
d. Fruiting head.
e. Achene.

34. ***Sagittaria calycina*** **(Mississippi arrowhead).**
a. Leaf.
b. Pistillate inflorescence.
c. Staminate inflorescence.
d. Fruiting head.
e. Achene.
f. Depauperate plant.

Annual or perennial, monoecious or dioecious, erect or lax and sprawling, the flowering stem 5 cm to 2 m tall, equaling or shorter than the leaves, drooping; emersed leaves sagittate, the floating leaves ovate, the submersed leaves reduced to linear phyllodia; petiole long, usually rather smooth with expanded base; emersed blade sagittate, prominently nerved, deltoid or deltoid-orbicular with more or less rounded apex, 1–10 cm long, 1–16 cm broad, basal lobes ovate-lanceolate, acute or acuminate, shorter than the length of the blade, 1–12 cm long, 1.0–7.5 cm broad; floating blades ovate, to 2.5 cm long, the submersed leaves reduced to phyllodia; scape with 3–12 whorls of flowers; peduncle inflated, smooth, with few flowers; flowers in whorls of 3, the upper either staminate or pistillate, the lower usually pistillate; bracts 3, fused basally to form a sheath around the stem, not prominently nerved, small and hyaline, ovate, acute, to 1 cm long, about 0.5 cm broad; pedicels inflated, terete, to 6 cm long, the lowermost usually becoming reflexed; sepals free, short, persistent, erect and closed, appressed to fruiting head, ovate to ovate-elliptic with obtuse apex, to 1.3 cm long, to 1.7 cm broad; petals free, equaling or slightly shorter than the sepals, ovate, obtuse, 0.7–1.5 cm long, 0.5–1.5 cm broad; stamens numerous, the filaments to 3 mm long, narrowly winged at margin, pubescent, equaling the anthers; fruiting heads globose, up to 2 cm in diameter when mature, very smooth in appearance due to horizontal beak of achenes; achenes narrowly cuneate-ovoid, to 3 mm long, to 2.5 mm broad, narrowly winged on both margins, the beak horizontal, straight, about 1 mm long. June–September.

Marshes, pond margins, shorelines, ditches, sloughs, often in standing water.

IA, IL, IN, KS, KY, MO, NE, OH (OBL).

Mississippi arrowhead.

This species is identified rather easily by the greatly enlarged peduncle and pedicels, the erect sepals, which are tightly appressed to the fruiting head, and the leaves, which have a wrinkled appearance. As in most of the Sagittarias, the achene is an important characteristic in identification and should be mature if used as a distinguishing feature. If immature, the achene can readily be confused with that of *S. latifolia*, but *S. calycina* may be separated by its bracts, peduncle, pedicel, and usually more rounded apex of the leaves.

Both gigantic (to 2 m tall) and dwarfed (to 5 cm tall) specimens occur.

5. **Sagittaria cristata** Engelm. Proc. Davenp. Acad. Nat. Sci. 4:29. 1883. Fig. 35.
Sagittaria graminea Michx. var. *cristata* (Engelm.) Bogin, Mem. N. Y. Bot. Gard. 9 (2):210. 1955.

Perennials with rhizomes but without stolons and corms; plants monoecious or dioecious, erect, 5–50 cm tall, the flower stalk equaling or slightly shorter than the leaves, rather slender in appearance; petiole long, slender, minutely ridged, with expanded base; blade either entire, hastate, or represented by bladeless phyllodia, prominently nerved, the unlobed blade linear-lanceolate, tapering to the acute apex, 2–17 cm long, 3–5 cm broad, the hastate blade with lobes usually unequal in both length and width, up to 2.5 cm long, about 2 cm broad, the phyllodia linear-lanceolate, tapering to the acuminate tip, to 19 cm long, to 2 cm broad; peduncle slender, minutely ridged, sparsely flowered, the flowers in whorls of 3, the upper

35. *Sagittaria cristata* (Crested arrowhead).

a. Habit of staminate plant.
b. Habit of pistillate plant.
c. Achene.

flowers either staminate or pistillate, the lower flowers pistillate; bracts 3, opposite the pedicel, basally connate to fused one-half their entire length, weak, prominently nerved, hyaline in appearance, inconspicuous, ovate, obtuse, 2 mm long, 1–2 mm broad; pedicels long, slender, terete, with staminate and pistillate being the same length, 1–3 cm long; sepals basally connate, short, with a hyaline margin, persistent, reflexed in fruit, ovate-lanceolate with obtuse apex, 3–6 mm long, 1–3 mm broad; petals free, greatly exceeding the sepals, ovate, narrowing to a slender claw at the base, obtuse, 1–2 cm long, 1–2 cm broad; stamens numerous, the filaments up to 3 mm long, inflated, scaly, shorter than the anther; fruiting heads globose, small, 4–8 mm in diameter when mature, smooth in appearance; achenes narrowly ovoid, 2 mm long, 1 mm broad, with both ventral and dorsal wings well developed, the dorsal one strongly scalloped, with 1–3 facial wings with one well developed, the beak usually oblique to horizontal and borne below the summit of the achene, 0.4–0.7 mm long. June–September.

Borders of lakes and ponds.

IA, IL, MO, NE (OBL).

Crested arrowhead.

This species of the northern Midwest is distinguished from the similar *S. graminea* by the longer beak of the achene and the scalloped dorsal ridge on the achene. Submersed leaves are very stiff in *S. cristata*, while they are flaccid in *S. graminea*.

6. **Sagittaria cuneata** Sheldon, Bull. Torrey Club 20:283. 1893. Fig. 36.
Sagittaria arifolia Nutt. ex J. G. Sm. Ann. Rep Mo. Bot. Gard. 6:32. 1894.

Plants monoecious or dioecious, erect, to 70 cm tall, the flowering stalk usually exceeding the leaves, slender; petiole long, slender, generally smooth, with a slightly expanded base; blade sagittate, prominently nerved, broadly rounded or slightly tapering to the acute, bristle-tipped apex, with median constrictions along each margin opposite the attachment of the petiole, 6–10 cm long, 1.0–4.5 cm broad, basal lobes broadly lanceolate, tapering to the acuminate apex, equaling or slightly exceeding the blade in length; peduncle slender, smooth; inflorescence sparsely flowered, the flowers in whorls of 3, unbranched, the upper predominantly staminate, the lower predominantly pistillate, occasionally both staminate and pistillate at the same node; bracts 3, opposite the pedicel, basally connate, firm, prominently nerved with a subhyaline margin, usually shorter than the immature flower head, erect, narrowly lanceolate, acute to acuminate, to 2 cm long, to 5 mm broad; pedicels long, slender, terete, with the staminate slightly longer than the pistillate, the staminate up to 2 cm long, the pistillate up to 1 cm long; sepals free, short, persistent and usually reflexed after flowering, ovate with obtuse apex, 8 mm long, 5 mm broad; petals free, equaling or exceeding the sepals, ovate and tapering to a slender claw, obtuse, 8–10 mm long, 5–8 mm broad; stamens numerous, the filaments to 5 mm long, slender, glabrous, 1 mm longer than the anthers; fruiting head globose, to 1.5 cm in diameter when mature; achenes ovoid, up to 2.8 mm long, up to 2 mm broad, with small ventral and lateral wings, with a small facial wing, the erect to curved terminal beak up to 0.4 mm long. June–September.

Sloughs, ditches, along streams.

IA, IL, IN, KS, NE, OH (OBL).

Wedge-leaved arrowhead.

This variable species may have sagittate leaves or unlobed leaves and a beak on the achene that may be straight or curved.

7. **Sagittaria graminea** Michx. Fl. Bor. Am. 2:190. 1803. Fig. 37.
Sagittaria acutifolia Pursh, Fl. Am. Sept. 2:397. 1814.

Plants with rhizomes, but without stolons and corms, monoecious or dioecious, erect, 5–50 cm tall, the flowering stalk equaling or slightly shorter than the leaves, rather slender in appearance; petiole long, slender, minutely ridged, with expanded base; blade either entire, hastate, or represented by bladeless phyllodia, prominently nerved, the unlobed blade linear-lanceolate, tapering to the acute apex, 2–17 cm long, 3–5 cm broad, the hastate blade with lobes usually unequal in both length and width, up to 2.5 cm long, 2 cm broad, the phyllodia linear-lanceolate, tapering to the acuminate tip, to 19 cm long, to 2 cm broad; peduncle slender, minutely ridged, sparsely flowered, the flowers in whorls of 3, the upper flowers either staminate or pistillate, the lower flowers pistillate; bracts 3, opposite the pedicel, basally connate to fused one-half their entire length, weak, prominently nerved, hyaline in appearance, inconspicuous, ovate, obtuse, 2 mm long, 1–2 mm broad; pedicels long, slender, terete, with staminate and pistillate being the same length, 1–3 cm long; sepals basally connate, short, with a hyaline margin, persistent, reflexed in fruit, ovate-lanceolate with obtuse apex, 3–6 mm long, 1–3 mm broad; petals free, greatly exceeding the sepals, ovate, narrowing to a slender claw at base, obtuse, 1–2 cm long, 1–2 cm broad; stamens numerous, the filaments to 3 mm long, inflated, scaly, shorter than the anthers; fruiting head globose, small, 4–8 mm in diameter when mature, smooth in appearance; achenes narrowly ovoid, 2 mm long, 1 mm broad, with both ventral and dorsal wings well developed, 1–3 facial wings with one well developed, the beak usually oblique to horizontal and borne below the summit of the achene, short, less than 0.3 mm long. June–September.

Swamps, ditches, sloughs, often in standing water.

IA, IL, IN, KS, KY, MO, NE, OH (OBL).

Grass-leaved arrowhead.

This species is one of three in the central Midwest with pubescent filaments. Of the other two, *S. rigida* has sessile pistillate flowers, and *S. cristata* lacks rhizomes but has stolons and corms. *Sagittaria graminea* also differs from *S. cristata* by its flaccid rather than stiff submersed leaves.

8. **Sagittaria latifolia** Willd. Sp. Pl. 4:409. 1806. Fig. 38.
Sagittaria simplex Pursh, Fl. Am. Sept. 2:397. 1814.
Sagittaria variabilis Engelm. Gray, Man. Bot. 461. 1848.
Sagittaria variabilis Engelm. var. *diversifolia* Engelm. Gray, Man. Bot. 439. 1856.
Sagittaria latifolia Willd. f. *diversifolia* (Engelm.) B. L. Robins. Rhodora 10:31. 1908.

Plants monoecious or dioecious, erect, 40–83 cm tall, the flowering stalk equaling or slightly exceeding the leaves, robust in appearance; petiole long, stout, ridged, with expanded base; blade hastate or sagittate, prominently nerved, usually acute at the apex, with median constrictions along each margin opposite the attachment of the petiole, 6–20 cm long, 5–19 cm broad; basal lobes broadly lanceolate, tapering

36. *Sagittaria cuneata* (Wedge-leaved arrowhead). a. Leaves and inflorescence. b. Fruiting heads. c. Achene.

37. ***Sagittaria graminea*** **(Grass-leaved arrowhead).**
a. Habit (left).
b. Habit (right).
c. Leaf.
d. Pistillate flower.
e. Achene.

38. ***Sagittaria latifolia*** **(Common arrowhead).**
a. Leaf.
b. Leaves of narrow-leaved form.
c. Inflorescence.
d. Staminate flower.
e. Achene.

to the acute or obtuse, bristle-tipped apex, usually equaling or rarely exceeding the blade in length, 5–23 cm long, 2–11 cm broad; peduncle stout, smooth, sparsely flowered, the flowers in whorls of 3, rarely branched at the first node, the entire stalk staminate, pistillate, or both; bracts 3, opposite the pedicels, free or slightly connate basally, thin, prominently nerved with hyaline margins, much shorter than the flower head, erect, ovate-lanceolate, obtuse, 1 cm or less long, 1 cm or less broad; pedicels long, slender, terete, with those of the staminate and pistillate flowers about equal in length, to 5 cm long; sepals free, short, with hyaline margins, persistent and reflexed after flowering, ovate, the apex obtuse or shallowly bilobed, 0.5–1.5 cm long, 0.5–1.0 cm broad; petals free, exceeding the sepals, orbicular with a short claw, obtuse, 1–2 cm long, 1–2 cm broad; stamens numerous, the filaments to 5 mm long, slightly inflated, glabrous, longer than the anthers; fruiting head globose, 2–3 cm in diameter when mature; achenes obovoid, 2–5–3.0 mm long, 3–4 mm broad, with a narrow ventral wing and a broad dorsal wing forming a low crest at apex of achene, without facial wings, with the beak of the achene horizontal and straight, to 2 mm long. June–October.

Swamps, ditches, sloughs, ponds, shorelines, often in standing water.

IA, IL, IN, KS, KY, MO, NE, OH (OBL).

Common arrowhead; broad-leaved arrowhead.

This is the most common as well as the most widespread species of *Sagittaria* in the central Midwest. Its most distinguishing feature is its horizontal beak on the achene. The beak is straight and 1–2 mm long. Leaves with extremely narrow blades and lobes are found throughout the Midwest.

9. **Sagittaria longiloba** Engelm. ex. J. G. Sm. N. Am. Sagittaria 16–17, t. 11. 1894. Fig. 39.

Rhizomatous plants with corms, monoecious or dioecious; flowering stems erect, to 1.5 cm tall, smooth; leaves erect or more commonly spreading, sagittate, 12–22 cm long, the basal lobes linear to lanceolate, to 22 cm long, 0.5–2.0 cm broad, distinctly longer than the terminal lobe, glabrous; flowering stems simple or with no more than one branch, glabrous; bracts connate at the base, ovate-lanceolate, attenuate at the apex, 0.8–1.3 cm long, nearly as broad, glabrous; pedicels up to 3 cm long; lower whorls of flowers pistillate, the upper whorls staminate; sepals free, 4–7 mm long, reflexed; petals 8–15 mm long, free; stamens numerous, the filaments glabrous, longer than the anthers; fruiting heads globose, 1.0–1.2 cm in diameter when mature; achenes obovate, with a prominent dorsal wing and a much narrower ventral wing, the facial wings absent or very small, the beak deltoid, short, attached to one side of the middle of the top of the achene, or the beak sometimes nearly absent. July–August.

Swamps, ditches, usually in shallow water.

KS, NE (OBL).

Long-lobed arrowhead.

This species is readily identified by its sagittate lobes of the leaf that are longer than the terminal lobe.

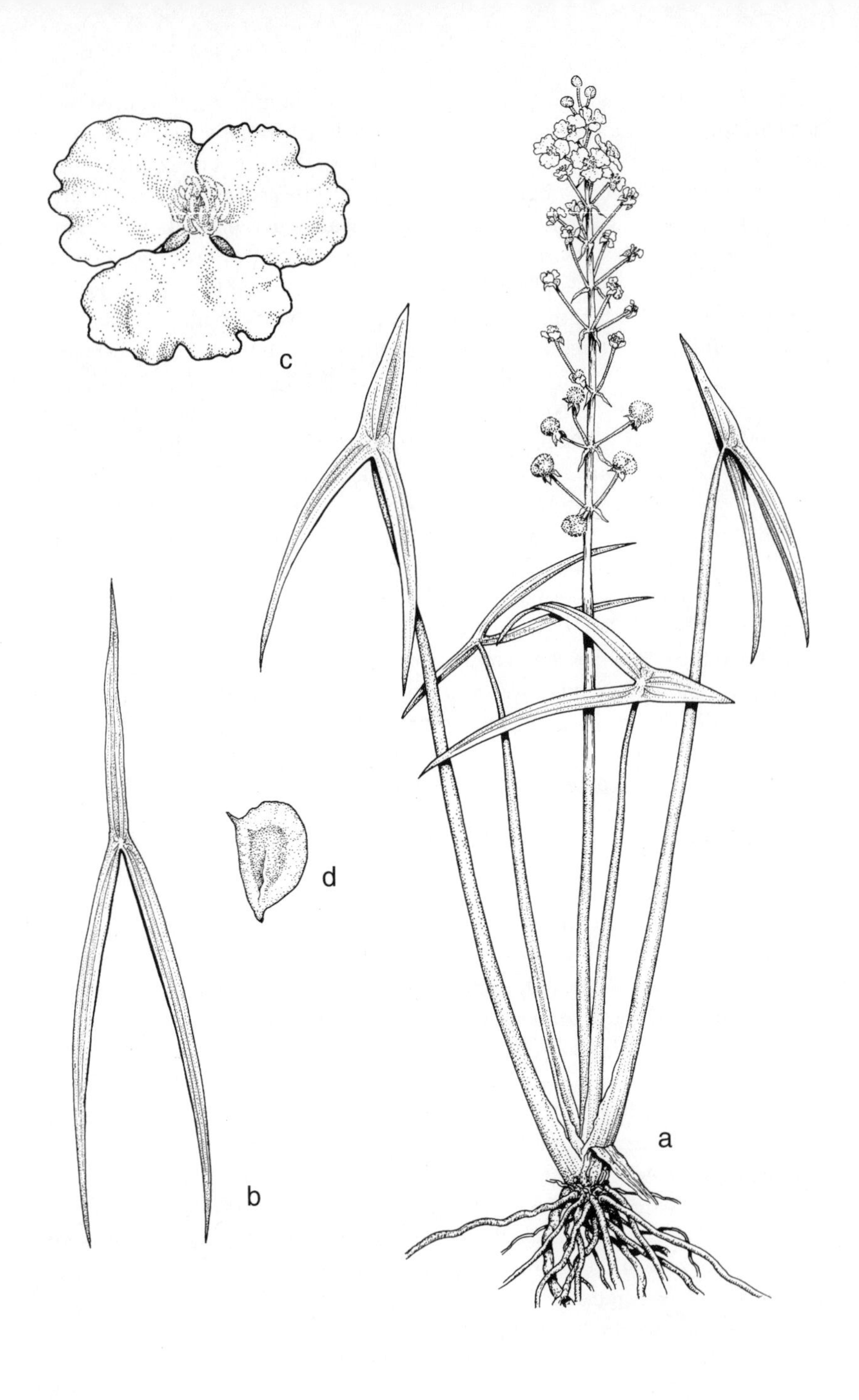

39. *Sagittaria longiloba* **(Long-lobed arrowhead).** a. Habit. b. Leaf. c. Flower. d. Achene.

10. **Sagittaria platyphylla** (Engelm.) J. G. Sm. N. Am. Sagittaria 29. 1894. Fig. 40.
Sagittaria graminea Michx. var. *platyphylla* Engelm. Man. Bot. N. U.S., ed. 5, 494. 1867.

Rhizomatous perennial, monoecious or dioecious; flowering stems to 75 cm tall, glabrous; leaves 10–70 cm long, glabrous, the petioles terete to flattened, the submerged leaves linear or absent, the emergent leaves lanceolate to ovate, occasionally with two basal lobes much shorter than the terminal lobe; flowering stems unbranched, erect; bracts connate about one-half their length, the apex obtuse, 3–7 mm long, about as broad; lower 1–3 whorls of flowers reflexed, pistillate, the pedicels thickened, glabrous, 15–30 mm long, the upper whorls of flowers staminate; sepals short, free, reflexed in fruit, 3–5 mm long; petals free, 5–10 mm long; stamens numerous, the filaments slender but somewhat swollen at the base, papillose, shorter than the anthers; fruiting heads globose, 0.9–1.2 cm in diameter when mature; achenes obovate, 1.5–2.0 mm long, glabrous, the beak curved, slender, 0.3–0.6 mm long. June–September.

Sloughs, ditches, around ponds, sometimes in standing water.

IL, KY, MO, OH (OBL).

Flat-leaved arrowhead.

Only *S. platyphylla* and *S. calycina* in the central Midwest have the pedicels of the fruits recurved and the sepals of the pistillate flowers mostly erect. *Sagittaria platyphylla* differs from *S. calycina* by its smaller fruiting heads up to 1.2 cm across and its nonsagittate leaves.

11. **Sagittaria rigida** Pursh, Fl. Am. Sept. 397. 1814. Fig. 41.
Sagittaria heterophylla Pursh, Fl. Am. Sept. 396. 1814, non Schreb. (1811).
Sagittaria heterophylla Pursh var. *rigida* (Pursh) Engelm. Gray, Man. Bot. 439. 1856.

Plants monoecious or dioecious, erect, 15–80 cm tall, the flowering stalk equaling or slightly shorter than the leaves, rather slender; petiole long, weak, slightly ridged, with an expanded base; blades ovate-elliptic, lanceolate, prominently nerved, acute, 2–16 cm long, 3–9 cm broad, or blade hastate, prominently nerved, acute, 5–15 cm long, 3–9 cm broad, the lobes either 1 or 2, if 1, poorly developed, unequal in length and width, 3–8 cm long, 0.5–1.5 cm broad; peduncle weak, slightly ridged, few-flowered; flowers in whorls of 2–8, the upper either staminate or pistillate, the lower pistillate; bracts opposite the pedicels, fused about three-fourths their entire length, weak, prominently nerved with a hyaline margin, rather inconspicuous, ovate, obtuse, 2–6 mm long, including the lobes up to 2 mm long, to 5 mm broad; pedicels of staminate flower long, slender, terete; pistillate flowers sessile; sepals free, short, with hyaline margins, persistent and reflexed after flowering, ovate-lanceolate with rounded or obtuse apex, 0.4–1.0 cm long, to 0.5 cm broad; petals free, greatly exceeding the sepals, ovate, narrowing to a slender claw at base, obtuse to acute, 1–2 cm long, nearly as broad; stamens numerous, the filaments to 4 mm long, inflated, densely pubescent or scaly, exceeding the length of the broad anther; fruiting head globose, 2–3 cm in diameter when mature, prickly in appearance due to the erect curved beaks of the achenes; achenes oblongoid, 2.5–7.0 mm long, 2.5–4.0 mm broad, with a well-developed crested dorsal wing, a small narrow ventral wing, usually

40. *Sagittaria platyphylla* (Flat-leaved arrowhead). Habit.

41. ***Sagittaria rigida*** **(Stiff arrowhead).**

a. Habit (shaded).
b. Leaves.
c. Inflorescence.
d. Achene.

one small facial keel, the beak of the achene usually terminal, erect, recurved, 1.0–1.5 mm long. May–October.

Swamps, margins of ponds, along waterways, sloughs, sinkhole ponds.

IA, IL, IN, KS, KY, MO, NE, OH (OBL).

Stiff arrowhead.

The distinguishing feature of this species is its sessile pistillate flowers.

14. ARACEAE—ARUM FAMILY

Herbaceous perennials from corms, rhizomes, or thick, fleshy roots; leaves sometimes net-veined, simple or compound; flowers unisexual or perfect, attached to a fleshy spadix; perianth parts 0, 4, or 6; stamens 2–6; staminodia sometimes present; ovary 1- to 3-celled, with 1–several ovules per cell; fruit dry, or fleshy and berrylike, 1- to 3-seeded.

Many genera of this primarily tropical family are cultivated as houseplants. Included among these are *Philodendron, Monstera, Aglaeonema, Pothos, Arum, Anthurium,* and *Dieffenbachia.*

This family consists of 110 genera and approximately three thousand species found throughout the world, but particularly abundant in the tropics. Three genera and three species may be emergent aquatics in the central Midwest.

1. Flowers perfect; spadix short-cylindric to globose; plants with rhizomes; leaves not arrowhead-shaped.
   2. Spadix short-cylindric; stamens 6; spathe open, white; berries red; perianth absent; plants not foul-smelling .... 1. *Calla*
   2. Spadix globose; stamens 4; spathe with inrolled margins, green with purple spots and stripes; fruit fleshy; perianth present; plants foul-smelling .... 3. *Symplocarpus*
1. Flowers unisexual; spadix elongated; plants with fleshy roots; leaves arrowhead-shaped .... 2. *Peltandra*

1. **Calla** L.—Water Arum

Characters of the species. This is the only species in the genus.

1. **Calla palustris** L. Sp. Pl. 2:968. 1753. Fig. 42.

Perennial with elongated rhizomes; leaves basal, ovate to nearly orbicular, short-acuminate at the apex, cordate at the base, entire, glabrous, 5–10 cm long, often nearly as broad, with curved-ascending, parallel veins; petioles rather stout, up to 20 cm long; spathe open, white, not concealing the spadix, ovate to elliptic, up to 6 cm long, narrowed to a linear, involute tip up to 10 mm long; spadix short-cylindric, 1.5–2.5 cm long, with a thickened stipe; flowers perfect; perianth absent; stamens 6, with flat, narrow filaments; berries red, 8–12 mm in diameter, with few seeds covered by a gelatinous material. June–August.

Swamps, bogs, lakes, usually in shallow water.

IL, IN, OH (OBL).

Wild calla.

The white spathes make this one of the most attractive native wetland plants in North America. The leaves are much more cordate than the leaves of *Alisma* species. *Caltha palustris,* the marsh marigold, has cordate leaves with teeth and net venation and yellow flowers.

42. *Calla palustris* (Wild calla). Habit.

2. **Peltandra** Raf.—Arrow Arum

Perennials with fleshy roots; leaves basal, sagittate; spathes and spadices elongated; flowers unisexual; perianth absent; stamens 4–5, embedded in the spadix; staminodia present; fruit a berry.

Two species, both in eastern North America, comprise the genus. Only the following occurs in the central Midwest.

1. **Peltandra virginica** (L.) Kunth, Enum. 3:43. 1841. Fig. 43.
Arum virginicum L. Sp. Pl. 2:966. 1753.
Peltandra undulata Raf. Journ. Phys. 89:103. 1819.

Perennial, with thick, fleshy roots; leaves basal, sagittate to hastate, acute at the apex, up to 30 cm long, up to 18 cm broad, expanding after flowering, pinnately veined and forming a submarginal nerve; petiole rather stout, up to 45 cm long;

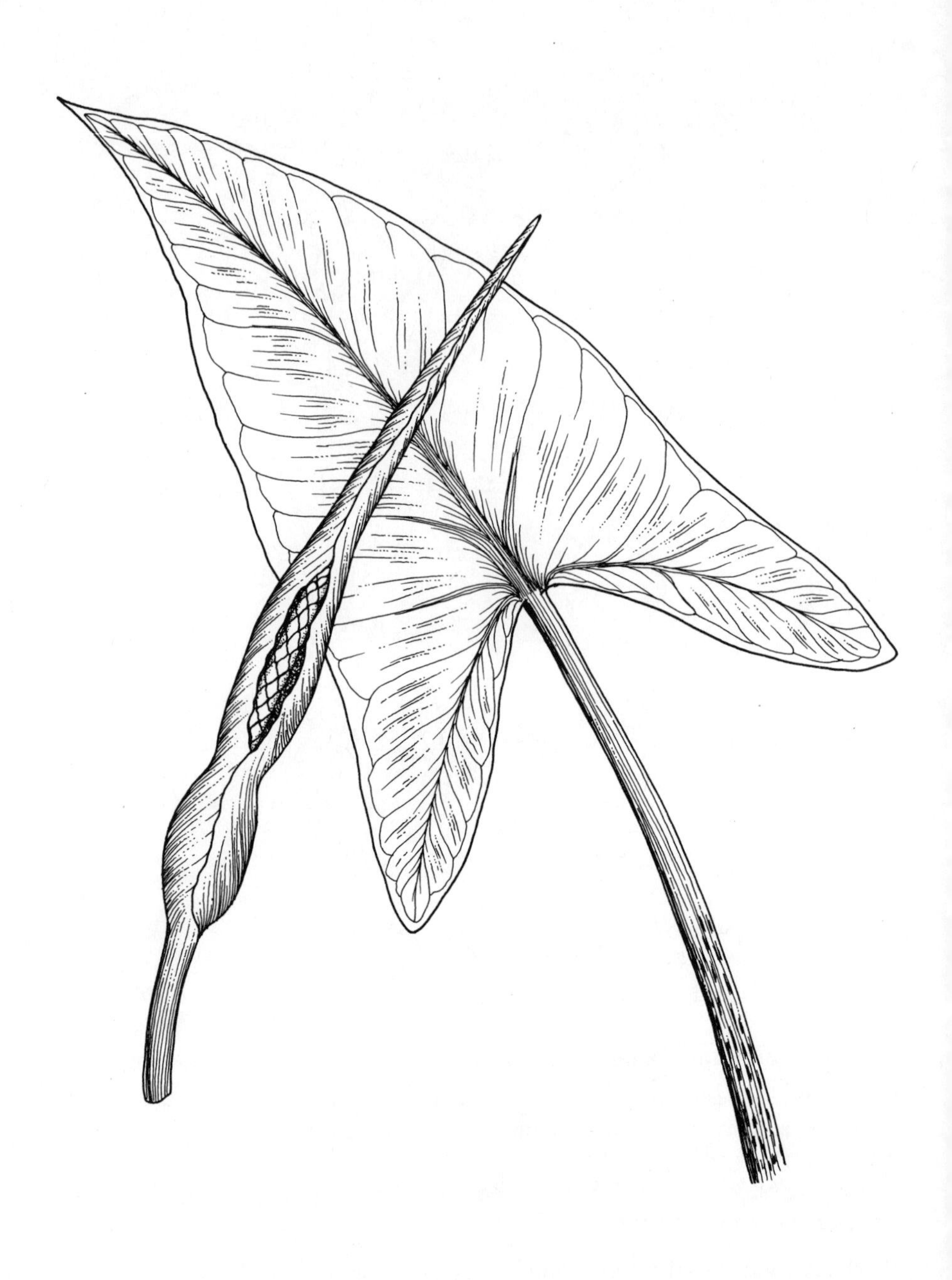

43. ***Peltandra virginica*** **(Arrow arum).** Leaf (right). Spathe and spadix (left).

spathes elongated, 10–20 cm long, convolute at both ends but open at the middle to expose the staminate flowers, green with a pale margin; spadix elongated, with a slender stipe, almost entirely covered by the flowers, the staminate flowers apical, soon falling away with the upper part of the spadix, the pistillate flowers basal; perianth absent; stamens 4–5, embedded in the spadix; berries greenish to brown, to 12 mm long, nearly as broad. May–July.

Swamps, ditches, sloughs, edge of lakes.

IA, IL, IN, KS, KY, MO, OH (OBL).

Arrow arum.

The stalks of the fruits bend downward so that the mature fruits develop beneath the surface of water, or on the ground if the plants are out of the water. The leaves, which resemble those of some species of *Sagittaria*, have a single large vein extending into each basal lobe instead of several veins that *Sagittaria* leaves have.

3. **Symplocarpus** Salisb.—Skunk Cabbage

Characters of the species. This is the only species in the genus.

1. **Symplocarpus foetidus** (L.) Nutt. Gen. 1:105. 1818. Fig. 44.
Dracontium foetidum L. Sp. Pl. 2:967. 1753.
Spathyema foetida (L.) Raf. Med. Repos. II 5:352. 1808.

Foul-smelling perennial from stout, more or less erect rhizomes; leaves appearing after the flowers, numerous, basal, ovate, obtuse at the apex, cordate at the base, net-veined, to 45 (–50) cm long, to 35 (–40) cm broad, the petiole rather stout, much shorter than the blade; spathe fleshy, ovoid, nearly completely enclosing the spadix, the margins inrolled, to 15 cm long, green with purple spots and stripes; spadix globose; flowers perfect; perianth parts 4, becoming fleshy, persistent on the fruit and strongly foul-smelling; stamens 4; ovary 1-celled, buried in the spadix; fruit fleshy, 8–12 cm in diameter, spongy, foul-smelling, the seeds embedded, 8–10 mm in diameter. February–March.

Swamps, wet woods, fens.

IA, IL, IN, KY, MO, OH (OBL).

Skunk cabbage.

This remarkable species is one of the first to flower in the central Midwest, often blooming while there is still snow on the ground. All parts of the plant have a foul-smelling odor, particularly the fleshy perianth and the spadix. After the plant flowers, the leaves emerge and often become huge, with each plant forming giant rosettes.

15. BUTOMACEAE—FLOWERING RUSH FAMILY

Only the following genus is in the Butomaceae.

1. **Butomus** L.—Flowering Rush

Perennials from rhizomes; leaves basal; inflorescence umbellate, many-flowered, bracteate; flowers perfect; perianth composed of 2 distinct series of 3 members each; stamens 9; ovaries 6, slightly coherent at the base; follicles many-seeded.

There is only one species of *Butomus*, native to Europe and sparingly established in the eastern and central United States.

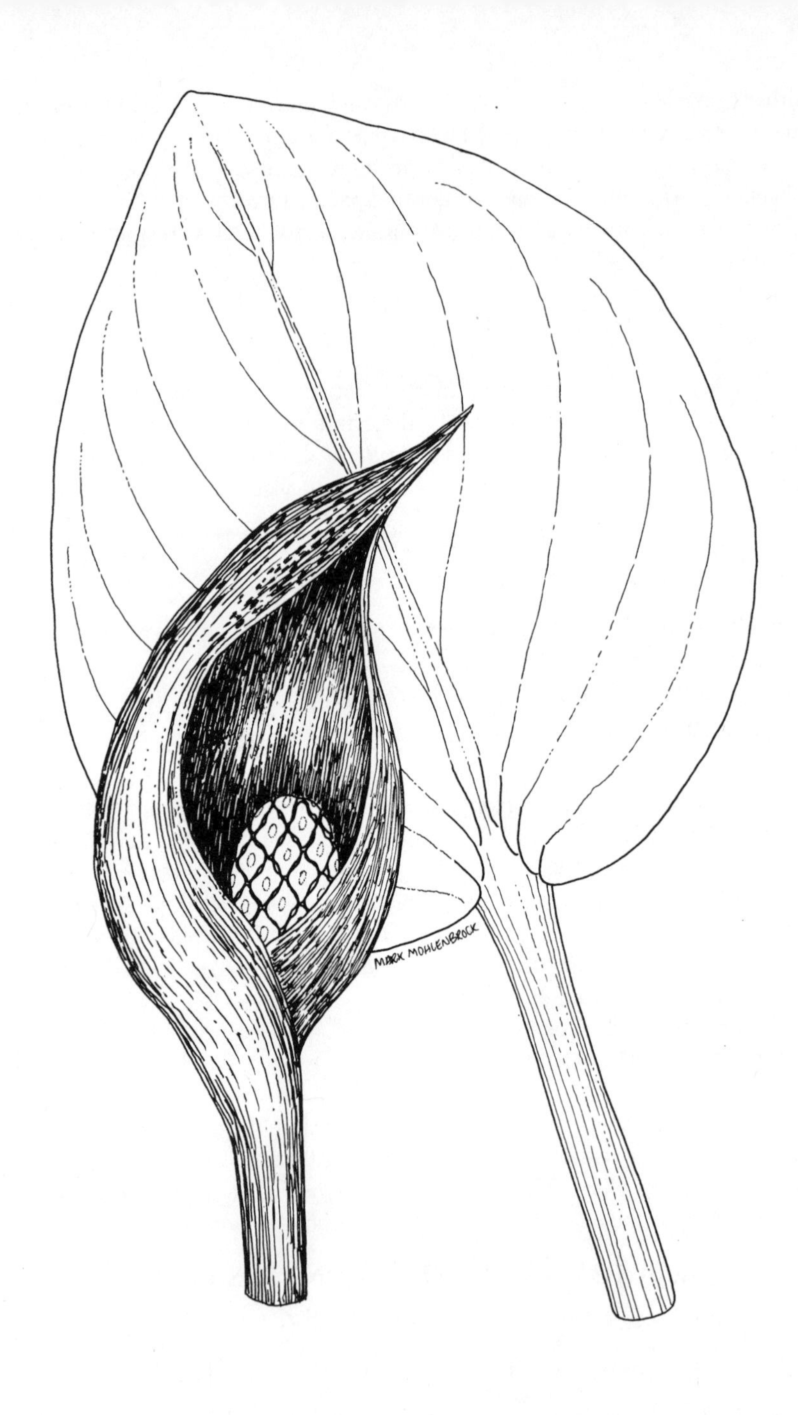

44. ***Symplocarpus foetidus*** **(Skunk cabbage).** Spathe and spadix (left). Leaf (right).

1. **Butomus umbellatus** L. Sp. Pl. 1:372. 1753. Fig. 45.

Perennial from a stocky rhizome; leaves basal, to 1 m long, 5–10 mm broad, glabrous; scape usually 1 m long, bearing a terminal umbel; umbel many-flowered; bracts 3, broadly lanceolate, acute, purplish; flowers pinkish, at least 18 mm across; pedicels arched-ascending, to 10 mm long; sepals 3, pinkish or greenish, persistent; petals 3, pinkish, persistent; stamens 9; follicles to 1 cm long, long-beaked. July–August.

Shallow water, wet ditches.

IL, OH (OBL), KS (not listed by the U.S. Fish and Wildlife Service).

Flowering rush.

This native of Europe is established in a few areas of the eastern and north-central United States. The showy, umbellate inflorescence gives this plant the appearance of an onion. Its nine stamens, six nearly free pistils, and a differentiated perianth distinguish this species.

16. CYPERACEAE—SEDGE FAMILY

This family is treated in a separate volume in this series.

17. ERIOCAULACEAE—PIPEWORT FAMILY

Annual or perennial herbs; leaves basal, parallel-veined, grasslike; flowers borne on scapes, the scape with a nearly bladeless sheath, unisexual, actinomorphic or slightly zygomorphic, borne in dense heads, each head bracteate; sepals 2–3, free or united at the base; petals 2–3, free or united at the base, sometimes absent; stamens 3–4, adnate to the petals, or borne at the tip of an androphore; ovary superior, sometimes borne on a gynophore, 2- or 3-locular; fruit a capsule.

Thirteen genera comprise this family, with about twelve hundred species. Only the following genus occurs in the central Midwest.

1. **Eriocaulon** L.—Pipewort

Perennial herbs from rather spongy roots; leaves basal, often translucent near the base with conspicuous cross-walls; flowers on scapes, crowded into heads, bracteate, unisexual, but both staminate and pistillate flowers in the same head; sepals 2, free or united at the base; petals 2, free or united at the base, each bearing a gland near the tip; stamens 4, borne at the tip of an androphore; ovary superior, borne on a gynophore with the petals; fruit a capsule.

Eriocaulon consists of about four hundred species, most of them in the tropics and subtropics. Only the following occurs in the central Midwest.

1. **Eriocaulon aquaticum** (Hill) Drake, Pharm. Journ. 83:200. 1909. Fig. 46.
Cespa aquatica Hill, Herb. Brit. 96. 1769.
Eriocaulon septangulare Withering, Bot. Arr. Veg. Gr. Brit. 784. 1776.
Nasmythia articulata Hudson, Fl. Anglica, Edit. Alt. 415. 1778.
Eriocaulon pellucidum Michx. Fl. Bor. Am. 2:166. 1803.
Eriocaulon articulatum (Hudson) Morong, Bull. Torrey Club 18:353. 1891.

Perennial from fleshy roots; leaves basal, linear, glabrous, to 10 cm long (longer when submersed); scapes very slender, up to 1 mm wide, with up to 7 ribs; flowers in heads, the heads white to gray, globose, up to 10 mm across, soft, with reflexed bracts, the bracts oblong to narrowly ovate, 1.0–1.5 mm long, entire; staminate flowers: sepals 2, gray, narrowly oblong, curved, acute or obtuse at the apex, up to 1.5 mm long, with club-shaped hairs on the back; petals 2, ciliate, about 0.5 mm long, whitish; stamens 4, borne on a club-shaped androphore; pistillate flowers: sepals 2, gray, oblong, curved, keeled, about 1.5 mm long, ciliate and with white hairs on the back; petals 2, narrowly oblong, ciliate, acute or obtuse at the apex, whitish; seeds brownish, ovoid to ellipsoid, about 0.5 mm long. July–August.

Bogs.

IN, OH (OBL).

Pipewort.

This delicate species inhabits bogs in the northern part of the central Midwest. The small white or grayish heads on scapes are distinctive. For many years, this species was known as *E. septangulare,* but the epithet *aquaticum* predates *septangulare.* This same species occurs in Europe where it appears to be somewhat different. If the North American plants are considered distinct from the European ones, then the correct binomial for our plant is *E. pellucidum.*

18. HYDROCHARITACEAE—FROG'S-BIT FAMILY

Annual or perennial, often aquatic, species with slender roots; leaves basal, alternate, opposite, or whorled, simple, sometimes stipulate; inflorescence axillary or terminal, the flowers solitary or in cymes and subtended by a spathe, unisexual, monoecious or dioecious, sometimes perfect, actinomorphic; sepals 3, free; petals 3, free; stamens 1–12; ovary inferior or superior, usually 1-locular; fruit berrylike.

This family consists of about seventeen genera and 715 species, found nearly throughout the world.

1. Leaves borne along the stem, opposite or whorled, the longest not more than 3 cm, the broadest not more than 5 mm.
    2. Leaves in whorls of 4 or 6, generally more than 2 cm long; petals 9–12 mm long .... 1. *Egeria*
    2. Leaves opposite or in whorls of 3, generally less than 2 cm long; petals minute or up to 5 mm long .... 2. *Elodea*
1. Leaves all basal, 2–200 cm long, 5–25 mm broad.
    3. Leaves ovate to orbicular; fertile stamens 6–12, the filaments united into a column .... 3. *Limnobium*
    3. Leaves narrow, elongate; fertile stamens 2, the filaments free .... 4. *Vallisneria*

1. **Egeria** Planch.—Waterweed

Aquatic perennials without rhizomes; stems branched or unbranched; leaves in whorls of (4–) 6, submersed, the blades linear; inflorescence 1-flowered, sessile, subtended by a spathe; flowers unisexual, the plants dioecious; staminate flowers: sepals 3, petals 3, stamens 9 or 10, the filaments free; pistillate flowers: sepals 3, petals 3, ovary apparently inferior, 1-locular, styles 3; fruit dry, dehiscing irregularly.

There are two species in the genus, both in warm regions of the world. They are often placed in the genus *Elodea.*

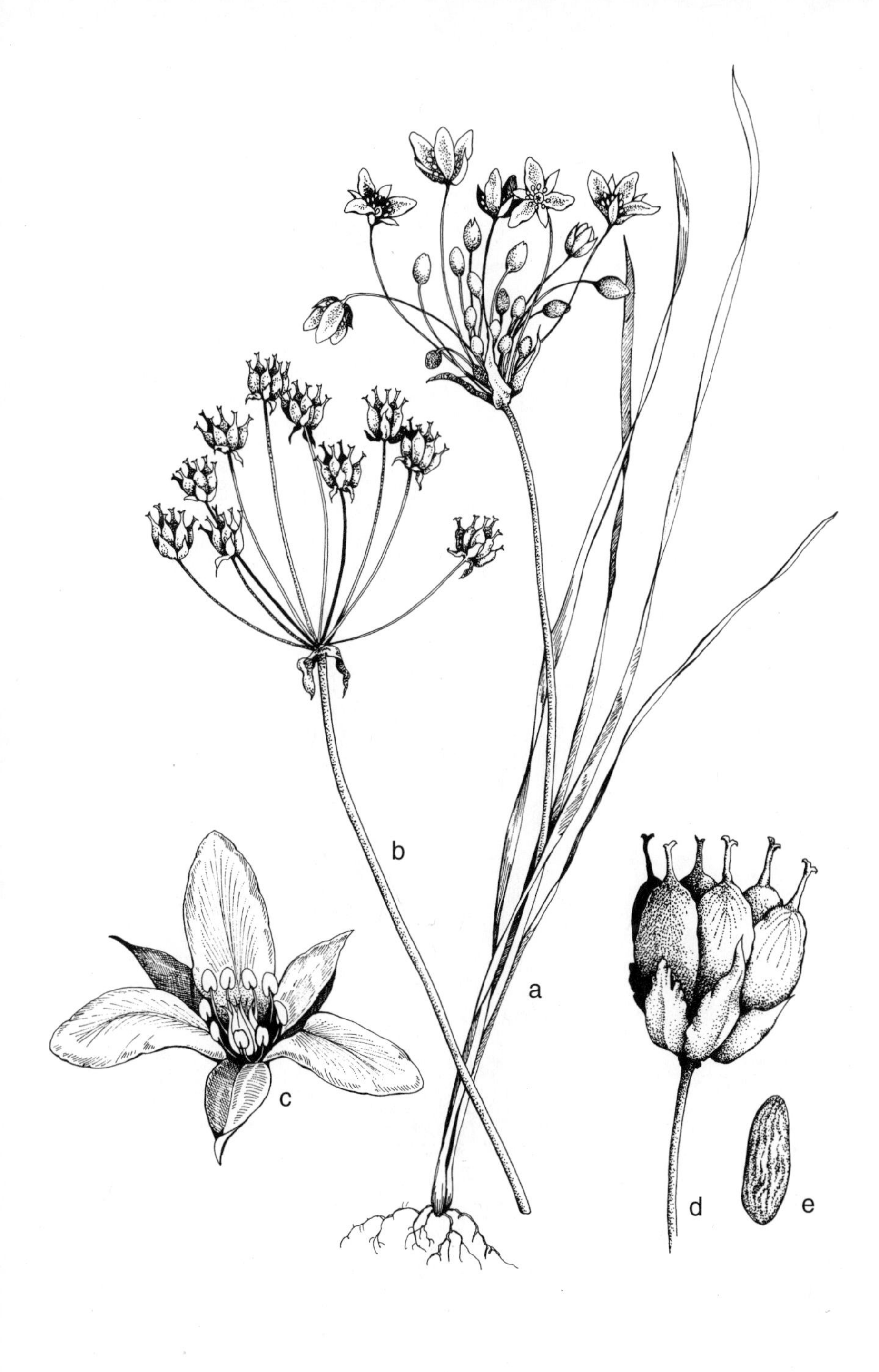

45. ***Butomus umbellatus*** **(Flowering rush).**
a. Habit.
b. Fruiting branch.
c. Flower.
d. Cluster of fruits.
e. Seed.

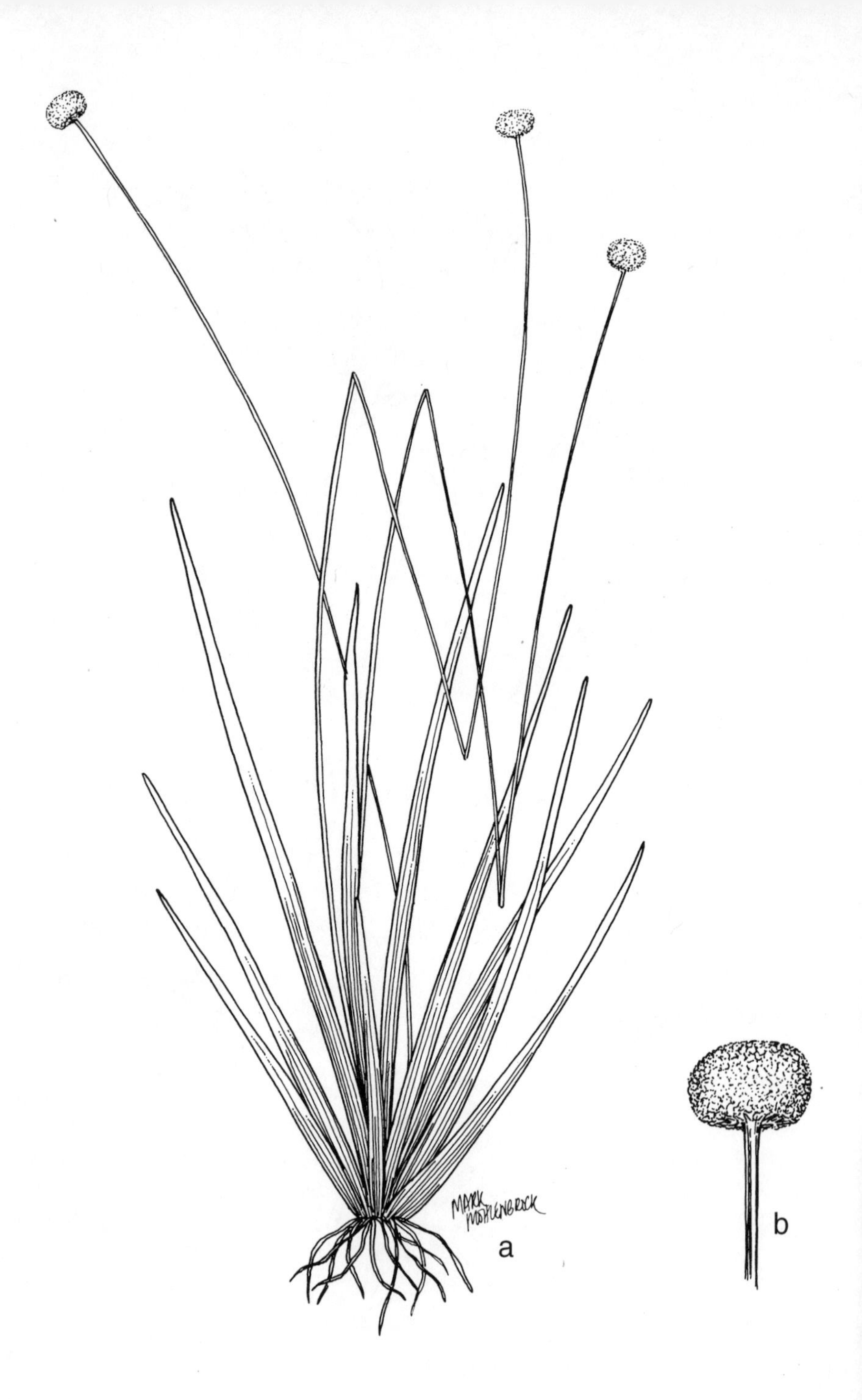

46. ***Eriocaulon aquaticum*** **(Pipewort).** a. Habit. b. Flowering head.

1. **Egeria densa** Planch. Ann. Sci. Nat. Bot. III, 11:80. 1849. Fig. 47.
Elodea densa (Planch.) Caspary, Monaster. Kgl. Preuss. Akad. Wissensch. 1857:48. 1857.
Anacharis densa (Planch.) Vict. Contr. Lab. Bot. Univ. Montreal 18:41. 1931.

Pistillate plants not observed in the wild; staminate plants: leaves linear-lanceolate, subacute, recurved, mostly in whorls of (4–) 6, over 2 cm long, 2–5 mm broad; sepals 3, 5–8 mm long; petals 3, 9–12 mm long, white; stamens apparently 9; pedicels up to 8 cm long. July–September.

Quiet water.

IL, IN, KY, MO (OBL).

Giant waterweed.

This species is native to Brazil, but it is a common aquarium plant in the United States. Its presence in the United States may be the result of aquaria having been discarded.

2. **Elodea** Michx.—Waterweed

Aquatic perennial, without rhizomes; stems branched or unbranched; leaves opposite or in whorls of 3 (rarely more), submersed, the blades linear; inflorescence solitary, in the axils of the leaves; flowers unisexual, subtended by a spathe, rarely bisexual, the plants usually dioecious; staminate flowers: sepals 3; petals 3, white; stamens 7–9, at least the innermost 3 and sometimes all of them connate; pistillate flowers: sepals 3; petals 3, white; ovary apparently inferior, 1-locular, styles 3; fruit dehiscing irregularly.

The perianth tube in the pistillate flowers becomes greatly elongated (to 30 cm) at anthesis so that the lobes of the perianth reach the surface of the water.

There are probably five species in the genus, although the variability of some of them has led to the naming of several other species.

1. Leaves, or most of them, opposite; seeds up to 4 mm long; pollen in monads .... 1. *E. bifoliata*
1. Leaves in whorls of 3; seeds 4.0–5.7 mm long; pollen in tetrads.
    2. Leaves at least 1.8 mm wide; spathes of staminate flowers 8.0–13.5 mm long; styles 2.5–4.0 mm long .... 2. *E. canadensis*
    2. Leaves up to 1.7 mm wide; spathes of staminate flowers up to 4 mm long; styles up to 2 mm long .... 3. *E. nuttallii*

1. **Elodea bifoliata** H. St. John, Res. Stud. State Coll. Wash. 30:23. 1962. Fig. 48.

At least some of the leaves opposite, spreading, linear to narrowly elliptic, acute, up to 20 mm long, 1.5–4.3 mm wide; staminate flowers: spathes cylindric, 10–40 mm long; sepals 3; petals 3, 2–4 mm long, white; stamens 7–9, the inner three filaments connate, forming a column, the pollen in monads; pistillate flowers: spathes linear, 10–65 mm long; sepals 3; petals 3, 1.4–3.0 mm long, white; styles 2.5–3.0 mm long; seeds ellipsoid, 2.8–3.0 mm long, long-hairy. July–September.

Rivers, lakes.

KS (OBL).

Opposite-leaved waterweed.

This western species just barely reaches the central Midwest in western Kansas. It differs from the other species of *Elodea* by having most of its leaves opposite and

47. *Egeria densa* (Giant waterweed).

a. Habit.

b. Flower (staminate).

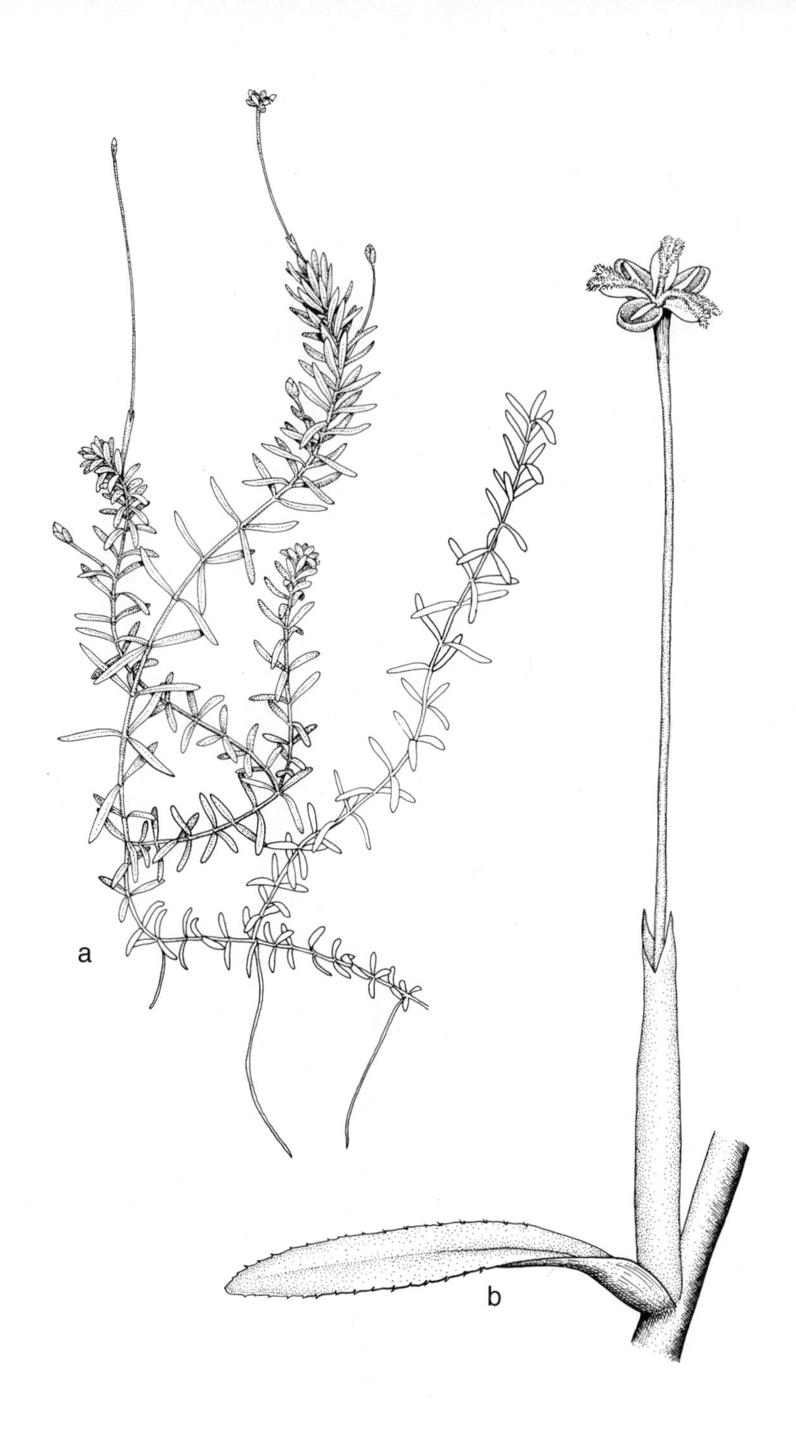

48. *Elodea bifoliata* (Opposite-leaved waterweed). a. Habit. b. Flower.

by having its pollen borne in monads. *Elodea laevivaginata* H. St. John may be the same species.

2. **Elodea canadensis** Rich. in Michx. Fl. Bor. Am. 1:20. 1803. Fig. 49.
Udora canadensis (Rich.) Nutt. Gen. N. Am. Pl. 2:242. 1818.
Anacharis canadensis Planch. Ann. & Mag. Nat. Hist. II, 1:86. 1848, not based on *E. canadensis* Rich. in Michx. (1803).

Upper leaves in whorls of three, lower leaves sometimes opposite, linear to broadly lanceolate, obtuse to acute at the apex, to 13 mm long, to 4 (–5) mm broad; staminate flowers: spathes cylindric, 8.0–13.5 mm long; sepals 3, 3.5–5.0 mm long; petals 3, 4–5 mm long, white; stamens 9, all connate, forming a column, the pollen borne in tetrads; pistillate flowers: spathes threadlike, 8.5–17.5 mm long; sepals 3, about 2 mm long; petals 3, 2.3–2.6 mm long, white; staminodia 3, very slender, less than 1 mm long; stigmas 3, bilobed; ovary 1, broadly lanceoloid, 2.5–3.0 mm long; fruit 5–6 mm long, ovoid. July–September.

Rivers, lakes.

IA, IL, IN, KS, KY, MO, NE, OH (OBL).

Waterweed; elodea; anacharis.

In most of the central Midwest, this is the common species of *Elodea*. It differs from *E. bifoliata* in that many of its leaves are in whorls of three, and from *E. nuttallii* by its larger leaves and larger flowers. The staminate plants are rarely seen and collected.

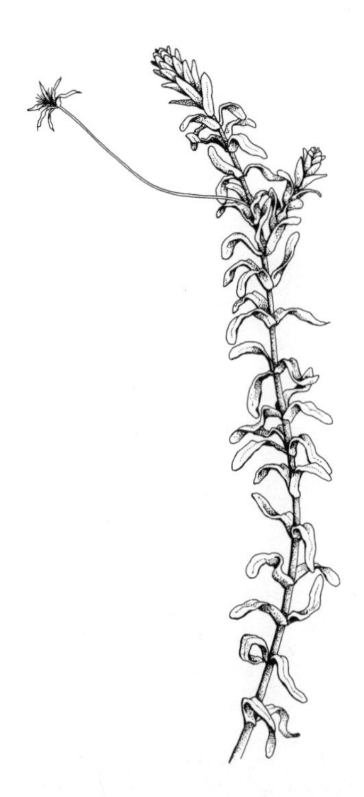

49. *Elodea canadensis* (Waterweed). Habit with flower.

3. **Elodea nuttallii** (Planch.) H. St. John, Rhodora 22:27. 1920. Fig. 50
Serpicula occidentalis Pursh, Fl. Am. Sept. 1:33. 1814, *nomen illeg.*
Anacharis nuttallii Planch. Ann. & Mag. Nat. Hist. II 1:86. 1848.
Elodea occidentalis (Pursh) H. St. John, Rhodora 22:27. 1920, based on illeg. basionym.
Anacharis occidentalis (Pursh) Vict. Contr. Lab. Bot. Univ. Montreal 18:40. 1931, based on illeg. basionym.

Upper leaves in whorls of 3, lower leaves sometimes opposite, linear to linear-lanceolate, acute, to 12 mm long, to 1.5 mm broad; staminate flowers: spathes subglobose to ovoid, 2.5–4.0 mm long, sepals 3, about 2 mm long, reddish, petals 3, minute, whitish, stamens 9, the three inner ones connate forming a column, the pollen in tetrads; pistillate flowers: spathes linear, 8.5–15.0 mm long, sepals 3, about 1 mm long, petals 3, 1.2–1.5 mm long, whitish, staminodia minute, about 0.5 mm long, stigmas 3, bilobed, ovary 1, lanceoloid, 1.5–2.5 mm long; fruit 2–4 mm long, lanceoloid. July–September.

Rivers, lakes.

IA, IL, IN, KS, KY, MO, NE, OH (OBL).

Smaller waterweed; smaller elodea; smaller anacharis.

This species is similar in appearance to *E. canadensis* but has smaller leaves and flowers.

50. *Elodea nuttallii* **(Smaller waterweed).** Habit with flower.

3. **Limnobium** Rich.—Sponge-plant

Aquatic perennials, without rhizomes but sometimes with stolons; leaves basal, emergent or floating, petiolate; inflorescence cymose; flowers unisexual, emersed, subtended by a spathe, the plants monoecious; sepals 3, united; petals 3 (–4), free or nearly so; stamens 9–12, the filaments united into a column; ovary 1, inferior, 6- to 9-locular; fruit fleshy, many-seeded.

There are two species, ours and one in Central and South America.

1. **Limnobium spongia** (Bosc) Rich. ex Steud. Nom. Bot. ed. 2, 2:45. 1841. Fig. 51.
Hydrocharis spongia Bosc, Ann. Mus. Paris 9:396. 1807.

Aquatic perennial with stolons; leaves floating, ovate, nearly as wide as long, subacute at the apex, rounded to cordate at the base, to 7 cm long, glabrous, green above, sometimes purplish below, the blade, at least when young, bearing spongy cells on the lower surface; spathes 1-leaved in the staminate flowers, 2-leaved in the pistillate flowers; peduncle of staminate flower 4–9 cm long, of pistillate flower 2.5–3.5 cm long; sepals 3, oblong, 4–7 mm long; petals 3, white, linear, 4–7 mm long; fruit globose, fleshy or soon becoming dry, bluish, about 5–6 (–10) mm in diameter. June–September.

Swamps, bayous, lakes, sometimes stranded on land.

IL, KY, MO (OBL).

Sponge-plant.

This remarkable plant sometimes bears patches of spongy cells on the lower surface of the leaves, presumably to keep the leaves buoyed in the water.

4. **Vallisneria** L.—Eelgrass, Water Celery

Aquatic perennials with rhizomes; erect stems rooted under water; leaves basal, submersed, ribbonlike; inflorescence cymose, the flowers unisexual, produced from a spathe, submersed or floating, the plants dioecious; sepals 3, unequal in size in the staminate flowers, equal in the pistillate flowers but united to form a tube; fertile stamens 2; ovary 1, 1-locular, style 1; fruit dehiscing irregularly.

Reproduction in this genus is curious. The staminate flowers, while in bud, are liberated and float to the surface of the water where the pistillate flowers have come due to rapid elongation of the peduncle. Following fertilization, the pistillate peduncle coils, pulling the fruit beneath the water where it matures.

There are two species in the genus, ours and one in the Old World.

1. **Vallisneria americana** Michx. Fl. Bor. Am. 2:220. 1803. Fig. 52.

Aquatic perennial with rhizomes; erect stems rooted under water; leaves basal, membranous, ribbonlike, up to 2 m long, 5–20 mm broad, obtuse at the apex, light colored down the center bordered by dark green on either side; staminate flowers on scapes, submersed, 1.0–1.5 mm wide, with 2 stamens; pistillate flower solitary on an elongated scape, the peduncle long enough so that the flower reaches the surface of the water, white, 2–3 cm long; fruit cylindrical, 5–12 cm long. July–October.

Rivers, streams, lakes.

IL, IN, KY, MO, OH (OBL).

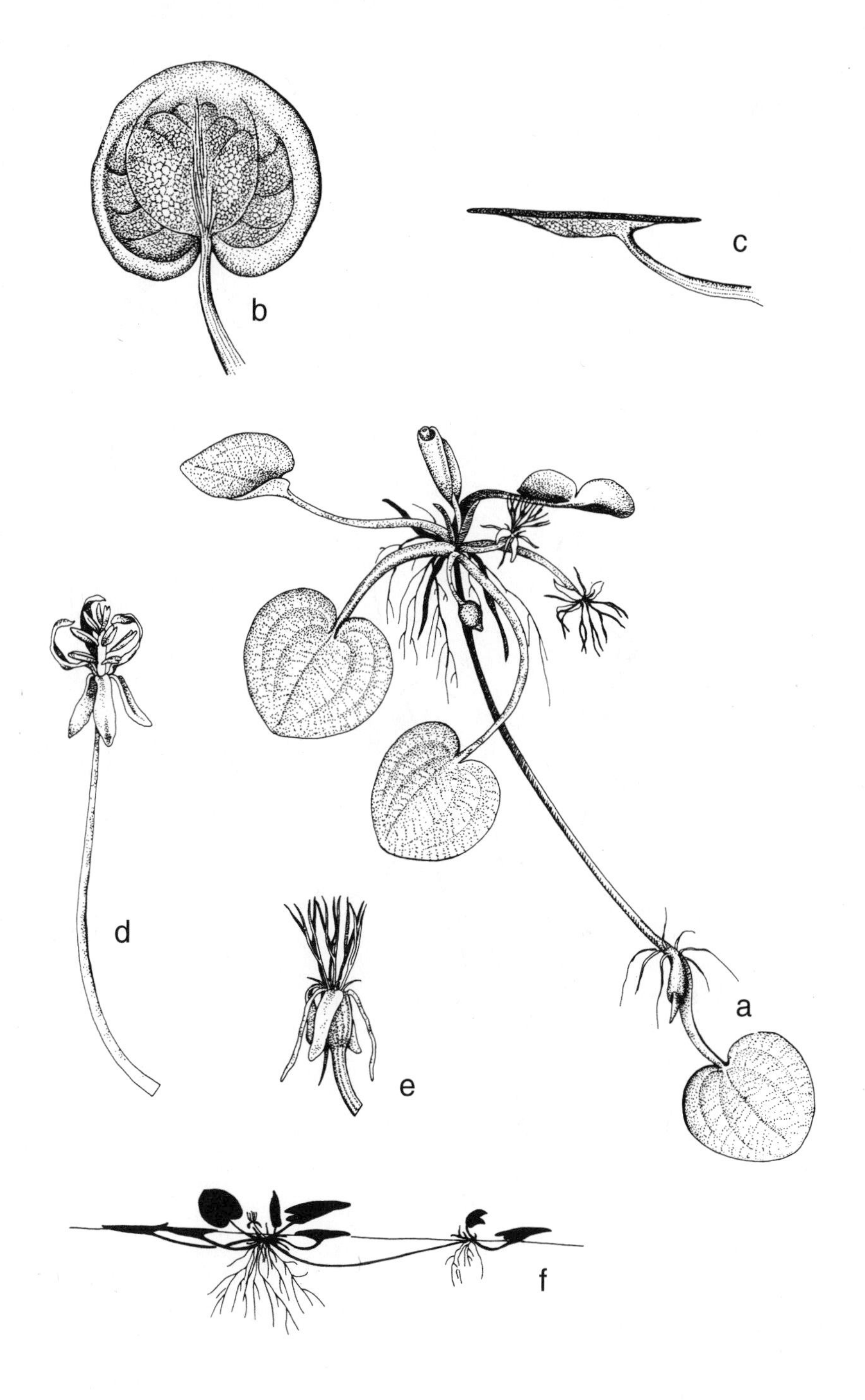

51. ***Limnobium spongia*** **(Sponge-plant).**
a. Habit.
b. Leaf.
c. Floating leaf.
d. Staminate flower.
e. Pistillate flower.
f. Habit (shaded).

Eelgrass; water celery; ribbongrass.

This is the only submersed species with long, ribbonlike leaves. It may occur at considerable depths below the surface of the water.

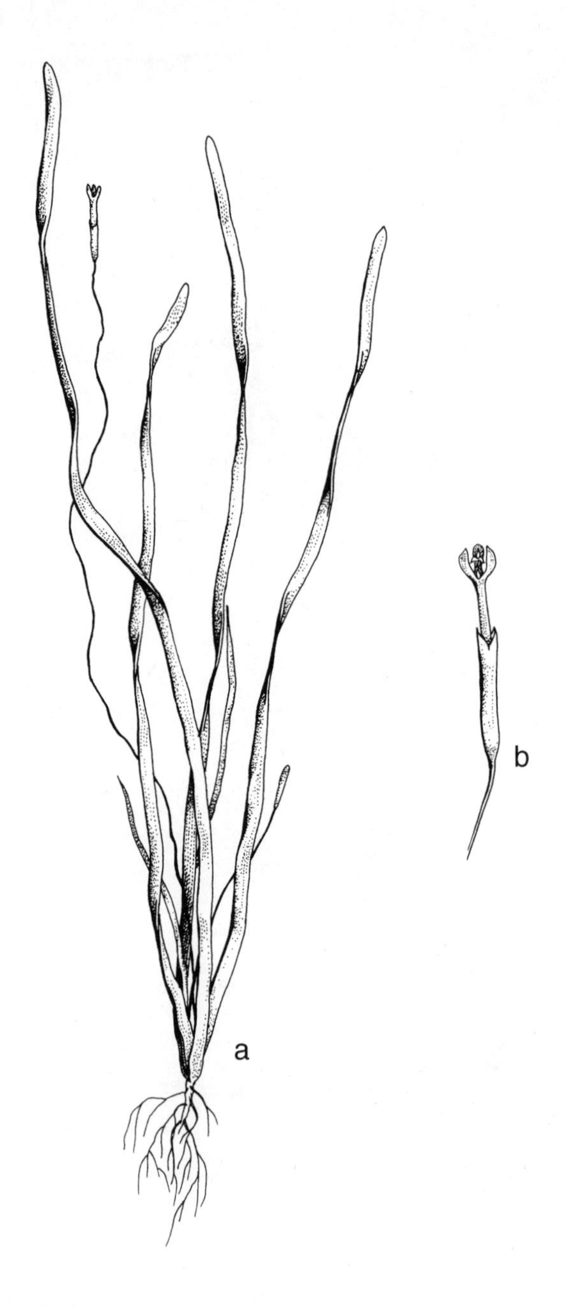

52. *Vallisneria americana* (Eelgrass). a. Habit. b. Flower.

19. IRIDACEAE—IRIS FAMILY

Perennial herbs with rhizomes or fibrous roots; leaves flattened, sometimes folded lengthwise, often sword-shaped; inflorescence 1- to several-flowered; flowers actinomorphic, perfect; perianth parts 6, united at least below, generally uniform in color; stamens 3; ovary inferior; fruit a capsule, 3-celled, loculicidal.

This family consists of about eighty genera and fifteen hundred species found throughout the world, but particularly abundant in South Africa.

Members of this family have the leaves and stems conspicuously flattened. The Iridaceae differs from the showy-flowered Liliaceae in the reduction of its stamens to three.

Besides the genus *Iris,* several genera of this family are grown as ornamentals, including *Gladiolus, Belamcanda* (blackberry lily), *Crocus, Freesia,* and *Tigridia.*

Only the genus *Iris* has aquatic species.

1. **Iris** L.—Iris

Perennial herbs from usually stout rhizomes; leaves sword-shaped, usually folded lengthwise, not confined to the base of the plant, flattened; inflorescence 1- to several-flowered, the flowers arising from spathes; perianth parts 6, of uniform color, clawed, the outer sepals generally slightly larger, more or less recurved, the inner petals spreading or erect; stamens 3, attached to the base of the sepals, concealed by the styles; styles petal-like, arching, bilobed; stigma flat, located between the lobes of the style; ovary 3- or 6-angled; capsule 3- or 6-lobed, with 1–2 seeds per cell.

There are about two hundred species of *Iris* worldwide, most of them in the Northern Hemisphere. The following may occur in shallow water in the central Midwest.

1. Flowers rusty-red (rarely yellow); capsule indehiscent; leaves rather soft .... 1. *I. fulva*
1. Flowers blue, violet, or yellow; capsule dehiscent; leaves firm.
    2. Flowers yellow; sepals 2-ridged on upper surface; perianth tube not constricted above the ovary .... 2. *I. pseudacorus*
    2. Flowers blue or violet; sepals not 2-ridged on the upper surface; perianth tube constricted above the ovary.
        3. Leaves usually brown at base; sepals with a bright yellow spot at base; seeds deeply pitted, dull .... 3. *I. shrevei*
        3. Leaves usually purple at base; sepals with a greenish yellow spot at base; seeds finely pitted, shiny. .... 4. *I. versicolor*

1. **Iris fulva** Ker, Bot. Mag. 36:pl. 1496. Fig. 53.
Iris cuprea Pursh, Fl. Am. Sept. 30. 1814.

Rhizome stout; leaves 40–80 cm long, 10–15 mm broad, green; flowering stem at least 50 cm tall, simple or branched, several-flowered; spathes unequal in length; flowers 7–9 cm broad, rusty red or rarely yellow; sepals obovate, recurved; petals obovate, slightly smaller than the sepals, spreading; ovary 6-angled; capsule 6-angled, 4.5–7.5 cm long, indehiscent. May–June.

Swamps, usually in shallow water.

IL, KY, MO (OBL).

Copper iris.

53. ***Iris fulva*** **(Copper iris).** a. Inflorescence and leaves. b. Capsule.

This beautiful species has migrated up the Mississippi Embayment into the southern part of the central Midwest. A very rare colony of yellow-flowered plants has occurred for years in Johnson County, Illinois. It may be distinguished from the yellow-flowered *I. pseudacorus* by the absence of a double ridge on the upper surface of the sepals, by its indehiscent capsules, and by its rather soft leaves.

2. **Iris pseudacorus** L. Sp. Pl. 1:38. 1753. Fig. 54.

Rhizome stout; leaves 50–100 cm long, about 1.5 cm broad, green; flowering stem at least 50 cm tall, branched or unbranched, several-flowered; spathes unequal in length; flowers 7–9 cm across, basically yellow, with dark brown markings; sepals obovate, 2-ridged on the upper surface, spreading; petals much smaller, narrowed at the middle, spreading or ascending; ovary usually 3-angled; capsule 3-angled or occasionally 6-angled, 5–8 cm long, dehiscent. June–August.

Wet ditches, along streams, around ponds, sometimes in shallow water.

IA, IL, IN, KS, KY, MO, NE, OH (OBL).

Yellow iris.

This species is native to Europe, but it occasionally escapes from cultivation into wetlands in the United States. It differs from the rare yellow-flowered form of *I. fulva* by the double ridge on the upper surface of the sepals, its dehiscent capsules, and its firm leaves. It is difficult to distinguish this species in the vegetative condition from either *Iris shrevei* or *I. versicolor*, except that the leaves of *I. pseudacorus* are usually narrower, up to 1.5 cm broad, while the leaves of the other two are usually 2–3 cm broad. In addition, the base of the leaves of *I. pseudacorus* are usually not purple as they are in *I. versicolor* and occasionally in *I. shrevei*.

3. **Iris shrevei** Small, Addisonia 12:13. 1927. Fig. 55.
Iris virginica L. var. *shrevei* (Small) E. Anders. Ann. Mo. Bot. Gard. 23:469. 1936.

Rhizome stout; leaves 40–100 cm long, up to 3 cm broad, green but often brown at base, rarely purple; flowering stem 60–100 cm tall, branched or unbranched, several-flowered; spathes unequal in length; flowers 6–8 cm across, basically bluish or violet; sepals spatulate, clawed, recurved, pubescent and with a bright yellow spot near the base; petals a little smaller, spreading to ascending; ovary 3-angled; capsule obtusely 3-angled, 6–9 cm long, dehiscent. May–June.

Marshes, sloughs, swamps, fens, around sinkholes.

IA, IL, IN, KS, KY, MO, NE, OH (OBL).

Blue flag; blue iris.

Some botanists and the U.S. Fish and Wildlife Service believe this plant to be a variety of the more southeastern *I. virginica*, or even the same as. The slightly thicker and somewhat more elongated, prominently beaked capsule is the chief distinction between *I. shrevei* and *I. virginica*.

Iris versicolor is very similar to *I. shrevei*, but the former has a greenish yellow spot at the base of the sepals rather than a bright yellow spot. It also has finely pitted, shiny seeds, rather than deeply pitted, dull seeds. In the vegetative state, *I. shrevei* usually has leaves with a brown base, while *I. versicolor* has leaves with a purplish base.

54. *Iris pseudacorus* (Yellow iris). Habit.

55. *Iris shrevei* (Blue flag). a. Inflorescence and leaves. b. Capsule.

4. **Iris versicolor** L. Sp. Pl. 1:39. 1753. Fig. 56.

Rhizome stout; leaves 40–100 cm long, up to 3 cm broad, usually purplish at base; stems 60–100 cm tall, branched or unbranched; inflorescence several-flowered; spathes unequal in length; flowers 6–8 cm across, blue or violet; sepals spatulate, clawed, recurved, pubescent and with a greenish yellow spot near the base; petals a little smaller, spreading to ascending; ovary 3-angled; capsule obtusely 3-angled, 3–6 cm long, dehiscent; seeds finely pitted, shiny. May–June.

Marshes, swamps, bogs.

IL, NE (OBL).

Northern blue iris; northern blue flag.

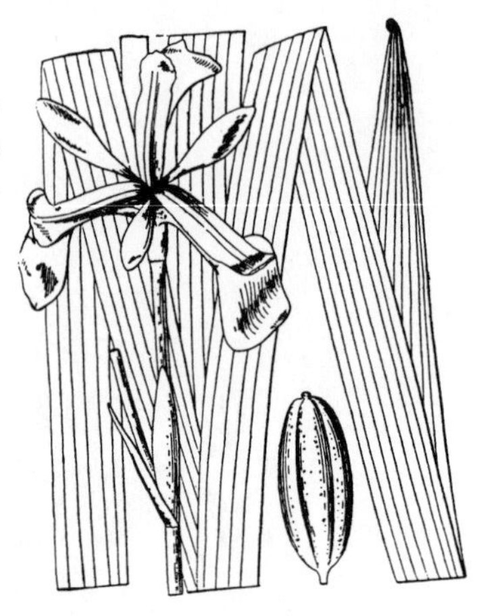

56. *Iris versicolor* (Northern blue iris). Habit. Petals (above left). Capsule (below).

Locations in Sarpy County, Nebraska, and in Jackson County, Illinois, are the only ones in the central Midwest. This species has purple leaf bases, finely pitted, shiny seeds, and a greenish yellow spot at the base of each sepal, thereby distinguishing it from the very similar *I. shrevei.*

20. JUNCACEAE—RUSH FAMILY

Annuals or usually perennials, with or without stolons and rhizomes; leaves flat or terete, often cross-septate; inflorescence paniculate, cymose, or umbellate; perianth 6-parted, green or brown; stamens 3 or 6; ovary superior, 1- to 3-locular; fruit a capsule; seeds often caruncùlate.

This family is distinguished from the often somewhat similar appearing Cyperaceae (sedges) and Poaceae (grasses) by the presence of a perianth. Those with terete leaves are distinctive because neither sedges nor grasses have terete leaves.

There are about 325 species in this family. In addition to *Juncus, Luzula* (the wood rushes), occurs in forested areas of the central Midwest. *Luzula* is readily distinguished from *Juncus* by the presence of cilia near the base of the leaves. Some species of *Juncus* in the central Midwest are not found in shallow water, and others are even upland species.

1. **Juncus** L.—Rush

Perennials, rarely annuals; sheath open, often auriculate; blades flat, involute, terete, or absent; inflorescence paniculate or cymose, bracteate; perianth 6-parted, scarious-margined, green or brown; stamens 3 or 6; ovary 3-celled or incompletely 3-celled; capsule with numerous seeds, the seeds apiculate or caudate.

Leafy proliferations in the inflorescences are not unusual in the septate-leaved species. They are usually formed through the activity of the homopteran insect *Livia juncorum* Lat.

For positive identification, flowers, fruits, and seeds are often necessary. However, many species of *Juncus* may be identified by their leaves or lack of leaves and size and shape of the flowering clusters.

There are approximately two hundred species of *Juncus* found throughout most of the world. A few species are not considered to be wetland species.

1. Leaves bladeless, reduced to sheaths, or with only 1–2 leaves near the base of the plant in *J. coriaceus*; flowers appearing lateral.
  2. Leaves 1–2 near the base of the plant .... 10. *J. coriaceus*
  2. Leaves bladeless, reduced to sheaths.
    3. Stems smooth, not ribbed.
      4. Culms in tufts, from short rhizomes .... 15. *J. effusus*
      4. Culms spaced along elongated rhizomes .... 17. *J. filiformis*
    3. Stems with longitudinal ribs.
      5. Culms in tufts, from short rhizomes; achenes not short-beaked .... 24. *J. pylaei*
      5. Culms spaced along elongated rhizomes; achenes short-beaked .... 3. *J. arcticus*
1. Leaves with blades, these either flat or terete.
  6. Flowers prophyllate, subtended by two small, opposite bracteoles.
    7. Leaves subterete; capsules shiny .... 12. *J. dichotomus*
    7. Leaves flat, sometimes channeled, but not subterete; capsules dull.
      8. Capsules less than twice as long as wide; perianth parts spreading at maturity .... 14. *J. dudleyi*
      8. Capsules twice as long as wide; perianth parts erect and appressed at maturity.
        9. Leaf blades channeled .... 29. *J. vaseyi*
        9. Leaf blades flat .... 8. *J. brachyphyllus*
  6. Flowers not prophyllate, not subtended by two small, opposite bracteoles.
    10. Leaves flat.
      11. Perianth segments 5–6 mm long; capsules long-beaked .... 18. *J. longistylis*
      11. Perianth segments 2–3 mm long; capsules apiculate but not long-beaked.
        12. Plants with 15–20 heads, each head 6–8 mm across; blades 1–3 mm wide .... 19. *J. marginatus*
        12. Plants with 20 or more heads, each head 4–6 mm across; blades 4–6 mm wide .... 5. *J. biflorus*
    10. Leaves terete or, if somewhat flattened in *J. validus* and *J. pelocarpus*, the leaves cross-septate.
      13. Leaves obscurely cross-septate, solid .... 23. *J. pelocarpus*
      13. Leaves, or some of them, obviously cross-septate, hollow.
        14. Capsules much longer than the perianth .... 13. *J. diffusissimus*
        14. Capsules shorter than, equaling, or barely longer than the perianth.
          15. Heads spherical or hemispherical.
            16. Heads hemispherical.
              17. Heads 6–10 mm in diameter.
                18. Seeds 0.3–0.4 mm long; flowers more than 10 per head .... 1. *J. acuminatus*
                18. Seeds 0.7–1.2 mm long; flowers usually 5 (–10) per head .... 26. *J. subcaudatus*
              17. Heads 10–20 mm in diameter .... 9. *J. canadensis*
            16. Heads spherical.
              19. Heads up to 10 mm in diameter.
                20. Some or all the leaves more or less flattened but somewhat hollow and cross-septate .... 28. *J. validus*
                20. All leaves terete and hollow and cross-septate.
                  21. Sepals 3.5–6.0 mm long; heads 30–100 per plant .... 6. *J. brachycarpus*
                  21. Sepals 2.0–3.2 mm long; heads 20–60 per plant. .... 25. *J. scirpoides*

19. Heads 10–20 mm in diameter.
22. Leaves 0.5–1.5 mm wide; heads 5–25 per plant .... 22. *J. nodosus*
22. Some leaves at least 2 mm wide; heads 25–100 per plant .... 27. *J. torreyi*
15. Heads wedge-shaped or top-shaped.
23. Heads 35–200 per plant.
24. Perianth about 2/3 as long as the capsules, obtuse to acute .... 7. *J. brachycephalus*
24. Perianth about equaling the capsules, acuminate to aristulate.
25. Inflorescence twice as long as wide or longer; heads 35–100 per plant .... 16. *J. elliottii*
25. Inflorescence less than twice as long as wide; heads 200–300 per plant .... 21. *J. nodatus*
23. Heads 3–35 (–50) per plant.
26. Cauline leaves 2 or more, up to 1.5 mm in diameter.
27. Capsules acute but not apiculate, pale cinnamon-brown; leaves 0.5–1.0 mm in diameter .... 11. *J. debilis*
27. Capsules apiculate, chestnut-brown; leaves 0.7–1.5 mm in diameter.
28. Heads 6–8 mm across; capsules short-apiculate .... 2. *J. alpinoarticulatus*
28. Heads 8–10 mm across; capsules long-apiculate .... 4. *J. articulatus*
26. Cauline leaf 1 (–2), 2–4 mm in diameter .... 20. *J. militaris*

1. **Juncus acuminatus** Michx. Fl. Bor. Am. 1:192. 1803. Fig. 57.
Juncus paradoxus E. Mey. Syn. Junc. 30. 1822.
Juncus acuminatus Michx. var. *legitimus* Engelm. Trans. Acad. St. Louis 2:463. 1868.
Juncus acuminatus Michx. var. *paradoxus* (E. Mey.) Farw. Am. Mid. Nat. 11:73. 1928.
Juncus acuminatus Michx. f. *sphaerocephalus* Hermann, Leaflets Western Botany 8:13. 1956.

Perennial from a short, inconspicuous rhizome; stems cespitose, 1.3–7.5 dm tall, 1–3 (–4) mm wide; leaves 3–4 dm long, 1–3 mm wide; involucral leaf shorter than the inflorescence; inflorescence compact to spreading, (0.5–) 2–14 (–18) cm long, the rays ascending to divergent; heads (1–) 2–82, hemispherical, 5–10 mm across, 5- to approximately 35-flowered; perianth segments lanceolate, acuminate; sepals 3–4 mm long; petals 0.7 mm shorter than to equaling the sepals; stamens 3; anthers shorter than the filaments; capsules ellipsoid, acute, mucronate, slightly shorter than the petals to exceeding the sepals by 0.3 mm; seeds ellipsoid, 0.5 mm long, minutely apiculate at both ends. May–August.

Ditches, around ponds and lakes, along rivers and streams, sloughs, fens, spring branches.

IA, IL, IN, KS, KY, MO, NE, OH (OBL).

Sharp-fruited rush.

This very common species is distinguished by its terete leaves and its hemispherical heads that are 5–10 mm across. It often has asexual shoots replacing flowers in the inflorescence, sometimes causing the plant to look like a *Rhynchospora*.

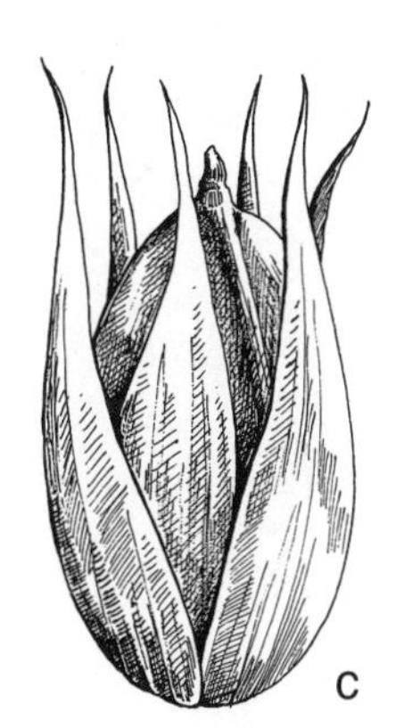

57. *Juncus acuminatus* (Sharp-fruited rush).

a. Habit.
b. Heads.
c. Capsule with perianth.

2. **Juncus alpinoarticulatus** Chaix, Hist. Pl. Dauphine 1:378. 1786. Fig. 58.
Juncus alpinus Vill. Hist. Pl. Dauphine 2:233. 1787.
Juncus alpinus Vill. var. *rariflorus* Hartm. Skand. Fl., ed. 7:140. 1868.
Juncus alpinus Vill. var. *fuscescens* Fern. Rhodora 10:48. 1908.
Juncus alpinoarticulatus Chaix ssp. *fuscescens* (Fern.) Hamat-Ahti, Bot. Fenn. 23: 280. 1986.

Cespitose perennial from slender rhizomes; stems (5–) 9–30 cm tall; leaves 1–2, to 16 cm long, 0.7–1.5 mm in diameter, terete, hollow, cross-septate; involucral leaf shorter than the inflorescence; inflorescence (2.5–) 3.5–24.0 cm long, with 2–38 heads; heads wedge-shaped or ellipsoid, 2–7 mm wide, with 2–9 flowers; flowers sessile or on pedicels to 5 mm long; perianth segments lanceolate; sepals 1.9–3.0 mm long, acuminate to acute to obtuse, usually mucronulate; petals obtuse to acute, occasionally acuminate, rarely apiculate, 0.2–0.5 (–1.0) mm shorter than to rarely equaling the sepals; stamens 6; anthers shorter than the filaments; capsules pale or dark brown, oblongoid or ellipsoid, acute or obtuse, with a mucro up to 0.5 mm long, slightly shorter than to exceeding the sepals by 1 mm; seeds ovoid to oblongoid, acute or acuminate, 0.5–0.6 mm long, dark apiculate at one end. July–September.

Marshes, sandbars, occasionally in shallow water.

IA, IL, IN, MO, NE, OH (OBL).

Alpine rush.

Three variations have sometimes been named in the United States for this species. The typical variety does not occur in the central Midwest. Specimens with one or several of the flowers elevated above the others on slightly elongated pedicels may be called var. *rariflorus*. Specimens with flowers sessile or equally short-pedicellate may be called var. *fuscescens*.

The U.S. Fish and Wildlife Service calls this plant *J. alpinus*.

3. **Juncus arcticus** Willd. Sp. Pl. 2 (1):206. 1799. Fig. 59.
Juncus balticus Willd. var. *littoralis* Engelm. Trans. Acad. St. Louis 2:442. 1866.
Juncus arcticus Willd. var. *balticus* (Willd.) Trautv. Tgrudy Imp. St.-Petersburgs Bot. Sada 5:119. 1875.
Juncus balticus Willd. var. *littoralis* Engelm. f. *dissitiflorus* Engelm. ex Fern. & Wieg. Rhodora 25:208. 1923.
Juncus litorum Rydb. Brittonia 1:85. 1931.

Perennial with single stems arising at intervals from a stout, elongated, occasionally branched rhizome with conspicuous internodes; stems 3.5–10.5 mm tall, 1.00–2.75 mm in diameter, finely many-ribbed; basal sheaths purplish to purplish brown to yellow, often glossy, the uppermost sheath 5–15 cm long, apiculate or with a mucro to 4 mm long; involucral leaf erect, terete, 5–28 cm long, exceeding the inflorescence and resembling a prolongation of the stem; inflorescence appearing lateral on the stem, condensed to diffusely spreading on unequal, compound rays, 1.1–20.0 cm long; perianth segments brown on both sides of a green mid-nerve, the margins scarious and hyaline; sepals lance-attenuate or acuminate, 3.2–5.0 mm long; petals acuminate or acute, 0.5 mm shorter than to nearly equaling the sepals;

58. *Juncus alpinoarticulatus* (Alpine rush).

a. Habit.
b. Head with capsules.
c. Seed.

59. *Juncus arcticus* (Baltic rush).

a. Habit.
b. Portion of ray.
c. Capsule with perianth.

stamens 6; anthers 1.5–2.0 mm long, three times longer than the filaments; capsules dark brown, ovoid, acuminate, with a distinct beak 0.5–1.0 mm long, longer than the petals, subequal to 1 mm longer than the sepals; seeds ovoid, 1 mm long, usually minutely apiculate at both ends. July–September.

Swamps, sandy shores, sandbars, mudflats, occasionally in shallow water.

IA, IL, IN, KS, MO, NE, (OBL), OH (FACW+).

Baltic rush.

This is one of four species of *Juncus* in the central Midwest that appears to have leafless stems and lateral inflorescences. Its ribbed stems distinguish it from *J. effusus* and *J. filiformis,* and the single stems borne along an elongated rhizome distinguish is from *J. pylaei. Juncus coriaceus* also has a lateral inflorescence, but it also has at least one leaf near the base of the plant.

4. **Juncus articulatus** L. Sp. Pl. 1:327. 1753. Fig. 60.

Perennials from course rhizomes; stems tufted, up to 60 cm tall; leaves 2–4, up to 15 cm long, 0.7–1.5 mm in diameter, terete, hollow, cross-septate; involucral leaf usually shorter than the inflorescence; inflorescence up to 30 cm long, nearly twice as long as wide, with several hemispheric clusters, each cluster 8–10 mm across, with 3–10 flowers; flowers sessile or subsessile; perianth segments lanceolate to lance-subulate; sepals 2.0–2.7 mm long, acute to acuminate; petals 2.2–3.0 mm long, obtuse to subacute; stamens 6; capsules chestnut-brown, narrowly ovoid, acute, 2.5–4.0 mm long; seeds ovoid to oblongoid, acute or acuminate, 0.5 mm long, short-apiculate. July– October.

Bogs, wet meadows.

IL, IN, KY, OH (OBL).

Jointed rush.

This species is similar to *J. alpinoarticulatus* and often confused with it. *Juncus articulatus* differs by having larger heads 8–10 mm across and long-apiculate capsules. On rare occasions this species may creep along the ground, rooting at the nodes.

5. **Juncus biflorus** Ell. Bot. S. C. & Ga. 1:407. 1817. Fig. 61.
Juncus marginatus Rostk. var. *biflorus* (Ell.) Wood, Class-Book Bot. 725. 1861.
Juncus marginatus Rostk. var. *aristulatus* (Michx.) Coville, Proc. Biol. Soc. Wash. 8:123. 1893.
Juncus biflorus Ell. f. *adinus* Fern. & Grisc. Rhodora 37:157. 1935.

Perennial from conspicuous, scaly rhizomes; stems solitary, approximately 3.5 cm apart, 4.9–8.8 (–10.2) dm tall; leaves 2.0–6.5 mm wide; auricles hyaline, pale brown, rounded; involucral leaf inconspicuous, shorter than the inflorescence; inflorescence 2.5–19.5 cm long, with (13–) 20–135 heads, the heads approximate to distant, 4–6 mm across, 2- to 10-flowered; sepals lanceolate, acuminate or mucronate to short-aristate; petals 2–3 mm long, ovate, obtuse or apiculate, longer than the sepals; stamens 3, nearly equaling the sepals to slightly exceeding the petals; anthers purplish brown; capsules obovoid, obtuse to truncate, beakless, slightly shorter than to exceeding the petals by 1 mm; seeds oblongoid, approximately 0.5 mm long, apiculate at both ends. June–August.

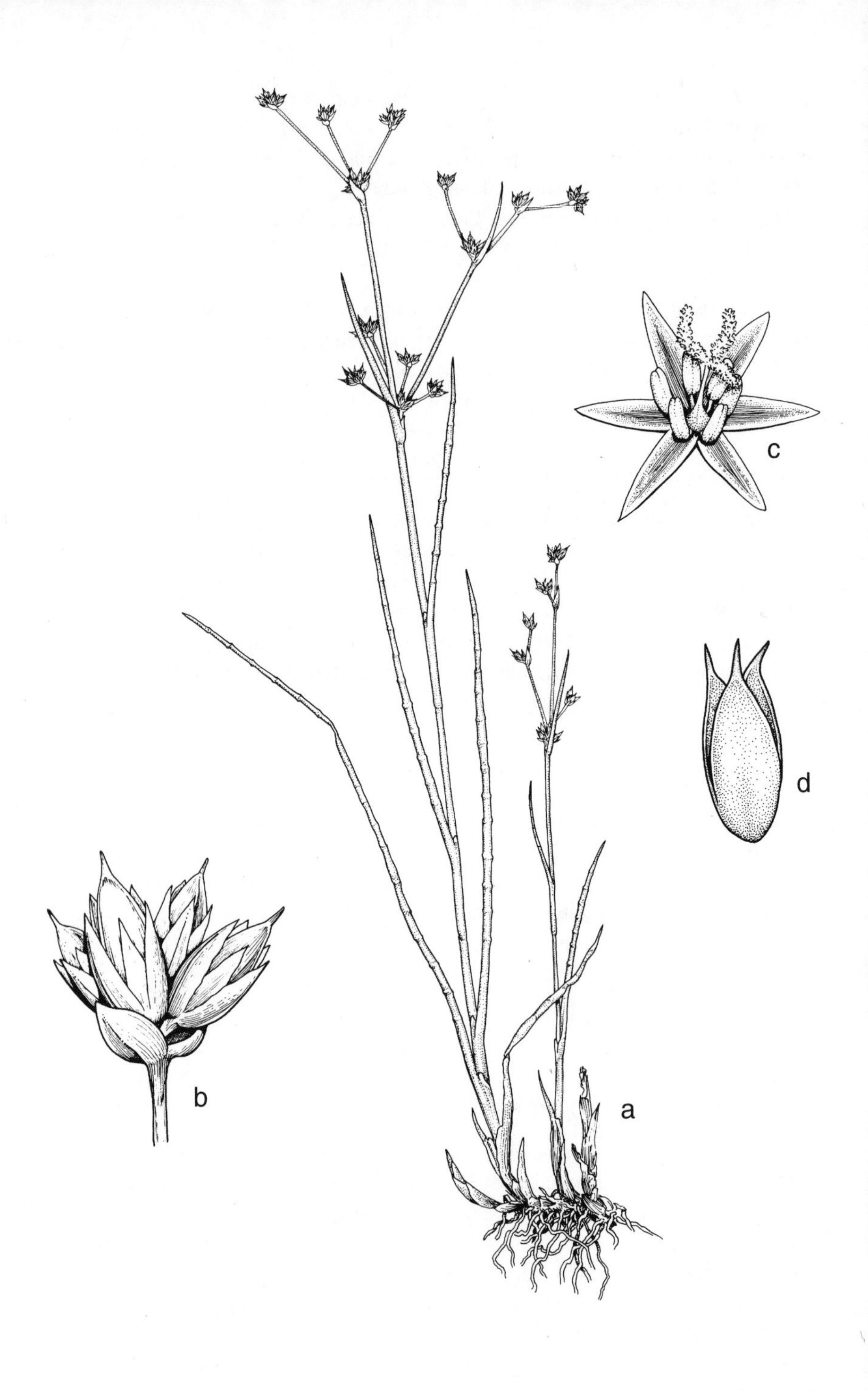

60. ***Juncus articulatus* (Jointed rush).** a. Habit. b. Flowers. c. Flower. d. Open capsule.

61. *Juncus biflorus* (Grass-leaved rush).

a. Habit.
b. Head with capsules.
c. Capsule with perianth.

Wet ditches, along streams, rarely in shallow water.

IL, IN, MO, (FACW), KY, OH (NI).

Grass-leaved rush.

This species is sometimes merged with *J. marginatus*, and both species have flat leaves and red-brown flower clusters. *Juncus biflorus* has 20 or more heads per plant, while *J. marginatus* has 3–18 (–32) heads per plant, and the heads are 4–6 mm across. The leaves of *J. biflorus* are 4–6 mm wide, while the leaves of *J. marginatus* are less than 3 mm wide. Plants with 6–10 flowers per head have been called f. *adinus*.

6. **Juncus brachycarpus** Engelm. Trans. Acad. St. Louis 2:467. 1868. Fig. 62.

Perennial from rhizomes 3–4 mm thick; stems 1–several, (1.5–) 2.3–9.8 dm tall; leaves (2–) 3 (–5), erect or slightly spreading, terete, hollow, cross-septate, 1–2 (–3) mm wide; auricle scarious, acute or truncate, prolonged 2–3 mm beyond point of insertion; involucral leaf shorter than to exceeding the inflorescence by 2 cm; inflorescence compact to open and branched, (1–) 2–9 (–11) cm long; heads 3–10 (–15), spherical, approximately 1 cm across, densely 50- to 80-flowered; perianth segments lanceolate, subulate; sepals 3–4 mm long; petals 2–3 mm long, distinctly shorter than the sepals; stamens 3; anthers shorter than the filaments; capsules drab brown, ellipsoid, acute, mucronate, usually shorter than the petals by 0.5 mm, rarely equaling the petals; seeds 0.4 mm long, apiculate at both ends. June–August.

Wet ditches, along streams, often in prairie areas.

IL, IN, KS, MO, OH (OBL), KY, OH (NI).

Short-fruited rush.

This is one of the species with terete leaves and spherical heads. It differs from *J. scirpoides* by its shorter sepals and fewer heads per plant, and from *J. nodosus* and *J. torreyi* by its smaller heads.

7. **Juncus brachycephalus** (Engelm.) Buch. Bot. Jahrb. 12:268. 1890. Fig. 63.

Perennial from fibrous roots; stems cespitose, 1.4–5.6 dm tall; leaves 3–5, terete, hollow, cross-septate, to 12 cm long, 1.0–3.5 mm wide; auricles membranaceous; involucral leaf much shorter than the inflorescence; inflorescence generally obpyramidal, ascending or widely spreading, 4–19 cm long; heads 13 to approximately 325, wedge-shaped, 2–7 mm wide, 2- to 5- (10-) flowered; perianth segments lanceolate, acuminate to obtuse; sepals 2.0–2.5 mm long; petals 2.0–2.5 mm long, from scarcely exceeding to exceeding the sepals by up to 0.5 mm; stamens 3 or 6; anthers shorter than the filaments; capsules lanceoloid to narrowly ellipsoid, acuminate, exceeding the petals by up to 1.5 mm; seeds fusiform, 0.8–1.2 mm long, caudate at both ends, the tails comprising one-fourth to two-fifths of the total length of the seed. June–August.

Marshes, along rivers and streams, around ponds and lakes, sometimes in shallow water.

IL, IN, KY, OH (OBL).

Short-headed rush.

Of the species of *Juncus* with 35 or more wedge-shaped heads per plant, this is the only one where the perianth is about two-thirds as long as the capsule.

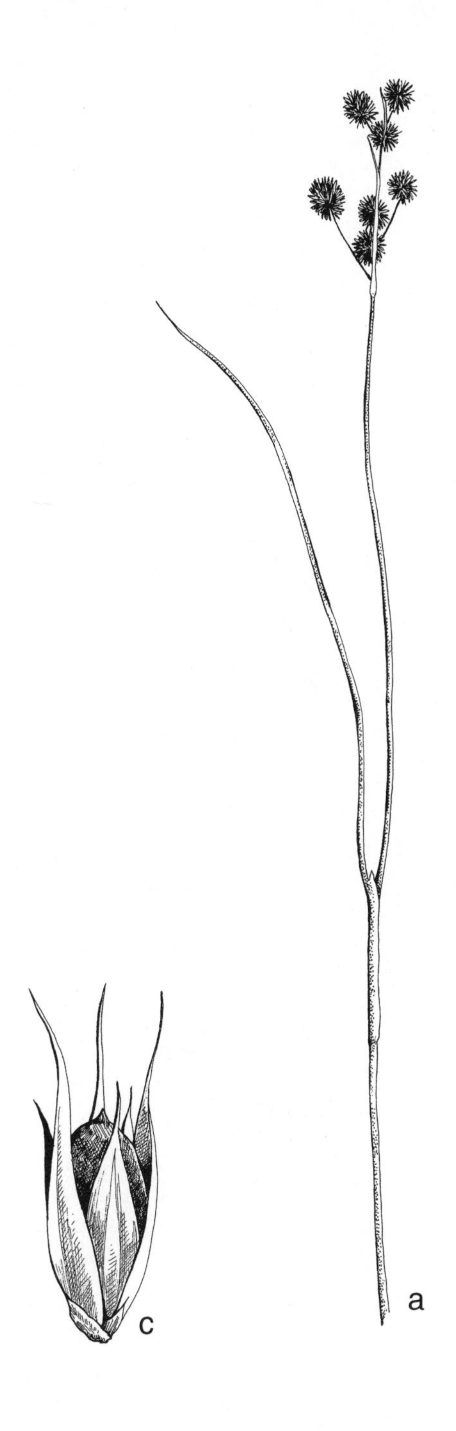

62. *Juncus brachycarpus* (Short-fruited rush).

a. Habit.
b. Flowering head.
c. Capsule with perianth.

63. ***Juncus brachycephalus*** **(Short-headed rush).** a. Habit. b. Head with capsules. c. Seed.

8. **Juncus brachyphyllus** Wieg. Bull. Torrey Club 27:519–520. 1900. Fig. 64.

Cespitose perennial; stems stout, stiff, up to 65 cm tall; leaves flat, 1–2 mm wide; auricles firm, membranaceous, up to 2 mm long; inflorescence a cyme, with many flowers, subtended by involucral bracts as long as or longer than the inflorescence; flowers prophyllate; perianth segments 3.5–6.0 mm long, the sepals longer than the petals, lanceolate-subulate, broadly scarious; stamens 6, anthers equaling the filaments; capsule oblongoid, about as long as wide, dull, slightly shorter than to about equaling the perianth, brownish; seeds oblongoid, 0.2–0.4 mm long, short-apiculate. June–August.

Wet prairies, salt flats, upland woods, sometimes in the depressions of the prairies and salt flats.

MO, NE.

This species is not listed by the U.S. Fish and Wildlife Service.

Small-leaved rush.

This species inhabits shallow depressions in wet prairies and salt flats. It is distinguished by its flat leaves and its dull capsules that are about as long as wide.

9. **Juncus canadensis** J. Gay ex Laharpe, Monogr. Junc. 134. 1825. Fig. 65.

Juncus canadensis J. Gay var. *longecaudatus* Engelm. Trans. Acad. St. Louis 2:474. 1868.

Juncus canadensis J. Gay var. *paradoxa* Farw. Pap. Mich. Acad. I, 30:60. 1944.

Juncus canadensis J. Gay var. *typicus* Fern. Rhodora 47:129. 1945.

Juncus canadensis J. Gay var. *typicus* Fern. f. *conglobatus* Fern. Rhodora 47:129. 1945.

Perennial; stems cespitose, 3.0–9.5 dm tall; leaves 3–4, to 3.2 dm long, 1–3 mm wide; involucral leaf shorter than the inflorescence; inflorescence compact to spreading, 2.5–18.0 cm long; heads 5–60 (to approximately 140), approximate to distant, hemispherical to spherical, 5- to 50-flowered; perianth segments lanceolate, acuminate to subulate; sepals 2.3–3.8 mm long; petals mostly exceeding the sepals by up to 0.2 mm, occasionally equaling the sepals; stamens 3, anthers shorter than the filaments; capsule ovoid to lanceoloid, acute or acuminate, slightly shorter than to exceeding the petals by 1.3 mm; seeds fusiform, caudate, 1.2–1.9 mm long, the tail comprising (one-third) one-half to five-eighths of the total length of the seed. June–September.

Bogs, swamps, wet meadows, ditches, marshes, around lakes and ponds.

IA, IL, IN, KY, MO, OH (OBL), NE (not listed for Nebraska by U.S. Fish and Wildlife Service).

Canada rush.

This species has the largest hemispherical head of any species in the central Midwest. The heads of *J. acuminatus* are about half as large. This species may form asexual plantlets, instead of flowers

10. **Juncus coriaceus** Mack. Bull. Torrey Club 56:28. 1929. Fig. 66.

Cespitose perennial; stems spreading, to 50 cm tall; leaves 1–2, terete, basal, persisting for up to two years, with brownish sheaths; inflorescence appearing lateral, more or less open, with an involucral bract extending beyond the inflorescence for 3 or more cm;

64. *Juncus brachyphyllus* **(Small-leaved rush).** a. Habit. b. Auricle. c. Capsule with perianth.

65. *Juncus canadensis* (Canada rush).

a. Habit.
b. Capsule with perianth.
c. Seed.

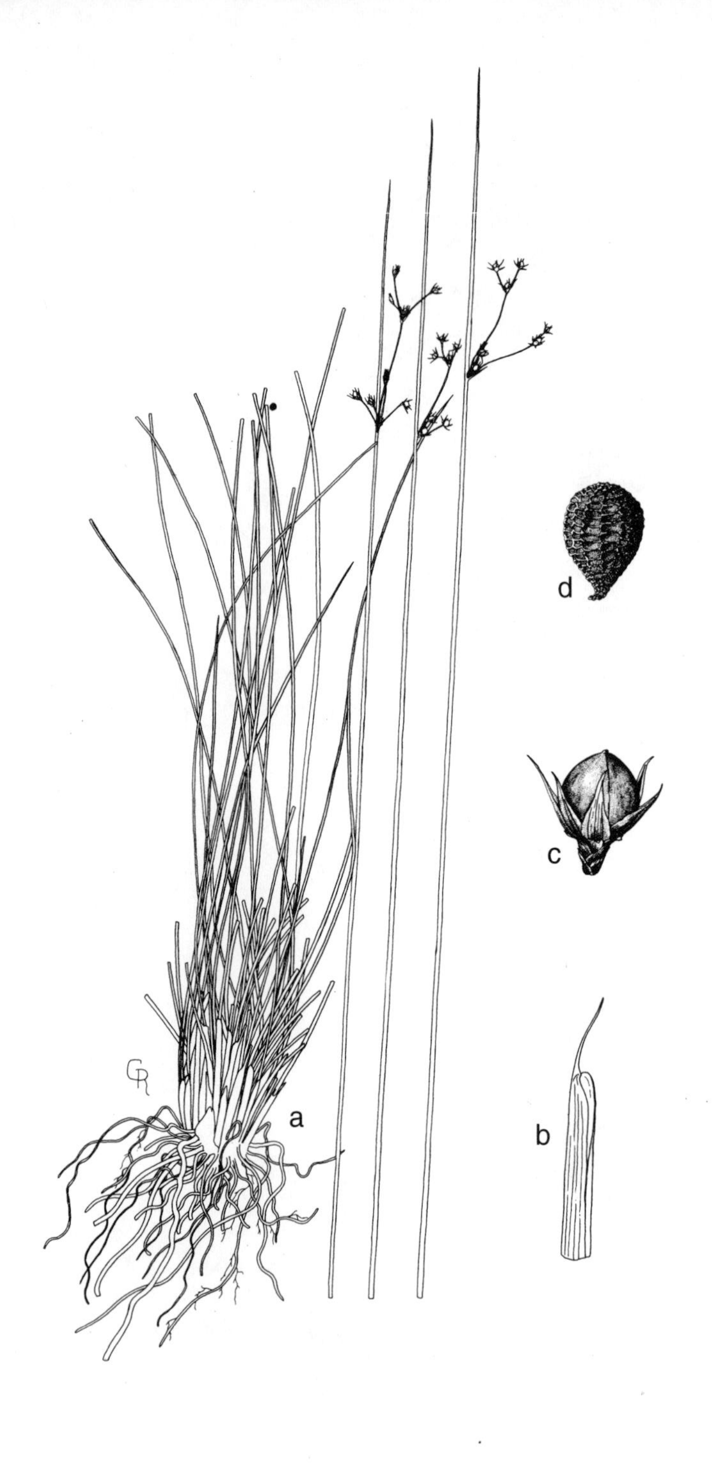

66. ***Juncus coriaceus*** **(Leather rush).** a. Upper part of plant. b. Auricle. c. Capsule with perianth. d. Seed.

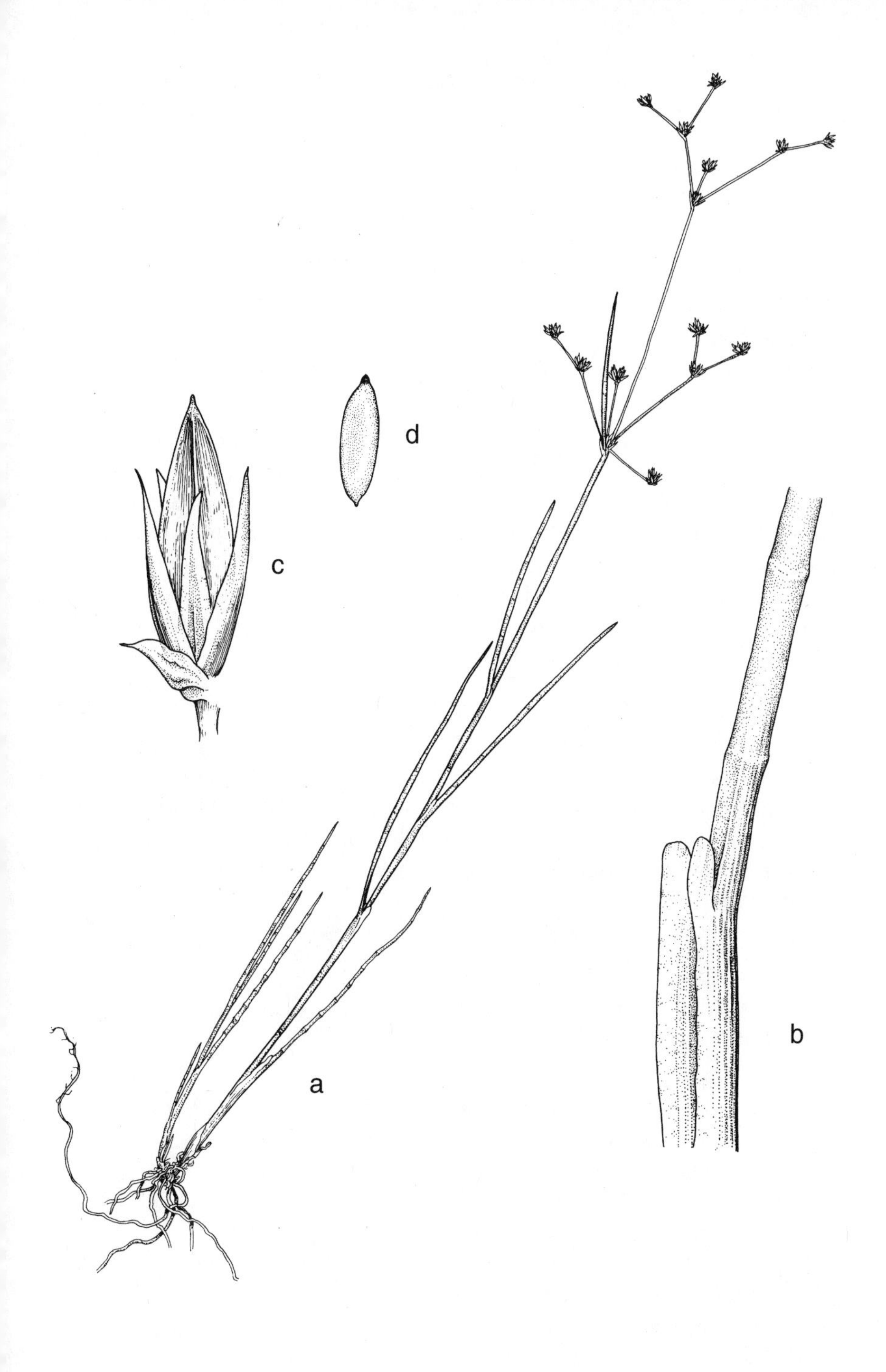

67. ***Juncus debilis*** **(Weak rush).**

a. Habit.
b. Auricle.
c. Capsule with perianth.
d. Seed.

sepals and petals about equal, lanceolate, 3.5–5.0 mm long, widely spreading; capsule brown, shiny, globose, a little shorter than the perianth segments; seeds angular, dark brown, distinctly reticulate, 0.5–0.6 mm long. June–September.

Swamps, wet ditches.

KY (FACW+).

Leather rush.

This is the only species of *Juncus* in the central Midwest with a lateral inflorescence and basal leaves.

11. **Juncus debilis** Gray, Man. Bot. N. U.S. 506. 1848. Fig. 67.
Juncus acuminatus Michx. var. *debilis* (Gray) Engelm. Trans. Acad. Sci. St. Louis 2:463. 1868.

Cespitose perennial from short rhizomes; stems slender, erect or spreading when in water, to 40 cm long; leaves terete, septate, 0.5–1.0 mm wide, the lowest leaf sometimes reduced to a sheath; involucral leaf shorter than the inflorescence; inflorescence diffuse, with several widely divergent rays; heads 3–35, 3–5 mm wide, wedge-shaped, 2- to 5- (to 10-) flowered; perianth segments linear-lanceolate, acuminate; sepals 1.2–2.8 mm long, about as long as the petals; stamens 3; capsules narrowly ovoid to ellipsoid, acute, longer than the perianth, pale cinnamon-brown; seeds narrowly ellipsoid, 0.3–0.4 mm long, apiculate. May–August.

In streams.

IL, KY, MO (OBL).

Weak rush.

This species has a diffuse inflorescence with up to 30 small clusters of flowers per plant. It differs from the similar appearing *J. alpinoarticulatus* and *J. articulatus* by the lack of an apiculus on the seed, by its pale cinnamon-brown seeds, and by its very narrow leaves. In Missouri streams, it often forms lax stems that produce leafy offshoots in the inflorescence.

12. **Juncus dichotomus** Ell. Sketch Bot. S. C. & Ga. 1:406. 1817. Fig. 68.
Juncus tenuis Willd. var. *dichotomus* (Ell.) A. Wood, Class-book Bot. 726. 1861.

Cespitose perennial; stems up to 1 m tall; leaves terete, rigid, up to 1 mm wide, sometimes slightly channeled; auricles rigid, rounded; inflorescence more or less open, subtended by involucral bracts, the lowest bract usually longer than the inflorescence; flower single, prophyllate; sepals lance-subulate, 3.0–4.5 mm long; petals similar but slightly shorter than the sepals; stamens 3; capsule ovoid to oblongoid, mucronate, shorter than the perianth; seeds dark brown, apiculate, 0.4–0.5 mm long. June–October.

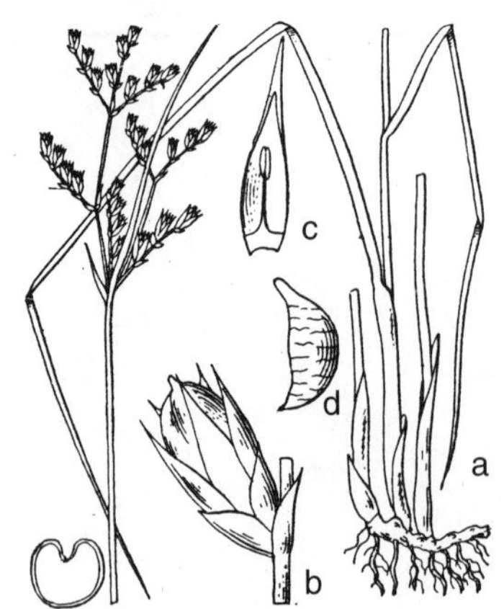

68. *Juncus dichotomus* **(Dichotomous rush).** a. Habit. b. Capsule with perianth. c. Stamen and perianth part. d. Seed.

Marshes.
OH (OBL).
Dichotomous rush.
This species has the widest leaves of any of the prophyllate species of *Juncus.*

13. **Juncus diffusissimus** Buckl. Proc. Acad. Phila. 14:9. 1862. Fig. 69.
Juncus acuminatus Michx. var. *diffusissimus* (Buckl.) Engelm. Trans. Acad. St. Louis 2:463. 1868.

Perennial; stems cespitose, (0.9–) 1.1–2.6 mm tall, 1–2 mm wide; leaves 2–3, 0.5–2.0 mm wide; involucral leaf shorter than the inflorescence; inflorescence diffusely branched, spreading, (6–) 8–16 cm long, comprising two-fifths to one-third the height of the plant in robust individuals; heads approximately 50–150, as few as 4–9 in small specimens, hemispherical or obpyramidal, 2- to 10-flowered; perianth segments lanceolate, subulate; sepals 2.2–3.0 mm long; petals equaling to often slightly shorter than the sepals; stamens 3, anthers shorter than the filaments; capsule linear-lanceoloid, acute, mucronulate, exceeding the perianth by at least 1.5 mm and usually 2 mm, rarely more; seeds narrowly ovoid, 0.4–0.5 mm long. May–September.

Ditches, sloughs, fens, around lakes and ponds, along rivers and streams.
IL, IN, MO, KY, OH (FACW), KS (FACW+).
Slimpod rush.
This species is readily identified by its long, slender capsules.

14. **Juncus dudleyi** Wieg. Bull. Torrey Club 27:524. 1900. Fig. 70.
Juncus tenuis Willd.. var. *dudleyi* (Wieg.) Hermann. Johnston, Journ. Arn. Arb. 25:56. 1944.

Perennial; stems cespitose, 3–9 dm tall; leaves flat or involute, 0.5–1.0 mm wide, 1–3 dm long; auricles cartilaginous, opaque, rigid, often slightly flaring, rounded at apex, yellow to orange-brown, less than 1 mm long and usually 0.75 mm long or less, not conspicuously prolonged beyond point of insertion; involucral leaves usually 2, exceeding the inflorescence by up to 13 cm; inflorescence rays erect to spreading; inflorescence 1–7 (–8) cm long; flowers approximate to distant, occasionally secund on spreading rays; prophylls ovate, obtuse to acute; perianth segments lanceolate, acuminate, spreading; sepals (3.5–) 4.0–6.0 mm long; petals slightly shorter than to equaling the sepals; stamens 6, anthers shorter than the filaments; capsule ovoid, obtuse or truncate, apiculate, usually three-fourths to seven-eighths the length of the sepals; seeds oblongoid or ovoid, 0.4–0.5 mm long, apiculate at both ends. June–September.

Along streams and rivers, around ponds and lakes, fens, seeps, meadows, some upland sites, rarely in standing water.
IA, IL, IN, KS, MO, NE (FAC), KY, OH (FAC-). The U.S. Fish and Wildlife Service considers this species synonymous with *J. tenuis.*
Dudley's rush.
Although this species is similar in appearance to *J. interior* and *J. tenuis,* it is distinguished by its yellow to orange-brown auricles of the leaves.

69. ***Juncus diffusissimus*** **(Slimpod rush).**

a. Habit.
b. Head with capsules.
c. Capsule with perianth.

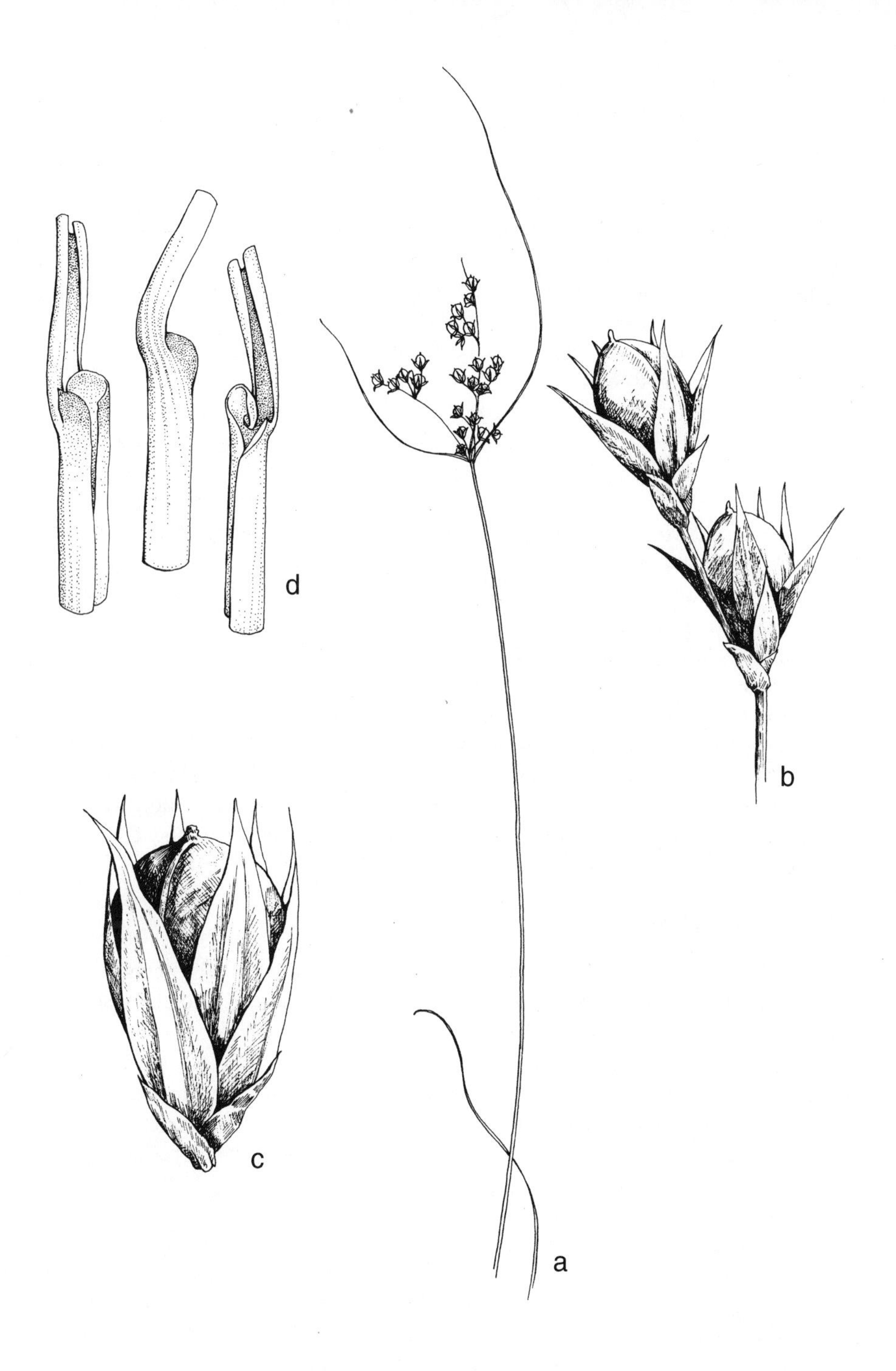

70. ***Juncus dudleyi***
(Dudley's rush).

a. Habit.
b. Portion of ray with capsules.
c. Capsule with perianth.
d. Sheaths, showing auricles (left auricle cut away in illustration on extreme right).

15. **Juncus effusus** L. var. **solutus** Fern. & Wieg. Rhodora 12:90. 1910. Fig. 71.
Juncus bogotensis HBK. var. *solutus* (Fern. & Wieg.) Farw. Pap. Mich. Acad. I, 23:127. 1937.

Perennial; stems densely cespitose from obscured rhizomes with inconspicuous internodes, smooth, (6–) 10–15 dm tall, 1.5–4.0 mm wide; basal sheaths 3–4, clasping, brown to purplish brown, obtuse, mucronate, the uppermost sheath 9–24 cm long with a mucro 2–4 mm long; involucral leaf erect, terete, appearing like the extension of the stem, 7–48 cm long, exceeding the inflorescence; inflorescence appearing lateral on the stem, irregularly spreading on unequal, compound rays, occasionally rounded and condensed, 1.5–9.0 cm long, 1.75–11.50 cm wide; perianth segments rigid, lance-attenuate, greenish, brown, or stramineous; sepals 2.0–3.5 mm long; petals 0.25–0.50 mm shorter than to equaling the sepals; stamens 3, anthers 0.5–0.8 mm long, approximately as long as the filaments; capsule brown, obovoid, truncate or retuse, beakless, 2.5–3.0 mm long, 0.5 mm shorter than to 0.75 mm longer than the sepals; seeds oblong-ovoid, 0.5 mm long, minutely apiculate at both ends. June–September.

Ditches, sloughs, swamps, meadows, around lakes and ponds, along rivers and streams, often in shallow water.

IA, IL, IN, KS, MO, NE (OBL), KY, OH (FACW).

Soft rush.

This and *J. filiformis* are the only species of bladeless rushes in the central Midwest that have smooth rather than ribbed stems. *Juncus effusus* grows in tufts, while *J. filiformis* grows singly along an elongated rhizome. The involucral leaf that appears like the continuation of the stem is at least 7 cm long, while the involucral leaf of the species of the sedge genus *Schoenoplectus* is always less than 7 cm long. *Juncus pylaei* is very similar but has distinctly ribbed stems.

16. **Juncus elliottii** Chapm. Fl. S. U.S. 494. 1860. Fig. 72.

Tufted perennial from short rhizomes with tuberous thickenings; stems slender, erect, to 70 cm tall; leaves terete, septate, 1–3 mm in diameter; involucral leaf shorter than the inflorescence; inflorescence diffuse with several widely divergent rays, twice as long as broad; heads 35–100, 3–5 mm wide, hemispherical, 2- to 7-flowered; perianth segments lanceolate, aristulate; sepals 2.2–3.0 mm long, about equaling the petals; stamens 3; capsule oblongoid-trigonous, abruptly short-acuminate to apiculate, dark purple-brown, 2.5–3.0 mm long, about equaling the perianth; seeds ellipsoid 0.5–0.6 mm long. May–September.

Wet areas, infrequent in shallow water.

KY (FACW).

Elliott's rush.

This species has small, hemispherical heads about 3–5 mm wide, while the hemispherical heads of *J. acuminatus* are usually 5–10 mm wide. *Juncus elliottii* has the general appearance of *J. nodatus* because of its open inflorescence, but it differs in having only 35–100 heads per plant and having dark purple-brown capsules.

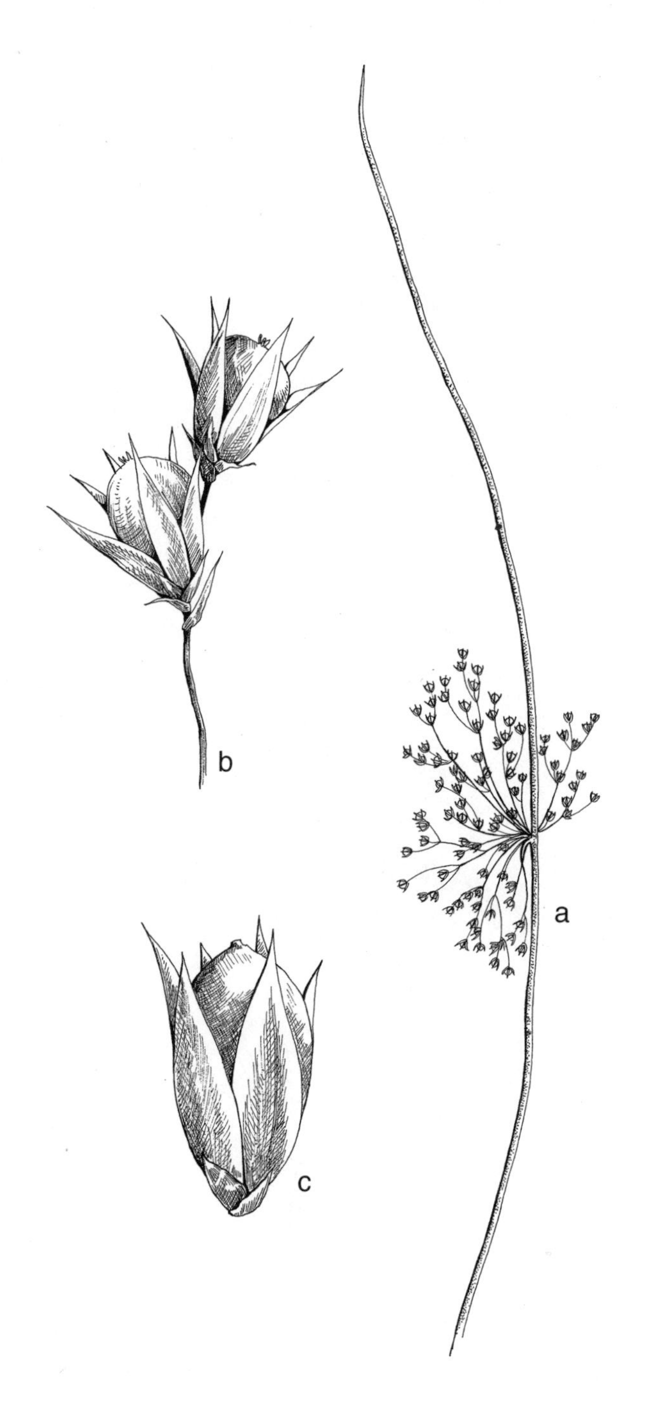

71. *Juncus effusus* var. *solutus* (Soft rush).

a. Habit.
b. Portion of ray with capsules.
c. Capsule with perianth.

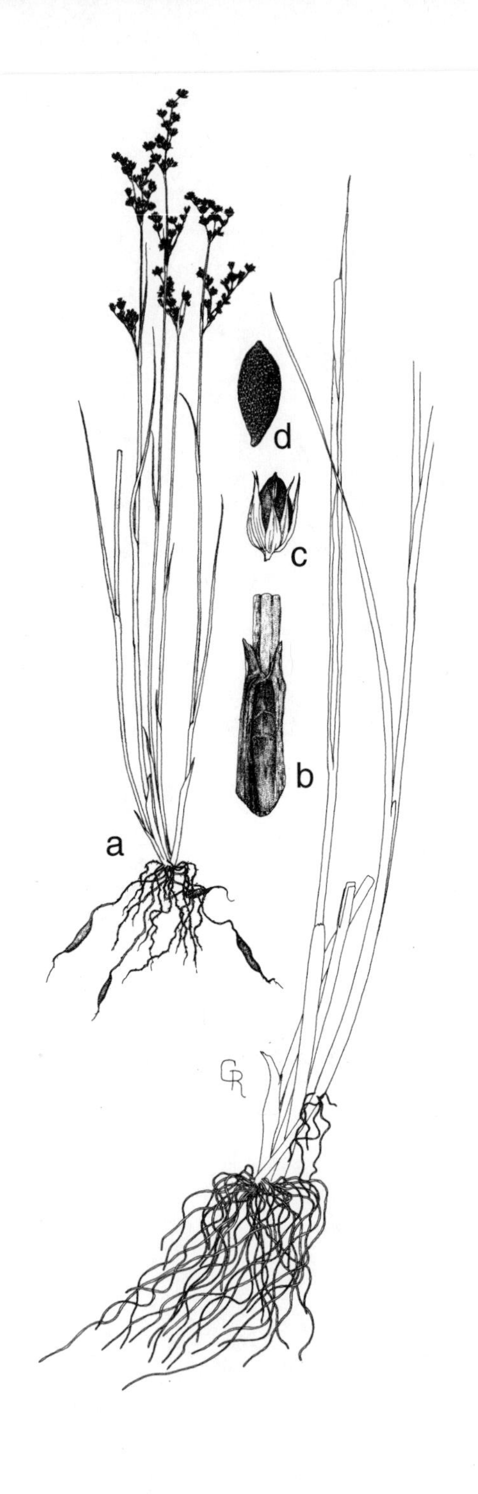

72. *Juncus elliottii* (Elliott's rush).

a. Habit.
b. Auricles.
c. Capsule with perianth.
d. Seed.

17. **Juncus filiformis** L. Sp. Pl. 1:326. 1753. Fig. 73.

Perennial from elongated rhizomes; stems arising singly along the rhizomes, slender, smooth, up to 1 m tall; leaves reduced to sheaths, the sheaths obtuse, mucronate; inflorescence appearing lateral, the involucral bract appearing like an extension of the stem, terete, 3–40 cm long; flowers on unequal, compound rays; sepals and petals lance-subulate, 2.5–3.5 mm long, about equal; stamens 3, the anthers and filaments about equal; capsule brown, ovoid, beakless, 2–3 mm long; seeds oblong-ovoid, about 0.5 mm long, apiculate at either end. June–August.

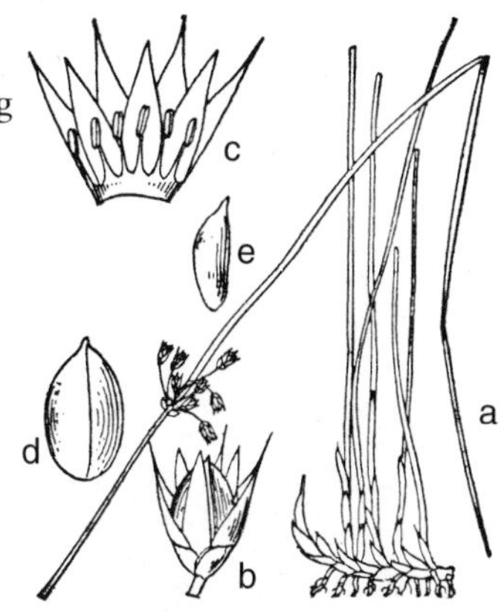

73. *Juncus filiformis* (Thread rush). a. Habit. b. Capsule with perianth. c. Perianth (open) and stamens. d, e. Seeds.

Lakes, ponds, depressions in meadows.

NE.

This species is not listed for Nebraska by the U.S. Fish and Wildlife Service.

Thread rush.

This species is distinguished by its slender stems with lateral inflorescences. It differs from *J. effusus* by its single stems arising from an elongated rhizome.

18. **Juncus longistylis** Torr. in W. H. Emory, Rep. U.S. Mex. Bound. 2 (1):223. 1859. Fig. 74.

Perennials with rhizomes; stems slender, flattened, to 70 cm tall; leaves mostly basal, shorter than the culms, flat, 1.5–3.0 mm wide; involucral bract usually shorter than the inflorescence; inflorescence cymose, with 2–15 heads; heads 2- to 12-flowered, without prophylls; perianth segments broadly lanceolate, acute, with scarious margins; sepals 5–6 mm long, equaling or slightly longer than the petals; stamens 6; capsule oblongoid, long-beaked, about as long as the perianth; seeds oblongoid, apiculate, 0.2–0.4 mm long. June–September.

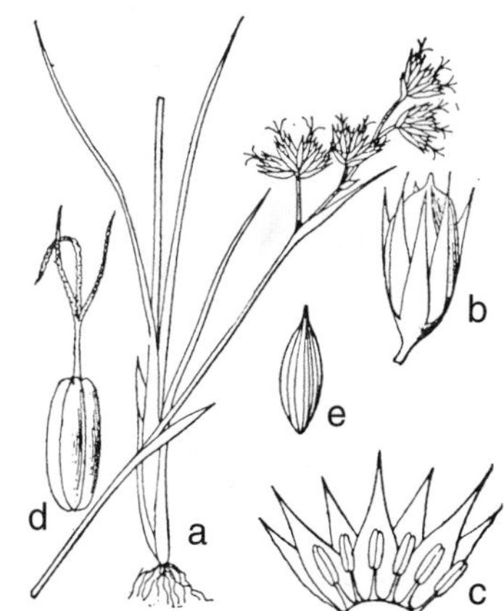

74. *Juncus longistylis* (Long-styled rush). a. Habit. b. Capsule with perianth. c. Perianth (spread) with stamens. d. Pistil. e. Seed.

Swamps, ditches, depressions in wet prairies, marshes, along streams.

NE (FACW).

Long-styled rush.

This western species is distinguished by its flat leaves and its long perianth segments that are 5–6 mm long.

19. **Juncus marginatus** Rostk. Monog. Junc. 38:pl. 2, f. 3. 1801. Fig. 75.
Juncus aristulatus Michx. Fl. Bor. Am. 1:192. 1803.
Tristemon marginatus Raf. Fl. Tellur. 4:32. 1836.

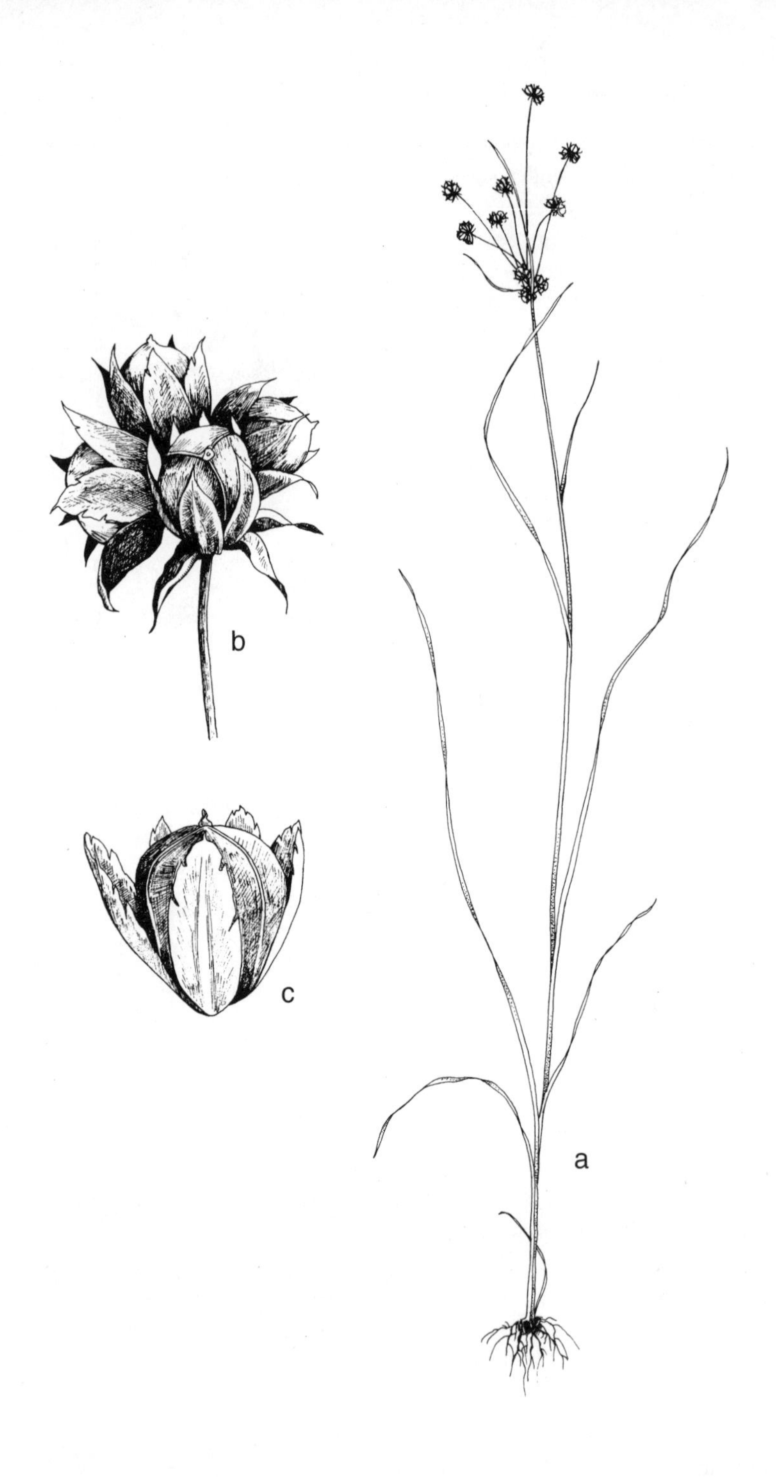

75. ***Juncus marginatus*** **(Flat-leaved rush).**

a. Habit.
b. Head with capsules.
c. Capsule with perianth.

Juncus marginatus Rostk. var. *vulgaris* Engelm. Trans. Acad. St. Louis 2:455. 1866.
Juncus marginatus Rostk. var. *paucicapitatus* Engelm. Trans. Acad. St. Louis 2:455. 1866.

Perennial; stems cespitose, from short, scaly, usually inconspicuous rhizomes; stems 0.45–5.90 dm tall; leaves 1.0–2.5 (–2.9) mm wide; auricles pale brown; involucral leaf inconspicuous, shorter than the inflorescence; inflorescence compact to spreading, the rays ascending to divaricate; inflorescence (1.5–) 2.5–9.3 (–11.0) cm long; heads 3–18 (–32), approximate to distant, 2- to 10- (15-) flowered; sepals lanceolate, acuminate or mucronate to short-aristate; petals 2.2–3.0 mm long, ovate, obtuse or apiculate, longer than the sepals; stamens 3, nearly equaling the sepals to equaling the petals, anthers purplish brown; capsule obovoid, obtuse to truncate, beakless, slightly shorter than to equaling the petals; seeds oblongoid, approximately 0.5 mm long, apiculate at both ends. June–September.

Wet, often sandy ground of pond borders, wet ditches, fields, wet prairies, marshes, seldom in shallow water.

IA, IL, IN, KS, KY, MO, NE, OH (OBL).

Flat-leaved rush.

The inconspicuous rhizomes of *J. marginatus* occasionally have offset rhizomes to 1.2 cm long. *Juncus biflorus,* in which the culms arise singly approximately 3.5 cm apart, often has offset rhizomes reaching a length of 6.5 cm. Since the flowers of the two species are essentially similar and the numbers of heads and flowers per head are variable, they are usually best separated by observation of the underground parts.

20. **Juncus militaris** Bigel. Fl. Bost. 2 ed., 139. 1824. Fig. 76.

Perennial with thick rhizomes; stems stout, to 1 m tall; leaves 1 (–2), to 30 (–50) cm long, terete, septate, 2–4 mm in diameter, usually longer than the stems; inflorescence branches ascending; inflorescence to 15 cm long, with 2–25 heads; heads hemispherical, 3–10 mm in diameter, 5- to 25- (30-) flowered; perianth segments lanceolate to lance-subulate; sepals 2.5–4.0 mm long, aristulate; petals equaling the sepals, acute; stamens 6; capsule ovoid-trigonous, beaked, slightly shorter than to a little longer than the perianth segments; seeds ovoid to oblongoid, apiculate, 0.5–0.6 mm long. July–September.

Shallow water.

IN (OBL).

Military rush.

This northern species grows in shallow water where it usually forms tufts of capillary leaves.

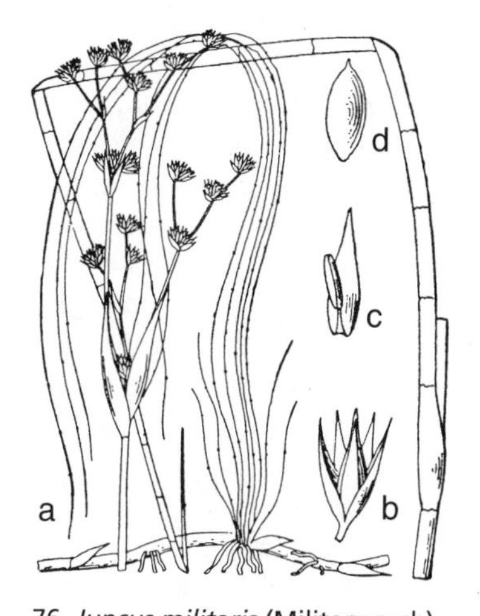

76. *Juncus militaris* (Military rush).
a. Habit. b. Capsule with perianth.
c. Stamen and one perianth part.
d. Seed.

21. **Juncus nodatus** Coville. Britt. & Brown, Illustr. Fl. N. U.S., ed. 2, 1:482. 1913. Fig. 77.
Juncus acuminatus Michx. var. *robustus* Engelm. Trans. Acad. St. Louis 2:463. 1868.
Juncus robustus (Engelm.) Coville. Britt. & Brown, Illustr. Fl. N. U.S., ed. 2, 1:395. 1896, non Wats. (1879).

Perennial; stems single or two together, 5.9–8.3 m tall, 3–10 mm wide, stout; leaves terete, strongly nodose-septate, 1.5–4.5 mm wide, to 5.1 dm long; involucral leaf much shorter than the inflorescence; inflorescence diffuse with numerous, widely divergent rays, 1.5–2.7 dm long; heads 150–280, 3–5 mm wide, wedge-shaped, 2- to 7- (8-) flowered; perianth segments lanceolate, acuminate; sepals 2.0–2.5 mm long; petals slightly shorter than to equaling the sepals; stamens 3; capsule ellipsoid, acute or obtuse, sometimes mucronulate, nearly equaling to exceeding the sepals by 0.75 mm; seeds narrowly ellipsoid, 0.5–0.6 mm long, minutely apiculate at both ends. June–September.

Ditches, around ponds and lakes, along rivers and streams, sloughs.

IL, IN, KS, MO (OBL), KY, OH (NI).

Diffuse rush.

This stout species with a much branched inflorescence may have as many as 280 heads per plant.

22. **Juncus nodosus** L. Sp. Pl. ed. 2:466. 1762. Fig. 78.
Juncus nodosus L. var. *vulgaris* Torr. Fl. N. Y. 2:326. 1843.
Juncus nodosus L. var. *genuinus* Engelm. Trans. Acad. St. Louis 2:471. 1868.
Juncus nodosus L. var. *proliferus* Lunell, Am. Midl. Nat. 4:238. 1915.

Perennial from rhizomes with numerous tuberous thickenings; stems to 6 dm tall, to 3 mm wide; leaves 2–5, erect or ascending, terete, septate, 1–2 dm long, 1.0–1.5 (–2.0) mm wide; auricles membranaceous, yellowish, prolonged 0.5–1.0 (–3.0) mm beyond point of insertion; involucral leaf exceeding the inflorescence; inflorescence compact to spreading, 1–5 (–7) cm long; heads (1–) 2–15, spherical, 8–11 (–12) mm wide, 9- to 35-flowered; perianth segments lanceolate, subulate; sepals 2.5–4.0 mm long; petals equaling to exceeding the sepals by 0.8 mm; stamens 6, anthers shorter than the filaments; capsule lanceoloid, subulate, equaling to exceeding the perianth by 1.5 mm; seeds ovoid or oblongoid, 0.5 mm long, apiculate at both ends. June–September.

Swamps, marshes, wet depressions in prairies and bottomland forests, bogs.

IA, IL, IN, KS, MO, NE, OH (OBL).

Round-headed rush.

Sometimes the flowers are replaced by proliferous tufts of leaves. These strange-looking plants have been called var. *proliferus.* The spherical heads are 8–11 mm in diameter, smaller than the spherical heads of *J. torreyi* and larger than the spherical heads of *J. brachycarpus* and *J. scirpoides.*

23. **Juncus pelocarpus** E. Meyer, Syn. Junc. 30. 1823. Fig. 79.

Colony-forming perennial with rhizomes; stems slender, solid, to 50 cm tall; leaves very narrow, more or less flat but septate; inflorescence much branched, to 1.5 cm long, the tip of each branch bearing 1 or 2 flowers, or the flowers sometimes replaced

77. *Juncus nodatus* (Diffuse rush).

a. Habit.
b. Heads with capsules.
c. Capsule with perianth.

78. *Juncus nodosus* (Round-headed rush).

a. Habit.
b. Flowering heads.
c. Capsule with perianth.

by bulblets; prophylls absent; perianth segments oblong, obtuse, scarious along the margins; sepals 1.6–2.3 mm long, a little shorter than the petals; stamens 6, anthers longer than the filaments; capsule narrowly ovoid-ellipsoid, acuminate-beaked, 2.5–3.0 mm long; seeds ovoid, 0.4 mm long. June–October.

Around ponds and lakes, often in shallow water.

IN (OBL).

Colonial rush.

This northern species often occurs around lakes and ponds where water has recently receded. In the autumn, the plants often become reddish. It and *J. validus* are our only species with flat, septate leaves. The stems of *J. pelocarpus* are solid, while those of *J. validus* are hollow.

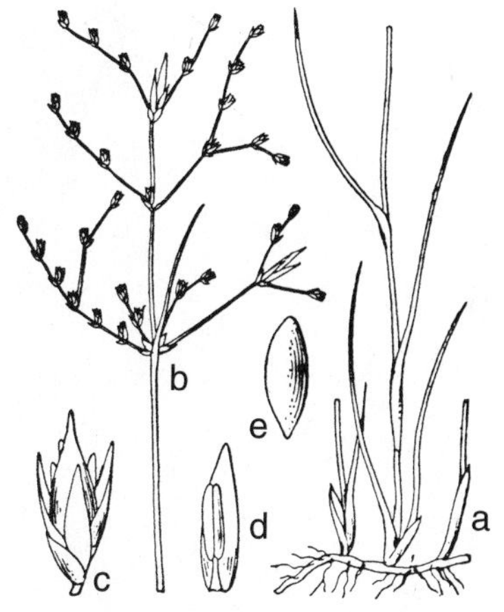

79. *Juncus pelocarpus* (Colonial rush). a. Habit. b. Upper part of plant. c. Capsule with perianth. d. Stamen and one perianth part. e. Seed.

24. **Juncus pylaei** Laharpe, Mem. Soc. Hist. Nat. Paris 3:119. 1827. Fig. 80.
Juncus effusus L. var. *pylaei* (Laharpe) Fern. & Wieg. Rhodora 12:92. 1910.

Perennial; stems densely cespitose, with 10–20 coarse longitudinal ridges, rather soft, terete, to 1 m tall, 2–4 mm wide; basal sheaths 3–4, clasping, dark red-brown to nearly black, up to 20 cm long; leaves absent; involucral leaf erect, terete, to 20 cm long, resembling a prolongation of the stem; inflorescence appearing lateral, irregularly spreading on unequal rays, occasionally rounded and condensed, 1.5–6.0 cm wide; perianth segments rigid, lanceolate, long-acuminate, pale brown; sepals 2.0–3.5 mm long, slightly spreading at the tip; petals a little shorter than to equaling the sepals; stamens 3; capsule brown, ovoid, obtuse to truncate to retuse at the tip, beakless, 2.5–3.0 mm long, shorter than to about equaling the perianth segments; seeds oblong-ovoid, 0.5 mm long, minutely apiculate. June–September.

Boggy areas.

IN.

The U.S. Fish and Wildlife Service has this species synonymous with *J. effusus*, which is OBL.

Rib-stemmed soft rush.

This species in the past usually has been either been combined with *J. effusus* or considered a variety of it. The strongly ribbed stems clearly distinguish it from *J. effusus*.

25. **Juncus scirpoides** Lam. Encycl. Meth. Bot. 3:267. 1789. Fig. 81.
Juncus scirpoides Lam. var. *genuinus* Buch. Bot. Jahrb. 12:323. 1890.

Perennial from short rhizomes; stems cespitose, (0.5–) 1.0–4.6 dm tall, 0.50–2.75 mm wide; leaves terete, septate, 2–3 in number, 1–2 mm wide, to 2.2 dm long; auricles membranaceous, pale, oblong, acute, prolonged 1–2 mm beyond point of

80. *Juncus pylaei* (Rib-stemmed soft rush). a. Habit. b. Capsule with perianth.

81. *Juncus scirpoides* (Sedgelike rush).

a. Habit.

b. Capsule with perianth.

insertion; involucral leaf shorter than to slightly exceeding the inflorescence, terete; inflorescence compact to spreading with several rays, (0.5–) 2–15 cm long; heads 1–34, spherical, 7–12 mm in diameter, 20- to approximately 70-flowered; perianth segments lanceolate, subulate, rigid; sepals (2.5–) 3.0–3.5 mm long; petals nearly equaling to approximately 1 mm shorter than the sepals; stamens 3, anthers shorter than the filaments; capsule oblongoid, subulate, exceeding the sepals by 0.75–1.00 mm; seeds ovoid, 0.4–0.5 mm long, apiculate at both ends. June–September.

Around ponds and lakes, wet prairies, usually in sandy areas.

IL, IN, KS, NE, MO (FACW+), KY, OH (FACW).

Sedgelike rush.

This is one of the species of *Juncus* with spherical heads. It differs from *J. brachycarpus* by its shorter sepals and fewer heads per plant, and from *J. validus* by its terete leaves.

26. **Juncus subcaudatus** (Engelm.) Coville & S. F. Blake, Proc. Bot. Soc. Wash. 31:45. 1918. Fig. 82.
Juncus canadensis J. Gay var. *subcaudatus* Engelm. Trans. Acad. Sci. St. Louis 2:474. 1868.

Cespitose perennial with few short rhizomes; stems to 80 cm tall; leaves terete, septate, up to 15 cm long, 1–2 mm wide, the lowest leaf sometimes reduced to a sheath; involucral bract shorter than the inflorescence; inflorescence with spreading branches, to 15 cm long; heads 6–35, hemispherical, 5–10 mm in diameter, 5- to 10-flowered; perianth segments lanceolate, acuminate; sepals 2.0–3.2 mm long, slightly shorter than the petals; stamens 3; capsule narrowly ovoid to ellipsoid, obtuse to acute, short-beaked, slightly longer than the perianth segments; seeds 0.7–1.2 mm long, with tail-like appendages at each end. July–October.

Along spring branches and streams, fens.

IL, IN, KY, MO, OH (OBL).

Short-tailed rush.

Because of its small hemispherical heads, this species resembles *J. acuminatus.* It differs from *J. acuminatus* by its larger seeds and fewer flowers per head.

27. **Juncus torreyi** Coville, Bull. Torrey Club 22:303. 1895. Fig. 83.
Juncus nodosus L. var. *megacephalus* Torr. Fl. N. Y. 2:326. 1843.
Juncus megacephalus A. Wood, Class-book Bot. ed. 2, 724. 1861, non Curtis (1835).
Juncus torreyi Coville var. *proliferus* Lunell, Am. Midl. Nat. 4:239. 1915.
Juncus torreyi Coville f. *longipes* Farw. Pap. Mich. Acad. I, 1:91. 1921.
Juncus torreyi Coville f. *brepipes* Farw. Pap. Mich. Acad. I, 1:92. 1921.
Juncus torreyi Coville var. *paniculata* Farw. Pap. Mich. Acad. I, 1:92. 1921.
Juncus torreyi Coville var. *globularis* Farw. Pap. Mich. Acad. I, 1:92. 1921.

Perennial from rhizomes with numerous tuberous thickenings; stems stout, to 10.7 dm tall, to 5 mm wide; leaves 2–5, terete, septate, often divaricate, 1.0–4.9 dm long, 1–3 mm wide, auricles membranaceous and yellowish or hyaline, prolonged 2.5–4.0 mm beyond point of insertion; involucral leaf often exceeding the inflorescence; inflorescence compact to spreading, 1–15 cm long; heads 1–21 (–45), spherical, 10–15 mm in diameter, (14-) 25- to approximately 90-flowered; perianth segments lanceolate, subulate; sepals 4–5 mm long; petals 1 mm shorter than to nearly equaling the sepals; stamens 6, anthers shorter than the filaments; capsule

82. *Juncus subcaudatus* (Short-tailed rush).

a. Habit.
b. Capsule with perianth.
c. Seed.
d. Seed (enlarged).

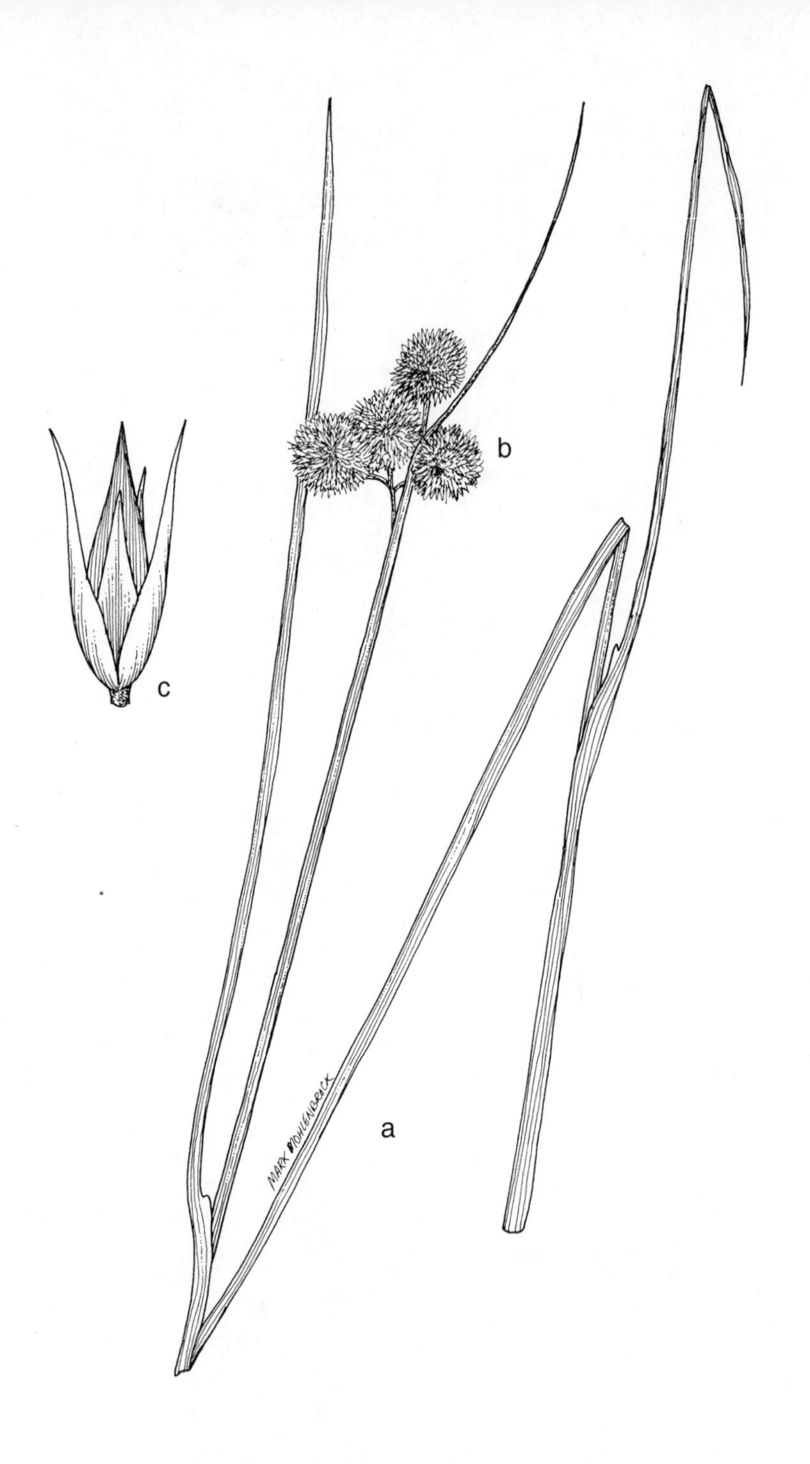

83. *Juncus torreyi* (Big round-headed rush).

a. Habit.
b. Flowering head.
c. Capsule with perianth.

lance-subuloid, slightly shorter than to exceeding the sepals by 1 mm; seeds ovoid or oblongoid, 0.5 mm long, minutely apiculate at both ends. June–September.

Swamps, marshes, sloughs, ditches, along streams, around ponds.

IA, IL, IN, KS, KY, MO, NE, OH (FACW).

Big round-headed rush.

This species has large, spherical heads. It differs from the similar *J. nodosus* by its larger heads, broader leaves, and by having more heads per plant. Several variations have been named for this rather variable species.

28. **Juncus validus** Coville, Bull. Torrey Club 22:305. 1895. Fig. 84.

Cespitose perennial from short rhizomes; stems stout, up to 1 m tall, hollow; leaves septate, more or less flat but tending to be a little terete and hollow, up to 45 cm long, 3–7 mm wide; involucral bract shorter than the inflorescence; inflorescence compact to open and branched, to 15 cm long; heads 15–75, mostly spherical, up to 10 mm in diameter, 25- to 70-flowered; perianth segments linear-lanceolate, acuminate; sepals 3.5–4.5 mm long, slightly longer than the petals; stamens 3; capsule lanceoloid, long-beaked, longer than the perianth segments; seeds 0.3–0.4 mm long, apiculate at both ends. July–September.

Ditches, along streams, marshes, wet prairies.

IL (not listed for IL by the U.S. Fish and Wildlife Service), KS (FAC), MO (FACW).

Stout rush.

This exceptionally stout species has leaves that are nearly flat but are still hollow and septate. It is the only *Juncus* with that type of leaf.

29. **Juncus vaseyi** Engelm. Trans. Acad. St. Louis 2:448. 1866. Fig. 85.

Perennial; stems cespitose from an inconspicuous rhizome, 1 mm wide, 3.7–6.9 dm tall, erect; leaves 1 or 2, terete throughout entire length of blade, septate, 2.1–4.0 dm long, 0.8 mm wide; auricles semimembranaceous, rounded, yellowish or brown, not prolonged beyond point of insertion; involucral leaf shorter than to exceeding the inflorescence by 4 cm; inflorescence compact with erect or ascending rays, (1.0–) 2.0–3.5 cm long, few-flowered; flowers mostly approximate, prophylls obtuse or acute; perianth segments lanceolate, acuminate or aristate, appressed; sepals 3.3–4.6 mm long; petals 0.5 mm shorter than to equaling the sepals; stamens 6, anthers as long as the filaments; capsule greenish or pale yellowish brown, ellipsoid-oblongoid, obtuse or truncate, retuse, slightly shorter than to exceeding the sepals by 1 mm; seeds fusiform, unequally caudate, 1.0–1.3 mm long, the bodies 0.5–0.8 mm long, the tails 0.2–0.4 mm long. June–September.

Bogs, around lakes, wet meadows, marshes.

IA, IL, IN (FACW).

Vasey's rush.

This is the only prophyllate species of *Juncus* to have hollow leaves that are septate but not particularly terete.

21. JUNCAGINACEAE—ARROW-GRASS FAMILY

Perennials with rhizomes; leaves basal, terete, ligulate, usually with open sheaths; inflorescence spicate or of spikelike racemes, without bracts; flowers usually

84. *Juncus validus* (Stout rush).

a. Habit.

b. Capsule with perianth.

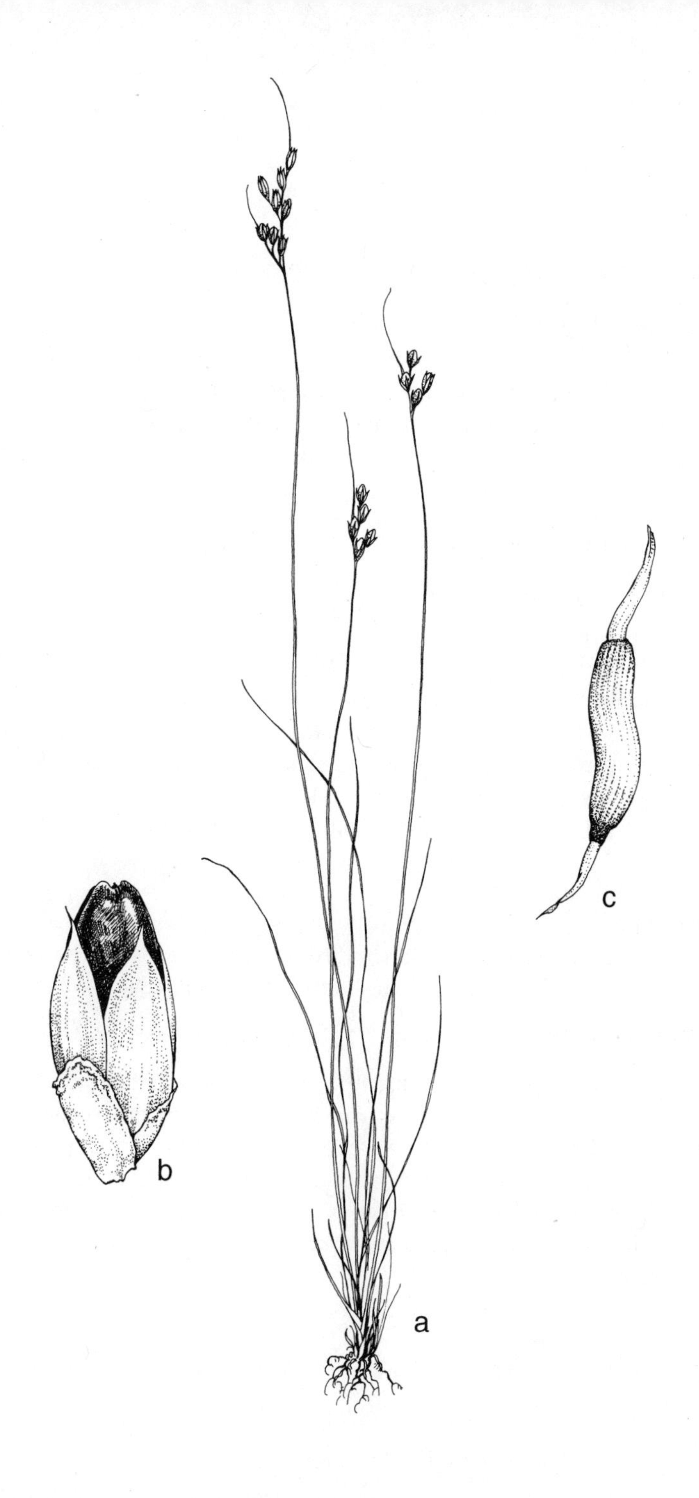

85. *Juncus vaseyi* (Vasey's rush).

a. Habit.
b. Capsule with perianth.
c. Seed.

perfect, small, inconspicuous; perianth composed of two scarcely distinguishable series of three members each; stamens usually 6; ovaries 3–6, adnate to a central axis; fruits composed of 3–6 follicles.

This family consists of five genera and twenty species. *Scheuchzeria,* often included in this family, in now placed in its own family, the Scheuchzeriaceae.

Only the following genus occurs in the central Midwest.

1. **Triglochin** L.—Arrow-grass

Flowers perfect; sepals 3, greenish, deciduous; petals 3, greenish, deciduous; stamens usually 6; carpels united at first, later separating and developing into 3 or 6 follicles; ovule 1; inflorescence a spikelike raceme, without bracts; leaves all basal.

This genus consists of 15 species. The following two occur in the central Midwest.

1. Carpels 6, the axis between them slender .... 1. *T. maritima*
1. Carpels 3, the axis between them broadly 3-winged .... 2. *T. palustris*

1. **Triglochin maritima** L. Sp. Pl. 1:339. 1753. Fig. 86.

Perennial from a stocky, nonstoloniferous rhizome; leaves basal, terete, linear, to 50 cm long, 1–3 mm wide, erect, glabrous; scape usually less than 10 dm tall, bearing a bractless, spikelike raceme, naked below; sepals and petals each 3, greenish, free from each other, deciduous, 1–2 mm long; stamens 6; carpels 6, rounded at base, sharply angled on the margins, the axis between them slender; follicles 6, linear, pointed at the base. April–August.

Marshes, bogs, ditches, swamps, around lakes and ponds, often in alkaline conditions.

IA, IL, IN, KS, NE, OH (OBL).

Arrow-grass.

The distinguishing characteristics of this species are the six sharply angled carpels borne on a slender axis and the linear follicles pointed at the base.

2. **Triglochin palustris** L. Sp. Pl. 1:338. 1753. Fig. 87.

Perennial from a stocky rootstock, with elongated, bulb-bearing stolons; leaves basal, terete, linear, to 30 cm long, 1–2 mm wide, erect, glabrous; scape usually less than 60 cm tall, bearing a bractless, spikelike raceme, naked below; sepals and petals each 3, greenish, free from each other, deciduous, 1.0–1.5 mm long; stamens 6; carpels 3, sharp-pointed at the base, the axis between them broadly 3-winged; follicles 3, narrowly linear. June–September.

Bogs, alkaline marshes.

IA, IL, IN, NE, OH (OBL).

Slender arrow-grass.

The three carpels and follicles are reliable characters that easily separate this species from *T. maritima.*

22. LEMNACEAE—DUCKWEED FAMILY

Free-floating thalloid plants with 0–several ventral rootlets originating at the node; fronds nerved or nerveless; vegetative buds with flowers produced at the node in

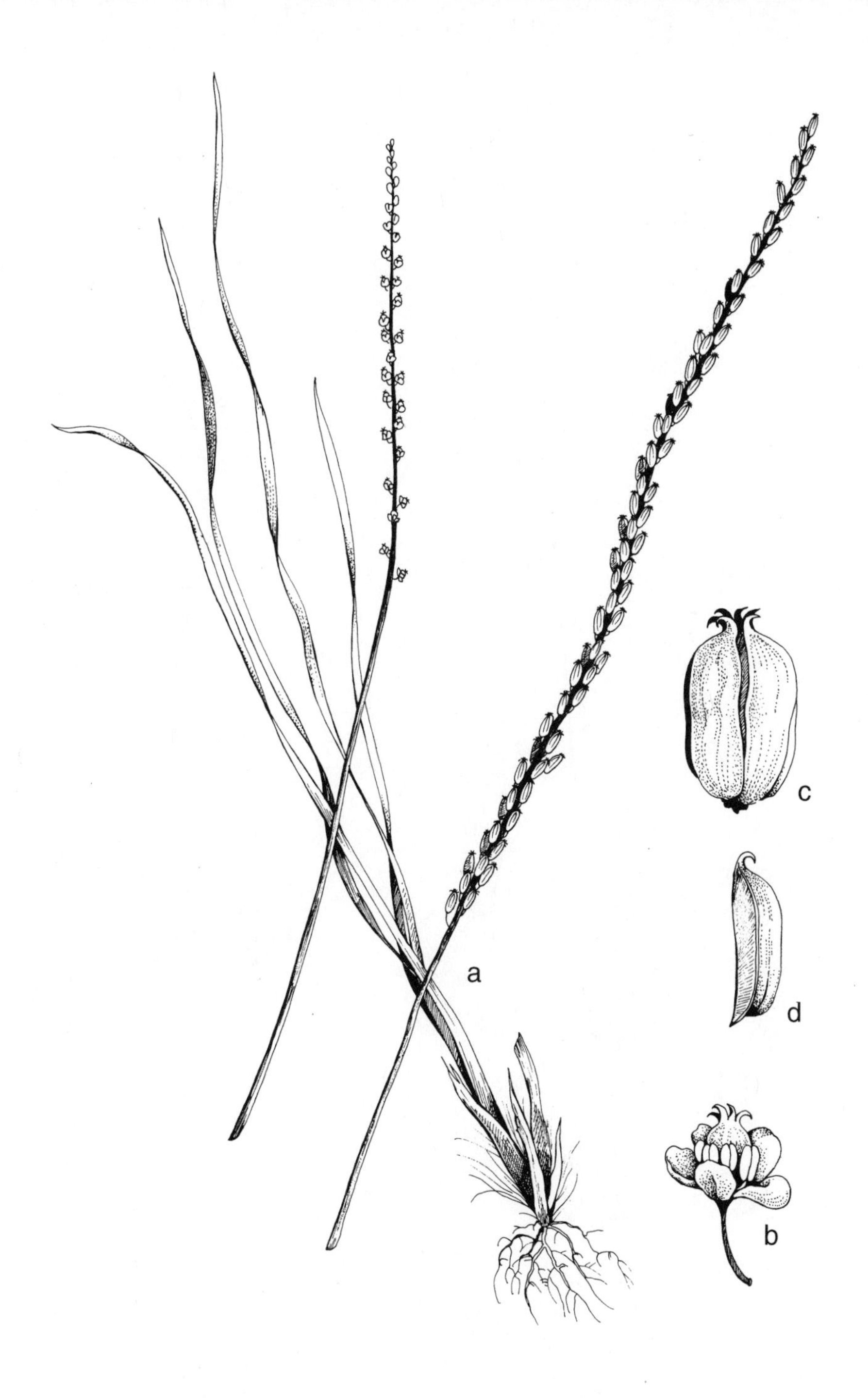

86. *Triglochin maritima* (Arrow-grass).

a. Habit, inflorescence, and fruiting branch.
b. Flower.
c. Cluster of fruits.
d. Single fruit.

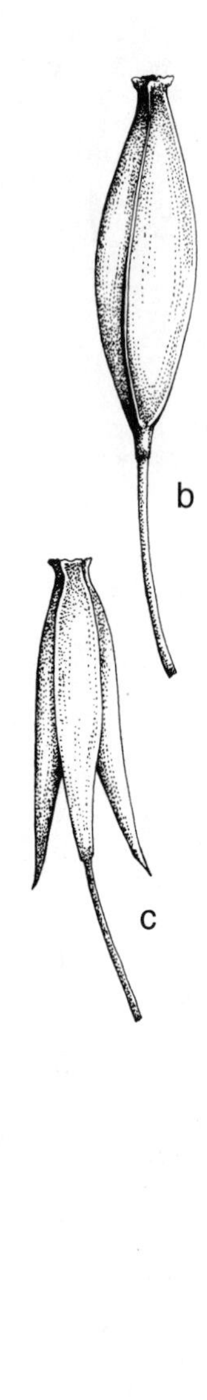

87. *Triglochin palustris* (Slender arrow-grass).
a. Habit.
b. Flower.
c. Follicles.

lateral, basal, or dorsal reproductive pouches; plants monoecious; flowers, in some species, surrounded by a spathe; staminate flowers 1–2 (–3), the anthers 1- to 2-celled; pistillate flower 1, the ovary 1-celled, with 1–several ovules; fruit a utricle, frequently ribbed; seed smooth or ribbed.

Flowering and fruiting specimens are extremely rare in collections and, when they do occur, are so reduced as to offer few characters of classificatory value. Thus, the delimitation of both the genera and the species rests heavily on vegetative features. It is partly because of this, since vegetative characters are variable, that taxonomic difficulties occur in the family.

The term frond is employed when speaking of the thalloid portion of the plant not including rootlets, flowers, and fruits. The term plant is more appropriately used when referring to the latter organs and the frond as a unit. The upper surface of the frond is referred to as the dorsal surface, while the lower surface is designated the ventral surface.

This family consists of four genera and about thirty-five species.

1. Rootlets 1–several per plant; reproductive pouches lateral, 2 per plant.
  2. Rootlet 1 per plant .... 1. *Lemna*
  2. Rootlets (1–) 2 or more per plant .... 2. *Spirodela*
1. Rootlets none; reproductive pouch basal, 1 per plant.
  3. Frond thick, globose or ellipsoid .... 3. *Wolffia*
  3. Frond thin, linear .... 4. *Wolffiella*

1. **Lemna** L.—Duckweed

Fronds flattened to strongly convex on the ventral surface, orbicular to spatulate, symmetrical to asymmetrical, occasionally long-stipitate, 1- to 3- (5-) nerved, the nerves frequently obscure; rootlet 1, the rootcap acute to obtuse; reproduction from two lateral pouches on either side of the node; flowers and fruit sparingly produced; ovules 1–6.

Lemna is the largest and most widespread genus of this family, with fifteen species worldwide.

1. Fronds spatulate with long, persistent stipes, often submerged in compact masses; rootlet frequently absent .... 8. *L. trisulca*
1. Fronds orbicular to elliptic, floating; rootlet present.
  2. Fronds producing rounded, rootless turions (specialized overwintering structures) late in the year; fronds with a distinct line of papillae near apex .... 9. *L. turionifera*
  2. Fronds not producing rounded, rootless turions late in the year; fronds without a papilla or with indistinct lines of papillae.
    3. Fronds 3- or 5-nerved.
      4. Rootlet sheaths cylindrical, unwinged.
        5. Lower surface of frond flattened or only weakly convex, apex rounded to acute; frond usually 3-nerved .... 4. *L. minor*
        5. Lower surface of frond moderately to strongly convex, apex broadly rounded or often obtuse; frond 3- or 5-nerved .... 2. *L. gibba*
      4. Rootlet sheaths winged.
        6. Fronds with a single apical papilla; seeds with 8–26 strong ribs .... 1. *L. aequinoctialis*
        6. Fronds with 2–3 papillae; seeds with 35 or more indistinct ribs.

7. Fronds firm, weakly to strongly asymmetrical, the main nerves often inconspicuous .... 6. *L. perpusilla*

7. Fronds membranous, symmetrical or nearly so, the main nerves prominent .... 7. *L. trinervis*

3. Fronds distinctly 1-nerved or obscurely nerved.

8. Fronds 1-nerved, the lower surface green.

9. Fronds ovate-elliptic, weakly to moderately asymmetrical, often floating in compact masses .... 10. *L. valdiviana*

9. Fronds ovate to orbicular, symmetrical, usually solitary .... 3. *L. minima*

8. Fronds obscurely nerved, the lower surface frequently reddish purple........ 5. *L. obscura*

1. **Lemna aequinoctialis** Welw. Apont. Phyto. 578. 1858. Fig. 88a.

Fronds single or connected in clusters of 2 or 3, ovate to nearly orbicular, symmetrical, firm, flattened, 3-nerved, with a single apical papilla; rootlet 1, the sheath narrowly winged, the tip acute; fruit 0.5–0.8 mm long, with 8–26 strong ribs.

Standing water.

IA, IL, IN, KS, KY, MO, NE (OBL).

Duckweed.

2. **Lemna gibba** L. Sp. Pl. 1:970. 1753. Fig. 89 f–l.

Fronds orbicular to cuneate, symmetrical to weakly asymmetrical, 2–6 mm long, 1–4 mm wide, commonly inflated and cavernous throughout, especially on the lower surface, obscurely 1- to 3- (5-) nerved; upper surface flat to weakly convex, the guard cells abundant, commonly with a median row of papillae and occasionally keeled, dark green or sometimes purple; lower surface flattened to strongly gibbous; rootlet one, the sheath cylindrical, prominent, the rootcap obtuse; plants solitary or commonly remaining attached in groups of 2–4 by short to somewhat elongated stipes; fruit winged, depressed-globose; seeds 2–4, compressed, asymmetrical, longitudinally ribbed.

Standing water.

IL, NE (OBL).

Duckweed.

3. **Lemna minima** Phil. Linnaea 33:239. 1864. Fig. 90 i–k.
Lemna minuta Kunth. HBK Nov. Gen. & Sp. 1:372. 1816, misapplied.
Lemna valdiviana Phil. var. *minima* (Phil.) Hegelm. Lemnac. 138. 1868.

Fronds obovate to elliptical, symmetrical to weakly asymmetrical, 1.5–1.7 mm long, 1.0–1.2 mm wide, weakly inflated, obscurely 1-nerved or apparently nerveless, cavernous only in the central region, thin-margined; upper surface convex, pale green, glossy, the guard cells present, with a median row of papillae commonly present; lower surface flattened or weakly convex, pale green or yellowish; rootlet one, the sheath cylindrical, thin, the rootcap acute or obtuse; plants frequently solitary or remaining attached in groups of 2 (–4); fruit elongated, symmetrical; seeds longitudinally ribbed and with cross-striae.

Standing water.

IL, IN, KS, KY, MO, NE, OH (OBL).

Duckweed.

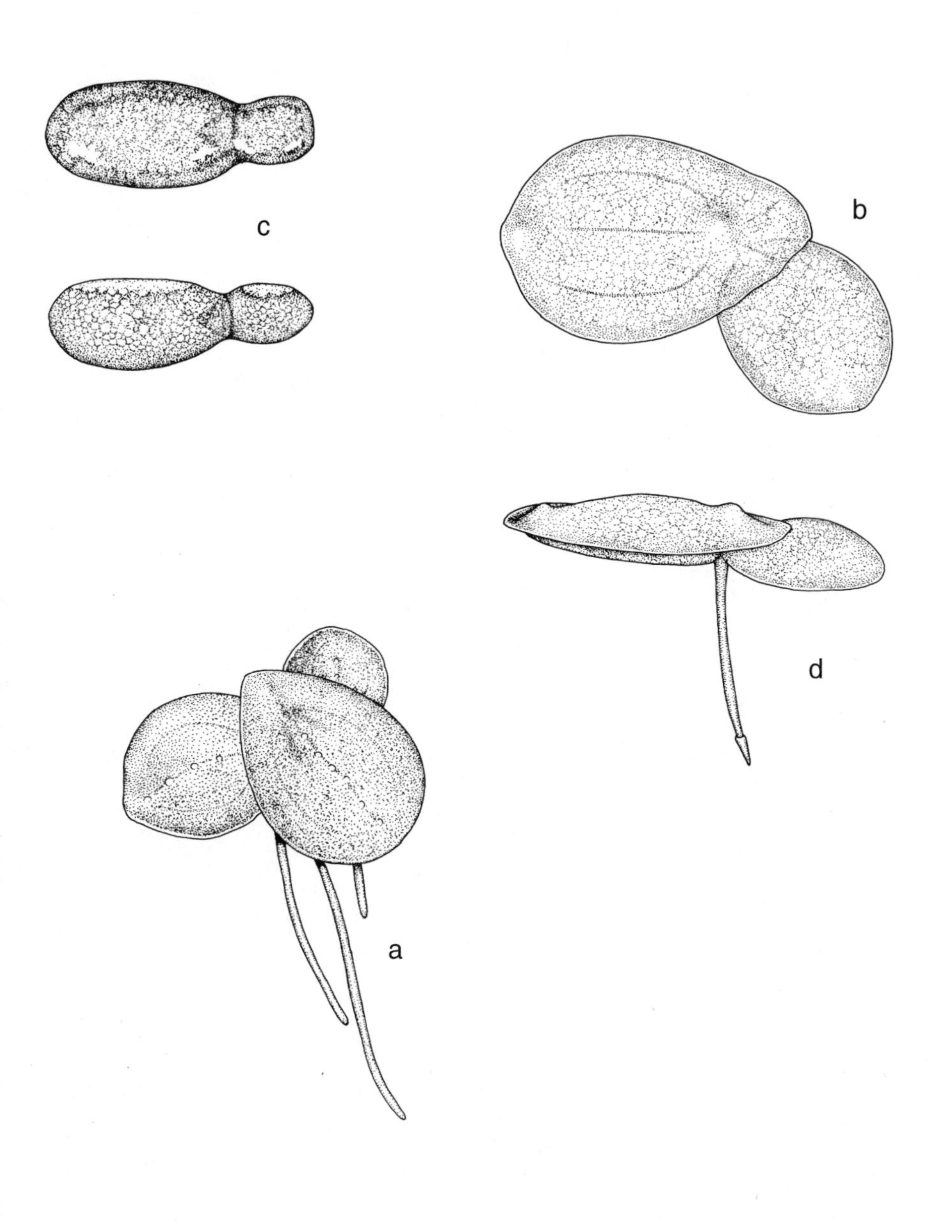

88. a. Habit. *Lemna aequinoctialis* (Duckweed). b. Habit. *Wolffia globosa* (Water meal). c. Habit. d. Habit. *Lemna turionifera* (Duckweed)

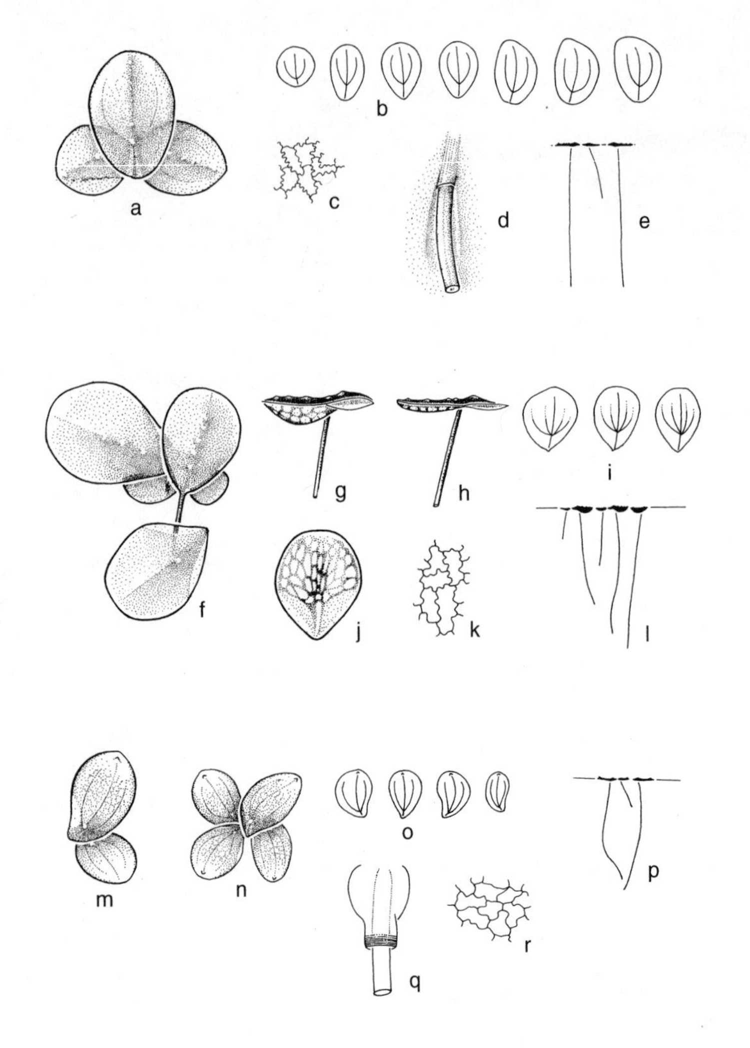

89. ***Lemna minor*** **(Duckweed).**
a. Habit, dorsal aspect.
b. Morphological variation among mature fronds.
c. Epidermal cell wall pattern, lower leaf surface.
d. Origin of single rootlet showing characteristic furrow formed on lower surface.
e. Floating habit, lateral aspect.
Lemna gibba (Duckweed).
f. Habit, dorsal aspect.
g. Lateral view of a frond showing strong development of gibbous condition on lower surface
h. Lateral view of a frond showing weak development of gibbous condition on the lower surface.
i. Morphological variation among mature fronds.
j. Lower surface of a frond showing development of air spaces in strongly gibbous condition.
k. Epidermal cell wall pattern, lower survace.
l. Floating habit, lateral aspect.
Lemna perpusilla (Duckweed).
m and n. Habit, dorsal aspects.
o. Morphological variation among mature fronds.
p. Floating habit.
q. Winged sheath surrounding rootlet at point of origin on lower surface.
r. Epidermal cell wall pattern, lower surface.

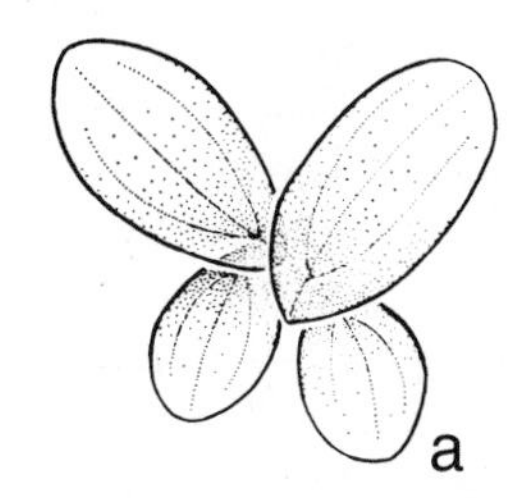

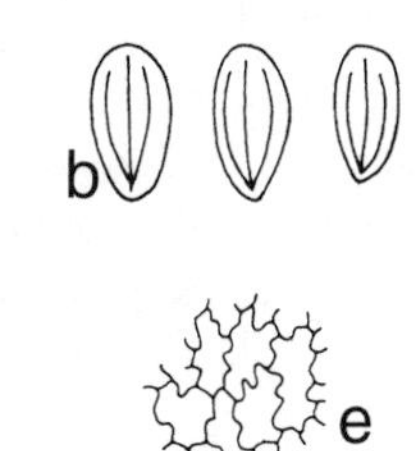

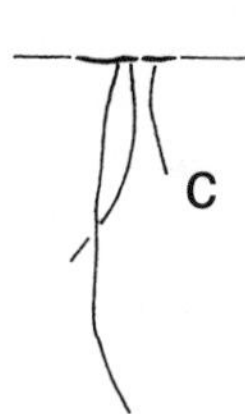

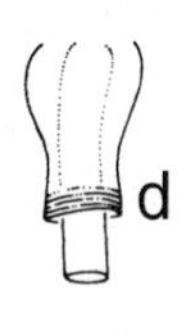

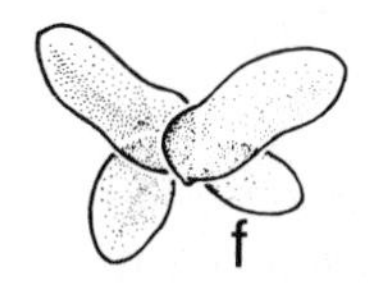

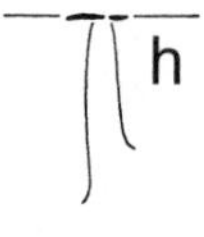

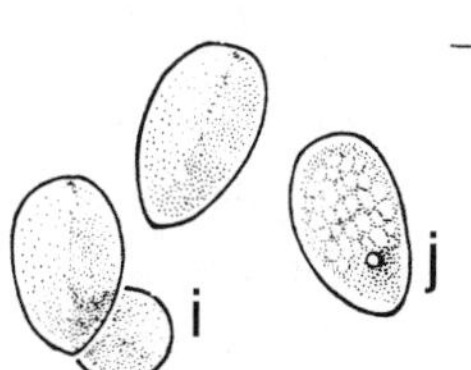

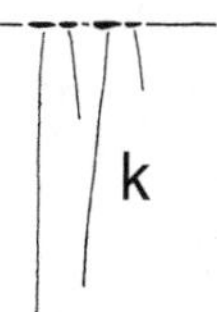

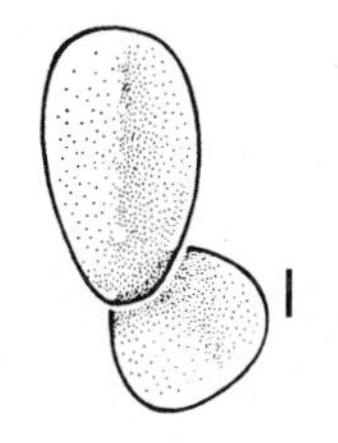

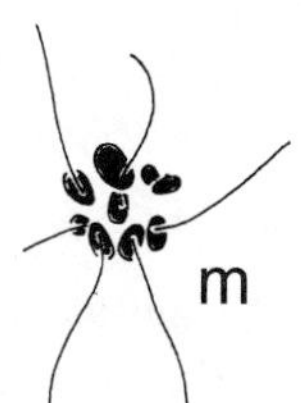

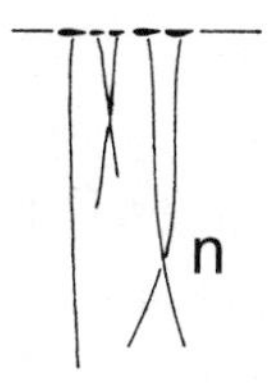

90. ***Lemna trinervis*** **(Duckweed).**
a. Habit, dorsal aspect.
b. Morphological variation among mature fronds.
c. Floating habit, lateral aspect.
d. Winged sheath surrounding rootlet at point of origin on lower surface.
e. Epidermal cell wall pattern, lower surface. *Lemna valdiviana* (Duckweed).
f. Habit, dorsal aspect.
g. Morphological variation among mature fronds.
h. Floating habit, lateral aspect. *Lemna minima* (Duckweed).
i. Habit, dorsal aspect.
j. Lower surface of a frond showing inflated central region.
k. Floating habit, lateral aspect. *Lemna obscura* (Duckweed).
l. Habit, dorsal aspect.
m. Lower surface of a frond showing dark pigmentation.
n. Floating habit, lateral aspect.

4. **Lemna minor** L. Sp. Pl. 1:970. 1753. Fig. 89 a–e.

Fronds obovate to elliptical (immature fronds orbicular), symmetrical or weakly asymmetrical, 2–5 mm long, 1.5–3.5 mm wide, flattened to weakly inflated, cavernous throughout, 3-nerved, the nerves occasionally obscure; upper surface usually slightly convex, with a thick, glistening cuticle, the guard cells abundant, a median row of papillae commonly present; lower surface flattened or weakly to moderately convex, pale green or occasionally reddish purple; rootlet one, arising obliquely beneath the node and lying in a narrow furrow, its length highly variable, the sheath frail, cylindrical, the rootcap obtuse; plants solitary or commonly remaining attached in groups of 2–4; fruit symmetrical, broadly ovoid; seeds longitudinally ribbed.

Standing water.
IA, IL, IN, KS, KY, MO, NE, OH (OBL).
Duckweed.

5. **Lemna obscura** (Austin) Daubs, Ill. Biol. Monogr. 34:20. 1965. Fig. 90 l–n.
Lemna minor L. var. *obscura* Austin. Gray, Man. Bot. 479. 1867.

Fronds obovate to elliptical, 1.5–2.0 mm long, 1.0–1.5 mm wide, obscurely nerved, moderately inflated and cavernous throughout, thin-margined; upper surface flattened to weakly convex, pale green or sometimes slightly reddish, the guard cells present, occasionally with a median row of small papillae; lower surface moderately to strongly convex, reddish purple; rootlet one, the sheath cylindrical, reduced, the plants solitary or remaining attached in groups of 2 (–4); rootcap obtuse; fruit obovoid, turbinate; seeds longitudinally ribbed, usually red pigmented.

Standing water.
IA, IL, IN, KS, KY, MO, NE, OH (OBL),
Duckweed.

6. **Lemna perpusilla** Torr. Fl. N. Y. 2:245. 1843. Fig. 89 m–r.

Fronds ovate to obovate, variable, asymmetrical, 1.5–3.0 mm long, 1.0–1.5 mm wide, inflated, obscurely 3-nerved or apparently nerveless, cavernous in the central region only; upper surface weakly convex, the guard cells abundant, with prominent nodal and apical papillae and occasional smaller papillae on median line between; lower surface flattened or weakly convex; rootlet one, with a winged sheath, the rootcap acute; plants solitary or commonly remaining attached in groups of 2–3; fruit asymmetrical, ovoid to oblongoid with a prominent oblique style directed toward the apex of the frond; seeds longitudinally ribbed and with numerous cross-striae.

Standing water.
IA, IL, IN, KS, KY, MO, NE, OH (OBL).
Duckweed.

7. **Lemna trinervis** (Austin) Small, Fl. S. E. U.S. 230. 1903. Fig. 90 a–d.
Lemna perpusilla Torr. var. *trinervis* Austin. Gray, Man. Bot. 479. 1867.

Fronds obovate to ovate-elliptic, symmetrical, 2.5–4.0 mm long, 1.5–2.0 mm wide, relatively thin and membranous, prominently 3-nerved, weakly cavernous in the central region only; upper surface flattened, the guard cells abundant, an apical papule sometimes present; lower surface flattened; rootlet one, with a winged sheath, the rootcap acute; plants solitary or commonly remaining attached in groups of 2–3; fruit asymmetrical, broadly ovoid to oblongoid with a prominent oblique style directed toward the apex of the frond; seeds longitudinally ribbed with numerous cross-striae.

Standing water.
IL, IN, OH (OBL).
Three-nerved duckweed.

8. **Lemna trisulca** L. Sp. Pl. 1:970. 1753. Fig. 91 g, h.

Fronds elliptic to oblanceolate, commonly falcate with long-tapering stipes, 4–10 mm long, 1.5–3.0 mm wide, membranous, obscurely 3-nerved, the apical margin serrulate; upper surface flattened, dull green, the guard cells absent; lower surface flattened; rootlets normally lacking, if present, the length highly variable, at times weakly coiled, the sheath winged, the rootcap long, obtuse; plants attached to one another at the nodes by elongated stipes, rarely solitary, generally submerged; fruit symmetrical; seeds longitudinally ribbed with numerous cross-striae.

Standing water.
IA, IL, IN, KS, KY, MO, NE, OH (OBL).
Star duckweed.

9. **Lemna turionifera** Landolt, Aquat. Bot. 1:355. 1975. Fig. 88d.

Fronds single to connected in clusters of 2–3, ovate to nearly orbicular, flattened to slightly convex below, sometimes with reddish margins, 1.5–4.0 mm long, with 2–4 papillae, producing rounded, rootless, olive to brown turions late in the season; rootlet 1, the sheath unwinged, the tip obtuse; fruit 1.0–1.4 mm long; seeds with 30–60 indistinct ribs.

Standing water.
IA, IL, IN, KS, MO, NE, OH (OBL).
Duckweed.
Turions are specialized, overwintering fronds.

10. **Lemna valdiviana** Phil. Linnaea 33:239. 1864. Fig. 90 f–h.
Lemna cyclostasa Ell. ex Thomp. Rep. Mo. Bot. Gard. 9:35. 1898.

Fronds oblong to oblong-elliptic, asymmetrical and often distinctly falcate, 1–5 mm long, 0.5–2.0 mm wide, obscurely 1-nerved or nerveless, cavernous throughout; upper surface convex, usually pale green and glossy, with guard cells sparingly produced, a median row of minute papillae occasionally present; lower surface flattened or weakly convex, pale green; rootlet one, the sheath long and cylindrical, the rootcap usually remaining attached in groups of 2–4 or more, frequently forming dense masses; fruit asymmetrical, oblongoid; seeds longitudinally ribbed and with numerous cross-striae.

Standing water.
IL, IN, KS, KY, MO, NE, OH (OBL).
Duckweed.

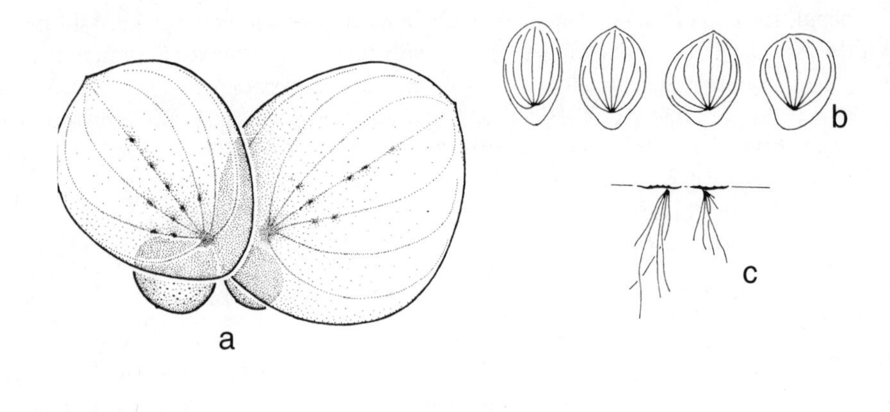

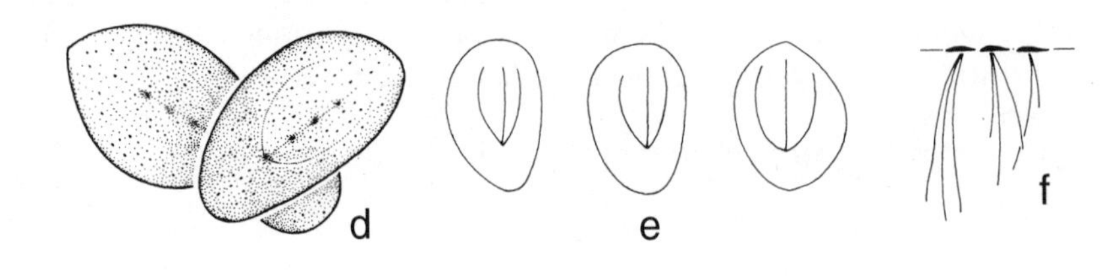

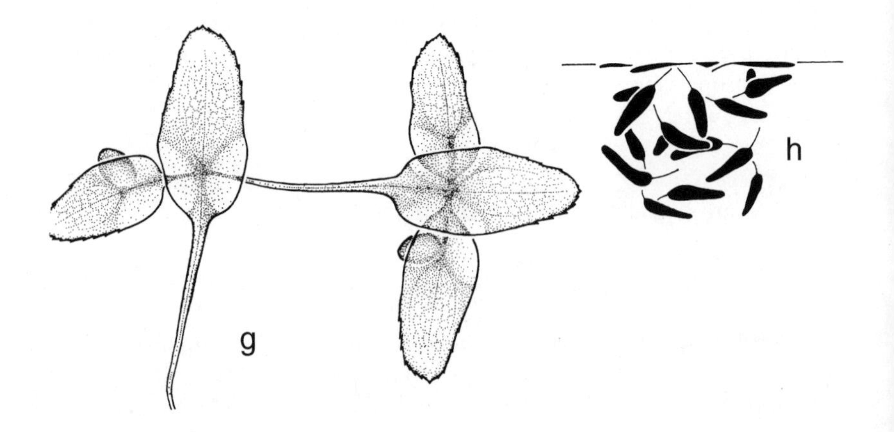

91. ***Spirodela polyrhiza*** **(Greater duckweed).**
a. Habit, dorsal aspect.
b. Morphological variation among mature fronds.
c. Floating habit, lateral aspect.
Spirodela punctata (Greater duckweed).
d. Habit, dorsal aspect.
e. Morphological variation among mature fronds.
f. Floating habit, lateral aspect.
Lemna trisulca (Star duckweed).
g. Habit, dorsal aspect.
h. Floating habit.

2. **Spirodela** Schleiden—Greater Duckweed

Fronds flattened, orbicular to reniform, usually asymmetrical, 3- to 8-nerved; rootlets (1–) 2–10, arising in a fascicle beneath the node; reproduction from two lateral pouches on either side of the node; flowers and fruit infrequently produced; ovules 2.

There are three species in this genus, two of them in the central Midwest.

1. Fronds orbicular, 5- to 8-nerved; rootlets 2–10 .... 1. *S. polyrhiza*
1. Fronds obovate to reniform, obscurely 3- (to 5-) nerved; rootlets (1–) 2–5 .... 2. *S. punctata*

1. **Spirodela polyrhiza** (L.) Schleiden, Linnaea 13:392. 1839. Fig. 91 a–c.
Lemna polyrhiza L. Sp. Pl. 1:970. 1753.
Lemna major Griff. Notul. 3:216. 1851.

Fronds orbicular to obovate (immature fronds orbicular), weakly to strongly asymmetrical, apiculate, 2–7 mm long, 2–6 mm wide, weakly inflated, cavernous throughout, prominently 5- to 8-nerved, the nerves originating palmately at the node, lateral nerves strongly incurved toward the central nerve after departure from the node; upper surface flattened, often olive-green, guard cells abundant, with a row of papillae frequently present along the central nerve and occasionally along lateral nerves; lower surface flattened or weakly convex, pale green or occasionally reddish purple; rootlets 2–10, arising in a fascicle directly beneath the node on the lower surface, up to 15 mm long; rootcaps acute; plants solitary or commonly remaining attached in groups of 2–5; vegetatively produced plants and infrequently observed flowers arising from two lateral, reproductive pouches; fruit a somewhat compressed and rounded utricle with winged margins; seed smooth or only slightly ribbed.

Standing water.

IA, IL, IN, KS, KY, MO, NE, OH (OBL).

Greater duckweed.

2. **Spirodela punctata** (G. Mey.) C. H. Thompson, Ann. Rep. Mo. Bot. Gard. 9:28. 1898. Fig. 91 d–f.
Lemna punctata G. Mey. Prim. Fl. Ess. 262. 1818.
Lemna oligorhiza Kurtz, Journ. Linn. Soc. London 9:267. 1867.
Spirodela oligorhiza (Kurtz) Hegelm. Die Lemnaceen 147. 1868.
Landoldtia punctata (G. Mey.) D. H. Les & D. J. Crawford, Nov. 9:530. 1999.

Fronds ovate to reniform, asymmetrical, subacute to obtuse at the apex, 2.2–6.0 mm long, 2.0–2.6 mm wide, inflated, cavernous throughout, obscurely 3-(5-) nerved; upper surface convex, sparsely punctate with pigment cells, the guard cells abundant; lower surface weakly convex, somewhat darker than the upper; rootlets (1–) 2–5, arising in a darkened fascicle directly beneath the node; rootcaps obtuse; plants solitary or commonly remaining attached in groups of 2–3; vegetatively produced plants and infrequently observed flowers arising from two lateral, reproductive pouches; fruit slightly winged, ovoid, longitudinally ribbed.

Standing water.

IL, KY, MO, OH (OBL).

Greater duckweed.

3. **Wolffia** Horkel—Water Meal

Plants reduced, globose or ellipsoid, nerveless; upper surface flattened or convex; lower surface strongly convex; pigment cells present in some species; rootlets absent; vegetative reproduction from a single, basal, reproductive pouch; flowers and fruit rarely observed, produced on the upper surface; ovule one.

Wolffia contains the world's smallest flowering plants. There are eleven species in the genus. Four species occur in the central Midwest.

1. Plants about as deep as wide or not as deep as wide, boat-shaped.
 2. Plants 1 1/2–2 times longer than wide, subacute at apex, without a papule .... 1. *W. borealis*
 2. Plants 1–1 1/2 times longer than wide, obtuse at apex, with a prominent central papule .... 2. *W. brasiliensis*
1. Plants deeper than wide, globose to ovoid.
 3. Plants up to 1.3 times longer than wide, 0.4–1.2 mm wide .... 3. *W. columbiana*
 3. Plants 1.3–2.0 times longer than wide, 0.3–0.5 mm wide .... 4. *W. globosa*

1. **Wolffia borealis** (Engelm.) Landolt, Ber. Geobot. Inst. Eidgen. Tech. Hoch. 44:137. 1977. Fig. 92 h–k.
Wolffia punctata Griseb. Fl. Brit. W. Ind. 512. 1864, misapplied.
Wolffia brasiliensis Wedd. var. *borealis* Engelm. Die Lemnaceen 127. 1868.

Plants narrowly ovoid to ellipsoid, 0.6–1.0 mm long, 1–1 1/2 times longer than wide, boat-shaped, nerveless, cavernous throughout with pigment cells abundant; upper surface flattened to slightly convex, the apex subacute to acute, without a central papule, the guard cells present; lower surface strongly convex; rootlets absent; plants solitary but usually remaining attached on a common axis in groups of 2; flowers and fruits produced on the upper surface but rarely observed.

Standing water.

IA, IL, IN, KS, KY, MO, NE, OH (OBL).

Water meal.

This boat-shaped species differs from *W. brasiliensis*, the other boat-shaped *Wolffia* in the central Midwest, by the absence of a central papule.

2. **Wolffia brasiliensis** Wedd. Ann. Sci. Nat. Bot. ser. 3, 12:170. 1849. Fig. 92 d–g.
Wolffia papulifera Thompson, Rep. Mo. Bot. Gard. 9:40. 1898.

Plants obovoid to ellipsoid, 1.0–1.5 mm long, 0.6–1.0 mm wide, 1–1 1/2 times longer than wide, nerveless, cavernous throughout, with pigment cells present; upper surface somewhat flattened but gradually rising in the center to form a single, distinct, conical papule, the apex obtuse to subacute, guard cells present; lower surface strongly convex; rootlets absent; plants solitary but usually remaining attached on a common axis in groups of 2; flowers and fruit produced on the upper surface, but rarely observed.

Standing water.

IA, IL, IN, KS, KY, MO, OH (OBL).

Water meal.

This species is distinguished from all others in the genus by its boat-shaped frond with a single central papule on the upper surface.

3. **Wolffia columbiana** Karst. Bot. Unters. 1:103. 1865. Fig. 92 l–o.

Plants globose to ellipsoid, 0.5–1.2 mm long, 0.4–1.2 mm wide, up to three times longer than wide, nerveless, conspicuously cavernous throughout, with pigment cells lacking; upper surface somewhat flattened, often with a median row of minute papules, guard cells present; lower surface broadly convex; rootlets absent; plants solitary but usually remaining attached on a common axis in groups of 2; flowers and fruits arising on the upper surface, but rarely observed.

Standing water.

IA, IL, IN, KS, KY, MO, NE, OH ((OBL).

Water meal.

This nearly globose species differs from the smaller *W. globosa* by its fronds 0.4–1.2 mm wide.

4. **Wolffia globosa** (Roxb.) Hartog & Plas, Blumea 18:367. 1970. Fig. 88 b, c.
Lemna globosa Roxb. Fl. Ind. 3:565. 1832.

Plants ovoid, 0.4–0.8 mm long, 0.3–0.5 mm wide, up to twice as long as broad, 1–1 1/2 times as deep as broad, rounded or slightly pointed at the tip; papilla 0; pigmented cells absent in vegetative tissue; rootlets absent; plants solitary but sometimes remaining attached in groups; flowers and fruits rarely observed.

Standing water.

IA, IL, IN, KS, KY, MO, NE, OH (OBL).

Water meal.

4. **Wolffiella** Hegelm.—Mud Midget

Fronds thin, linear, nerveless; rootlets absent; vegetative reproduction from a single, basal, reproductive pouch, the flowers and fruit rare, produced on the upper surface; ovule one; young plants remaining attached to form stellate colonies. The species of this genus appear to be green roots floating in the water. There are ten species in the genus.

1. **Wolffiella gladiata** (Hegelm.) Hegelm. Bot. Jahrb. Syst. 21:304–305. 1895. Fig. 92 a–c.
Wolffia gladiata Hegelm. Die Lemnaceean 133. 1868.
Wolffiella gladiata (Hegelm.) Hegelm. var. *floridana* J. D. Smith, Bull. Torrey Club 7:64. 1880.
Wolffiella floridana (J. D. Smith) Thompson, Rep. Mo. Bot. Gard. 9:37. 1898.

Fronds thin, linear, attenuate, with an acute or obtuse apex, often falcate and curving downward when floating, 3–12 mm long, 0.5–1.0 mm wide, cavernous throughout, nerveless, the guard cells rarely present; plants mostly submerged, solitary or commonly remaining attached in 2- to 4- or more-membered groups, often forming stellate colonies where many members are coherent; new plants arising from a ring, the basal node situated in a wedge-shaped, often spreading, membranous, basal reproductive pouch; flowers and fruits produced on the dorsal surface but rarely observed.

Standing water.

IL, IN, KY, MO, OH (OBL).

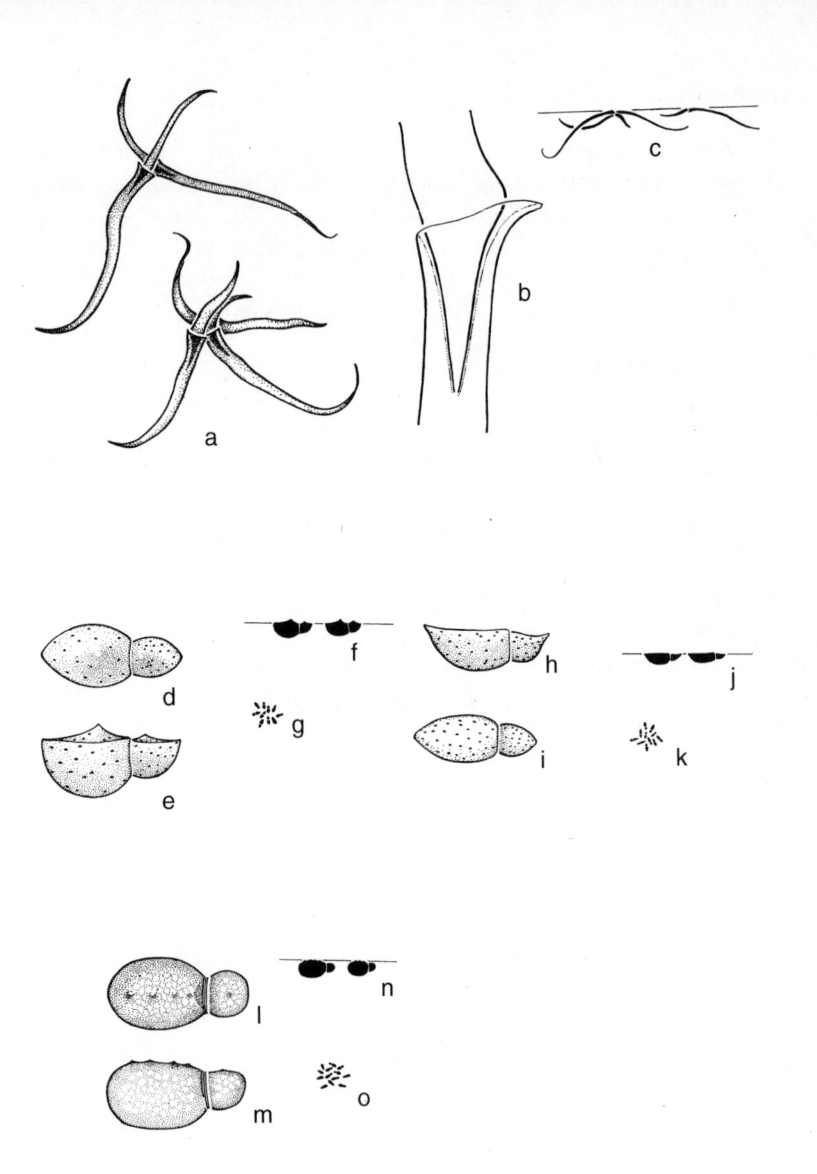

92. ***Wolffiella gladiata* (Mud midget).**
a. Habit, dorsal aspect.
b. Basal reproductive pouch showing vegetative origin of new plant.
c. Floating habit, lateral aspect. *Wolffia brasiliensis* (Water meal).
d. Lateral aspect, habit.
e. Habit, dorsal aspect.
f. Floating habit, lateral aspect.
g. Floating habit, dorsal aspect. *Wolffia borealis* (Water meal).
h. Habit, lateral aspect.
i. Habit, dorsal aspect.
j. Floating habit, lateral aspect.
k. Floating habit, dorsal aspect. *Wolffia columbiana* (Water meal).
l. Habit, lateral aspect.
m. Habit, dorsal aspect.
n. Floating habit, lateral aspect.
o. Floating habit, dorsal aspect.

Mud midget.

This is the only duckweed with a linear frond.

23. MARANTHACEAE—ARROWROOT FAMILY

Perennials herbs, often coarse, with rhizomes; leaves distichous, basal or nearly so, the veins pinnate from the midvein, with a pulviniform ligule; inflorescence a spike or panicle surrounded by bracts; flowers perfect, zygomorphic; sepals 3, attached at base; corolla 3-lobed, tubular below; fertile stamen 1 (–2), with 1 anther sac and with 1–2 petaloid staminodia; pistil 1, the ovary inferior, 2-locular, with 1 ovule per locule; fruit a capsule or fleshy, berrylike fruit, dehiscent; seeds often arillate.

This family consists of about thirty genera and about five hundred species, most of them tropical or subtropical.

1. **Thalia** L.—Thalia

Rhizomatous perennials; leaves large, parallel-veined; inflorescence paniculate, covered with a white powder; flowers perfect, bracteate; sepals 3; petals 3, united at base; staminodia 3, petaloid; stamen 1, two-lobed, one lobe with a fertile anther, one lobe petaloid; ovary inferior; style 1; fruit a utricle, the perianth and stamens persistent; seed 1 per fruit.

There are seven species in the genus, all subtropical or tropical, with one of them extending into the southern part of the central Midwest.

1. **Thalia dealbata** Fraser ex Roscoe, Trans. Linn. Soc. London 8:340. 1807. Fig. 93.

Perennials with stout rhizomes; leaves up to 3 m long, with petioles longer than the blades, with two sheathing stipules at base, the sheaths so arranged as to appear as a stem, the blades ovate, acute, up to 45 cm long, sometimes covered with a whitish powder; panicles arching, covered with white powder, on peduncles up to 2 m long, bearing flowers in pairs on zigzag branches; flowers perfect, surrounded by two leathery bracts; sepals 3, thin, papery, 1.0–1.5 mm long; petals 3, united at base, 7–9 mm long, purple-blue, the upper petal broader than the other two; staminodia 3, purple, petaloid, 7–12 mm long, two of them hoodlike and appearing as lips; stamen 1, 2-lobed, the upper lobe with a fertile anther, the lower petaloid; style curled at tip; utricles globose, 10–15 mm in diameter, purple to brown. July–August.

Swamps, wet ditches, lakes.

IL, MO (OBL).

Powdery thalia.

This is a robust species whose flowers are showy because of their petaloid staminodia. The blades of the leaves and the branches of the panicles are covered by a powdery substance. It is the only member of its family that is native to the United States.

24. NAJADACEAE—NAIAD FAMILY

There is only one genus in the family.

1. **Najas** L.—Naiad

Monoecious or dioecious, slender, branching aquatics, rooting from lower nodes; leaves linear, dilated at base, opposite, sessile, entire, serrulate, or coarsely toothed; flowers unisexual, reduced, usually solitary in leaf-sheath axils; perianth 0; staminate flowers borne near terminal nodes, a minute spathe enclosing a single stamen; pistillate flowers borne near middle and lower nodes, with a single 1-locular ovary and 1 ovule; stigmas 2–3, awl-shaped; achenes oblongoid, with reticulate surface, enclosed in a closely adhering membranaceous epicarp.

Najas consists of forty species and is nearly worldwide in distribution.

1. Leaves seemingly entire or serrulate; plants monoecious; achenes 1.5–3.5 mm long.
  2. Leaves linear, 0.5–1.4 mm wide, with slightly dilated base; marginal spinules microscopic.
    3. Achenes lustrous, with 30–50 rows of obscure, minute, often square areolae; style and 2 stigmas 0.8–1.8 mm long .... 1. *N. flexilis*
    3. Achenes dull, with 16–24 rows of distinct, large, hexagonal or rectangular areolae; style and 2–3 stigmas less than 1 mm long .... 3. *N. guadalupensis*
  2. Leaves filiform, 0.1–0.3 mm wide, with abruptly dilated base; marginal spinules macroscopic.
    4. Achenes with roughened appearance, with 22–40 rows of longer than broad, rectangular areolae; style and 2–3 stigmas 0.8–1.2 mm long .... 2. *N. gracillima*
    4. Achenes with ribbed appearance, with 10–18 rows of broader than long, rectangular areolae; style and 2 stigmas 1.0–1.4 mm long .... 5. *N. minor*
1. Leaves coarsely toothed; plants dioecious; achenes 4.0–7.5 mm long .... 4. *N. marina*

1. **Najas flexilis** (Willd.) Rostk. & Schmidt, Fl. Sed. 384. 1824. Fig. 94.
Caulinia flexilis Willd. Abh. Akad. Berlin 95. 1803.
Naias canadensis Michx. Fl. Bor. Am. 2:220. 1803.

Long and slender to closely tufted annuals; stems narrow, densely or sparsely branched, the terminal nodes crowded, the internodes dark green; leaves linear or sublanceolate, setaceous, light to dark green, 1–3 cm long, 0.5–1.4 mm wide, the apex usually acute and slightly recurved, gradually sloping to the base, ovate, slightly dilated, unlobed, each leaf margin with 16–35 microscopic spinules less than 0.5 mm long; persistent style and 2 stigmas 0.8–1.0 mm long; epicarp yellowish to purplish; achenes fusiform to slender, highly lustrous, yellowish to purplish brown, 0.8–3.0 mm long, one-third as thick, marked with 30–50 longitudinal rows of obscure, usually squarish areolae. June–September.

In lakes, streams, stripmine ponds.

IA, IL, IN, KY, MO, OH (OBL).

Northern naiad.

In *Najas flexilis* and *N. guadalupensis*, the surface of the achenes is covered by a pattern of nearly square or 6-sided areas, termed areolae. The squarish areolae in *N. flexilis* are arranged in 30–50 rows, while in *N. guadalupensis*, there are 16–24 rows.

2. **Najas gracillima** (A. Br.) Magnus, Beitr. 23. 1870. Fig. 95.
Naias indica L. var. *gracilllima* A. Br. Journ. Bot. Brit. & For. 2:277. 1864.

Usually moderately compact annuals; stems quite slender, often highly branched, most nodes somewhat crowded, the internodes purplish; leaves filiform, linear, serrulate, light to dark green, 1.5–3.0 cm long, 0.1–0.3 mm wide, the apex acicular

93. *Thalia dealbata* (Powdery thalia).

a. Inflorescence.
b. Flower.
c. Leaf.
d. Fruit.

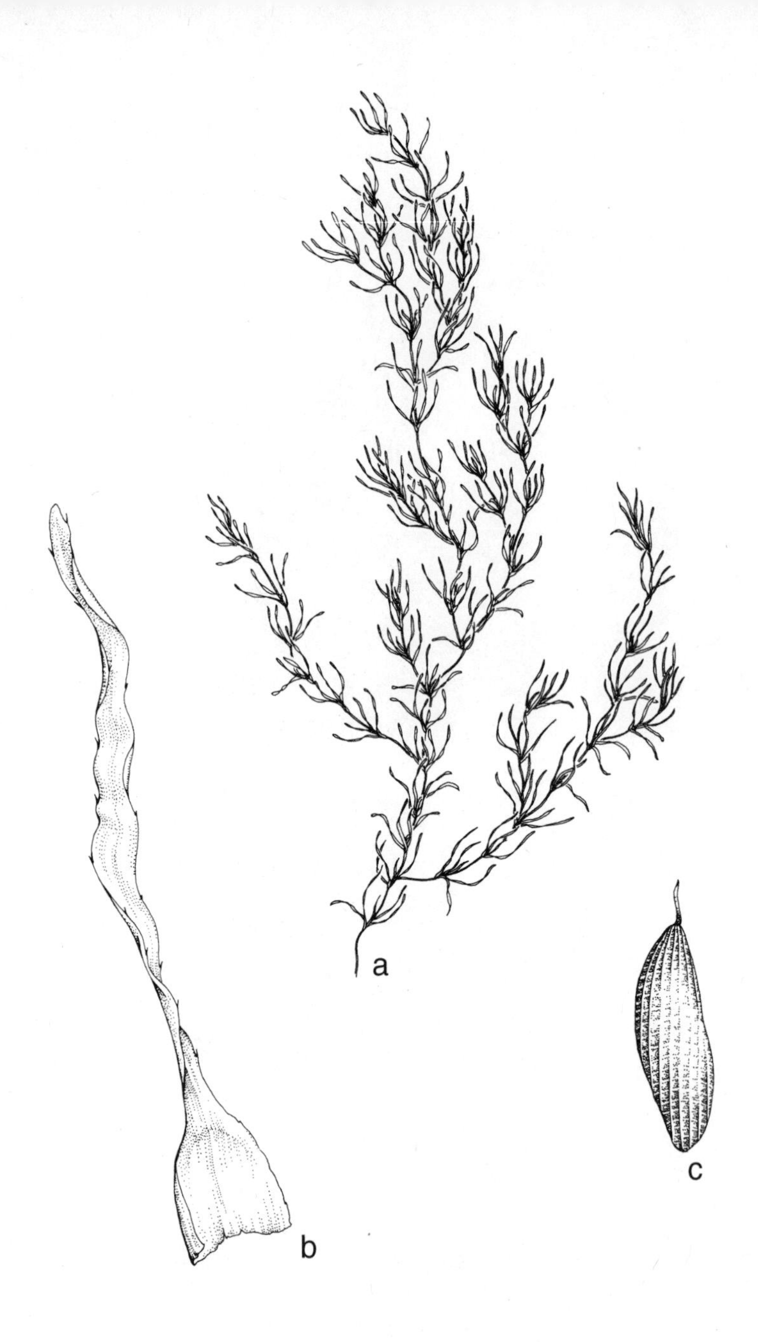

94. *Najas flexilis* (Northern naiad).

a. Habit.
b. Leaf.
c. Fruit.

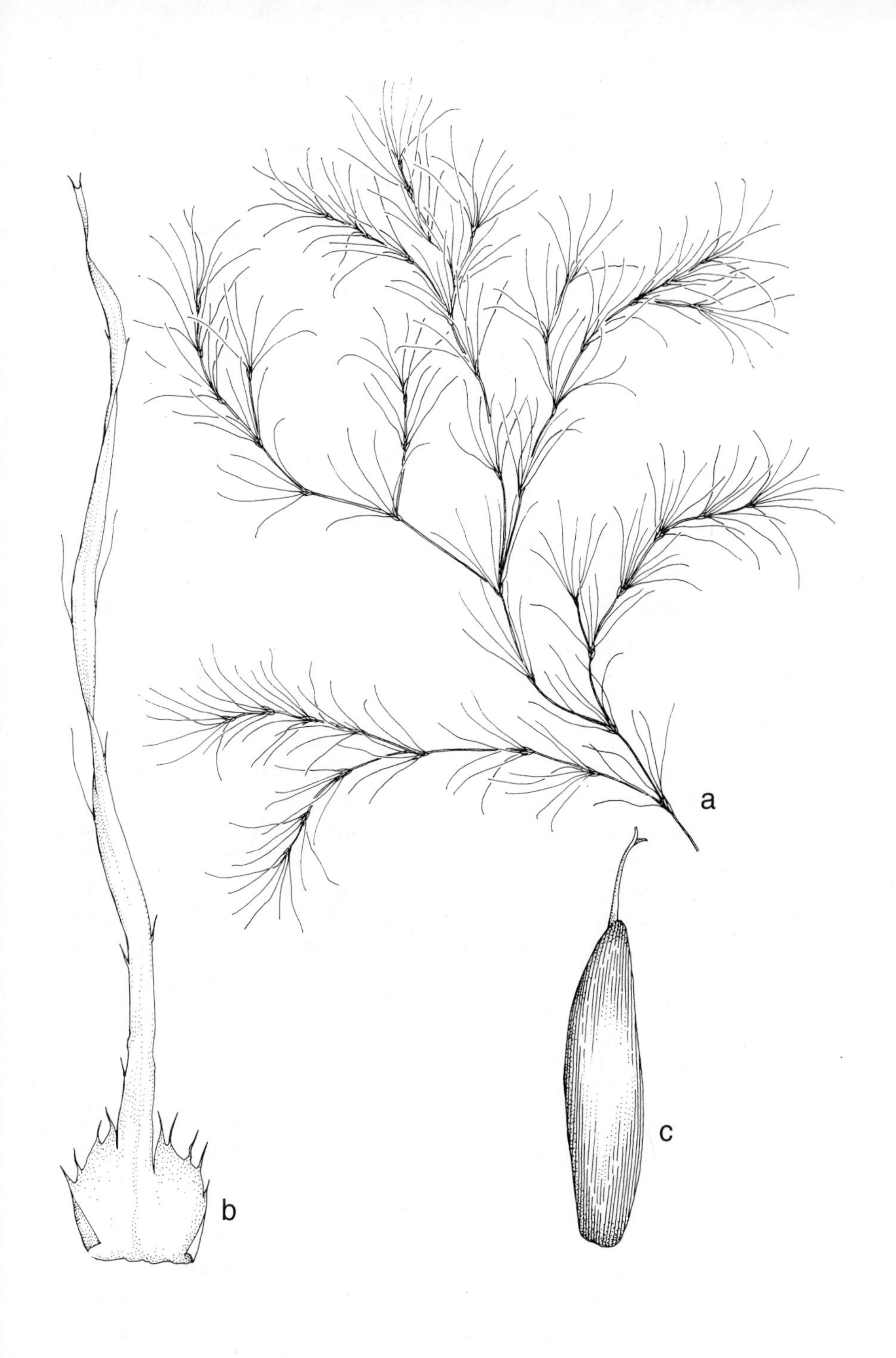

95. *Najas gracillima* (Slender naiad).

a. Habit.
b. Leaf.
c. Fruit.

but not recurved, the base scarious, abruptly dilated, auriculate with 4–7 antrorse, denticulate auricles, each leaf margin with 8–20 macroscopic spinules more than 0.5 mm long; persistent style and 2–3 stigmas 0.8–1.2 mm long; epicarp purplish; achenes usually subfalcate to linear with oblique sides, dull to slightly lustrous, yellowish to grayish, 2.5–3.5 mm long, one-quarter as thick, marked with 22–40 vertical rows of longitudinal rectangular areolae. June–September.

In ponds and lakes.

IL, IN, KY, MO, OH (OBL).

Slender naiad.

This is the most slender species of *Najas* in the central Midwest. It differs from *N. minor*, the other slender species, by the roughened ribbed appearance of the achenes with 22–40 rows of longer than broad areolae.

3. **Najas guadalupensis** (Spreng.) Magnus, Beitr. Gatt. 8. 1870. Fig. 96.
Caulinia guadalupensis Spreng. Syst. 1:20. 1825.
Naias flexilis (Willd.) Rostk. & Schmidt var. *guadalupensis* (Spreng.) A. Br. Journ. Bot. Brit. & For. 2:276. 1864.
Najas olivacea C. Rosend. & Butters. Rhodora 37:347–348. 1935.
Najas guadalupensis (Spreng.) Magnus var. *olivacea* (C. Rosend. & Butters) Haynes, Sida 8:43. 1979.
Najas guadalupensis (Spreng.) Magnus ssp. *olivacea* (Haynes) Haynes & Hellq. Novon 6:371. 1996.

Usually long and slender annuals; stems slender, rarely 1–2 mm thick, terete, rather sparsely branched, the terminal nodes not very crowded, the internodes dark green or purplish; leaves narrowly linear, setaceous, light green, 0.6–2.0 mm long, 0.5–1.2 mm wide, the apex acute or obtuse but not recurved, gently sloping to the base, slightly dilated, the shoulders rounded, unlobed, each leaf margin with 10–18 (–40) microscopic spinules less than 0.5 mm long; persistent style and 2–3 stigmas less than 1 mm long; epicarp purplish brown; achenes fusiform, ellipsoid, dull, light yellow to brownish, 1.2–2.4 mm long, one-third as thick, marked with 16–24 rows of distinct areolae, hexagonal at first, generally becoming more rectangular when dried at maturity. June–September.

In ponds, lakes, and streams.

IA, IL, IN, KS, KY, MO, NE, OH (OBL).

Southern naiad.

Plants with leaves having 20–40 teeth on each side of the leaf and stems 1–2 mm thick may be called var. *olivacea*. This variety is known from Illinois, Indiana, and Iowa.

4. **Najas marina** L. Sp. Pl. 1:1015. 1753. Fig. 97.

Rather slender, delicate, dioecious annuals; stems branching, to 5 (–10) cm long; leaves linear to linear-oblong, 1–3 cm long, 0.8–2.5 mm wide, each margin with 3–12 coarse teeth, the sheath distended, entire, or 1- to 2-toothed on each margin; staminate flower 3–4 mm long, rupturing from spathe; pistillate flower 3.0–3.5 mm long; achenes ellipsoid, 4.0–7.5 mm long (excluding the beak), 1.5–3.0 mm broad, the surface pitted, the slender recurved beak 1.0–1.5 mm long. June–September.

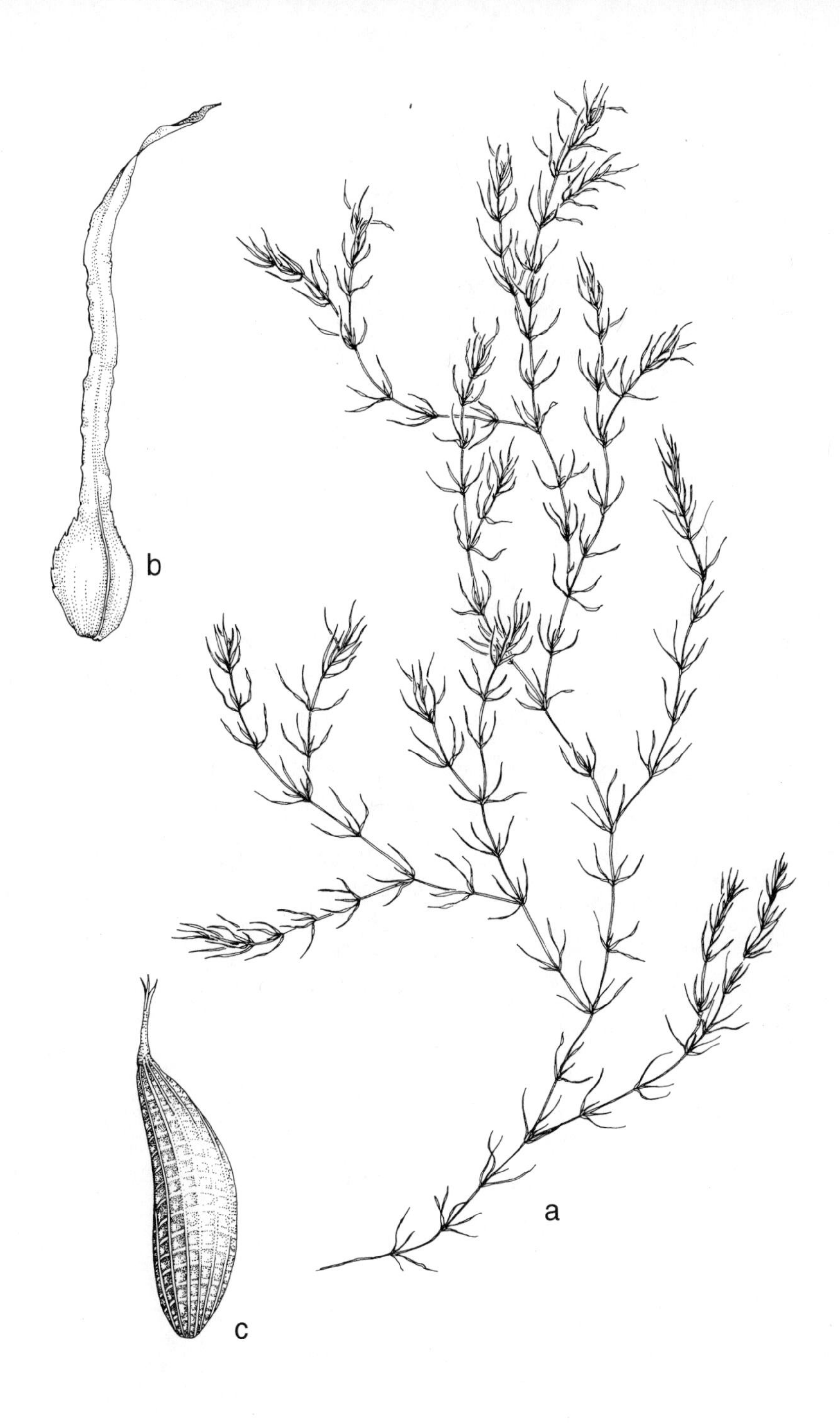

96. *Najas guadalupensis* (Southern naiad).
a. Habit.
b. Leaf.
c. Fruit.

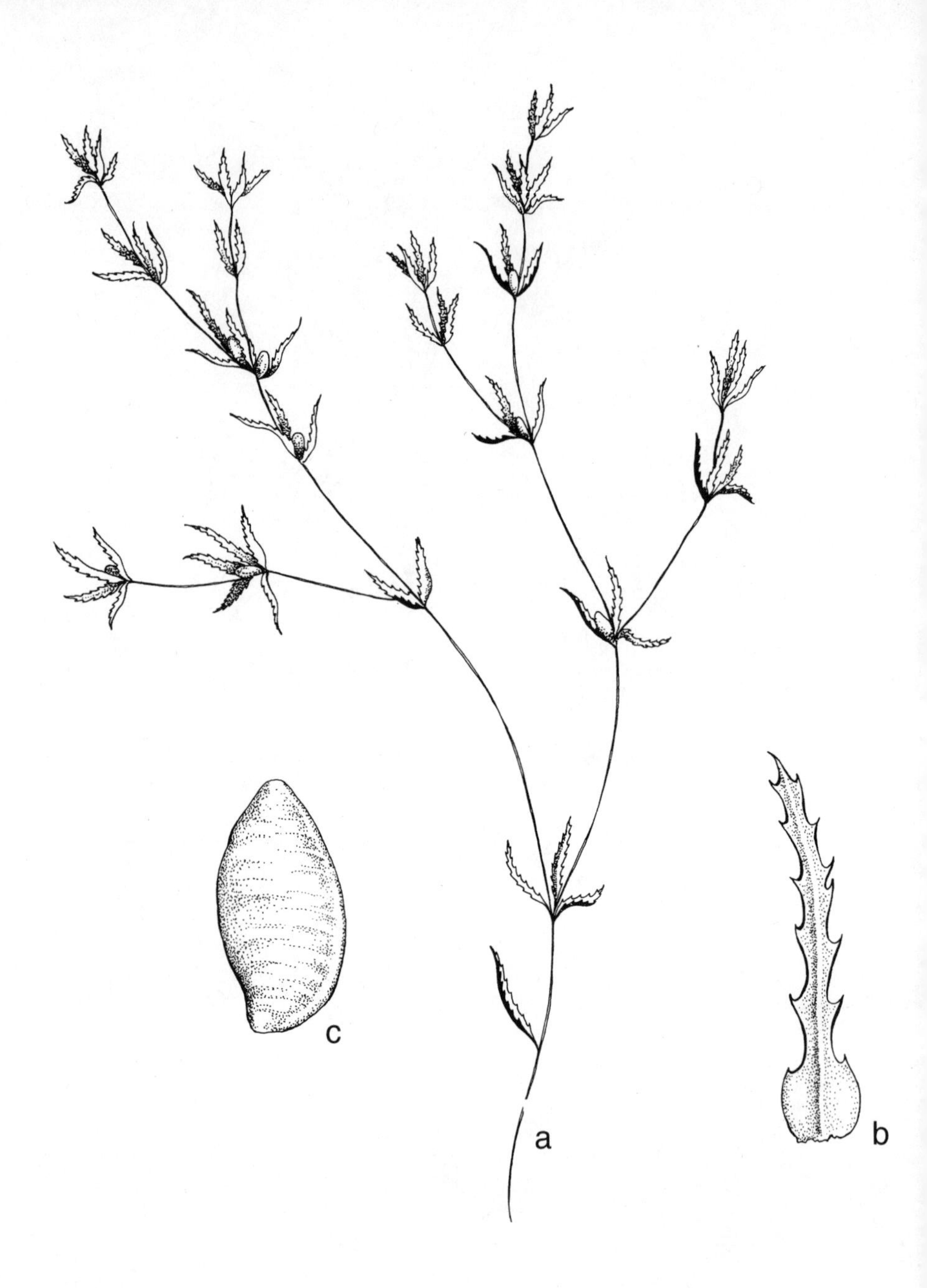

97. *Najas marina* (Toothed naiad). a. Habit. b. Leaf. c. Achene.

In ponds, lakes, and streams.

IL, IN, OH (OBL).

Toothed naiad.

This is the only species of *Najas* with coarsely toothed leaves. It also has the largest fruits of any species of the genus.

5. **Najas minor** All. Fl. Pedem. 2:221. 1875. Fig. 98.

Slender to terminally compact annuals; stems slender, but slightly stouter than *N. gracillima*, moderately branched, the terminal nodes crowded, with lime-green internodes; leaves filiform, colored the same as the internodes, 2.0–3.5 cm long, 0.2–0.3 mm wide, the distal end spinescent, recurved, the sheath abruptly distended, truncate, serrulate, each leaf margin with 8–16 barely macroscopic spinules more than 0.5 mm long; style and 2 stigmas 1.0–1.4 mm long; epicarp purplish; achenes falcate, slender, oblique, 2.6–3.6 mm long, one-quarter as wide, marked with 10–18 regular vertical ribs of broad, rectangular reticulations. June–September.

In ponds, lakes, and streams.

IL, IN, KY, MO, OH (OBL).

Brittle naiad.

Najas minor differs from *N. gracillima* by its ribbed achenes with 22–40 rows of areolae that are broader than long.

25. ORCHIDACEAE—ORCHID FAMILY

Mostly perennials; inflorescence various, or sometimes the flower solitary; flowers perfect, irregular, greatly modified; perianth parts (5–) 6, in 2 series; stamens 1 or 2, united with the style to form a central column; ovary 1, inferior, 1-celled; seeds minute.

This huge family consists of eight hundred genera and possibly as many as thirty-five thousand species, the great majority of them tropical and subtropical. Some species entirely lack chlorophyll, although none of the aquatic ones is like this. Most all species are mycorrhizal. Pollen frequently coheres in waxy masses known as pollinia.

None of the orchids in the central Midwest is a true aquatic, but several of them, particularly those that live in bogs and fens, may occur in standing water for periods of time during the year.

Key to the Genera Based Mainly on Floral Characters

1. Lip slipper-shaped .... 3. *Cypripedium*
1. Lip variously shaped, but not slipper-shaped.
    2. Leaves absent or senescent at flowering time.
        3. Flower solitary; solitary leaf formed after flowering .... 1. *Arethusa*
        3. Flowers several; leaves senescent at flowering .... 8. *Spiranthes*
    2. Leaves present at flowering time.
        4. Flowers with a spur .... 6. *Platanthera*
        4. Flowers without a spur.
            5. Flowers pink.
                6. Flower one per stem .... 7. *Pogonia*
                6. Flowers 3 or more per stem .... 2. *Calopogon*

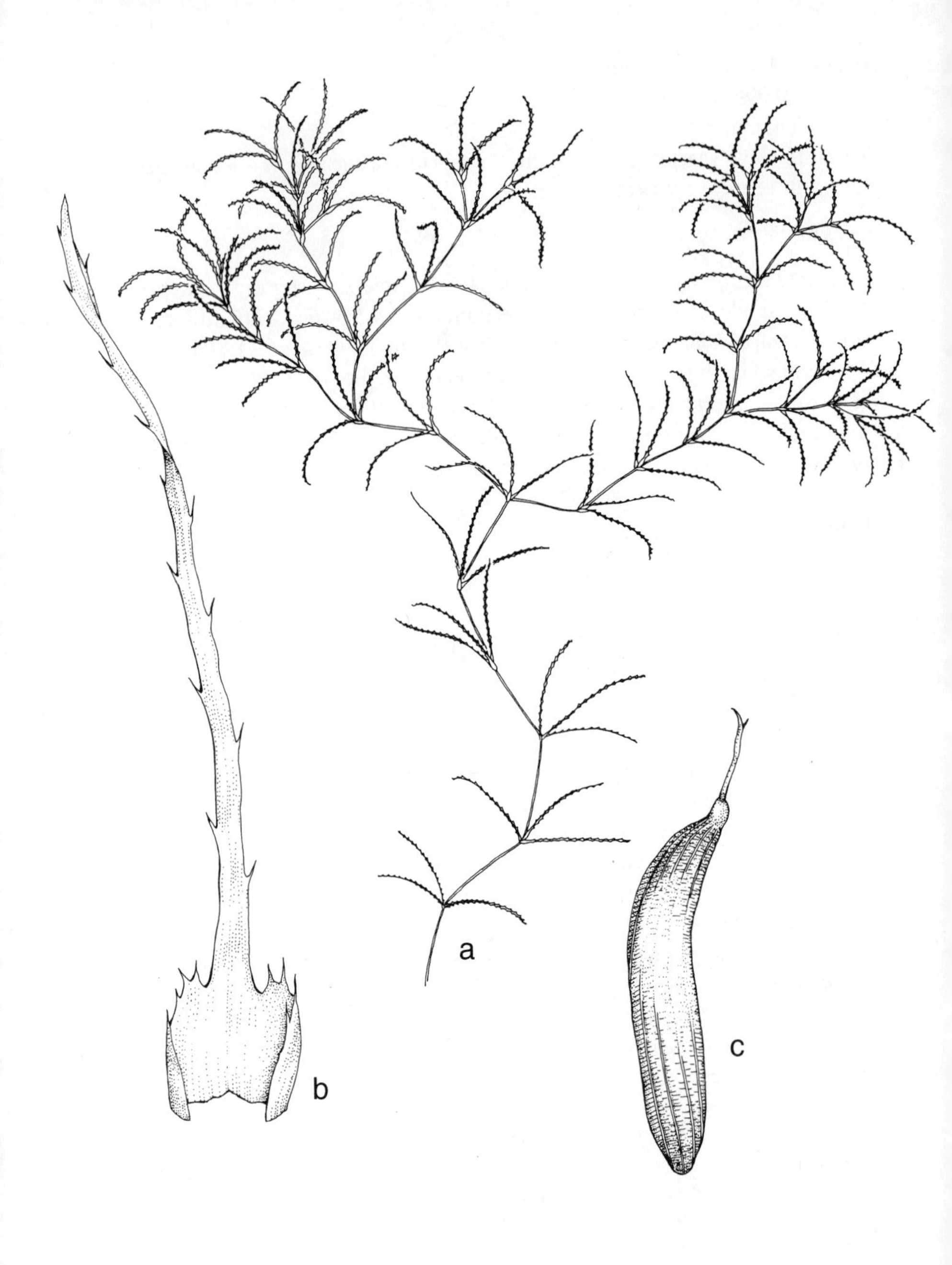

98. ***Najas minor*** **(Brittle naiad).** a. Habit. b. Leaf. c. Fruit.

5. Flowers not pink.
  7. Flowers spirally arranged .... 8. *Spiranthes*
  7. Flowers not spirally arranged.
    8. Leaf one at flowering time .... 5. *Malaxis*
    8. Leaves two at flowering time .... 4. *Liparis*

Key to the Genera Based on Stem and Leaf Characters

1. Leaves absent at flowering time, or at least senescent.
  2. No senescent leaves present at base of plant at flowering time .... 1. *Arethusa*
  2. One or more senescent leaves at base of plant at flowering time .... 8. *Spiranthes*
1. Leaves present at flowering time.
  3. Stems, or at least the upper part, pubescent.
    4. Stems pubescent throughout; leaves strongly ribbed, 5–12 cm wide ...... 3. *Cypripedium*
    4. Stems pubescent only in the upper part; leaves not strongly ribbed, 0.9–1.2 cm wide .. .. 8. *Spiranthes*
  3. Stems glabrous.
    5. Leaf one at flowering time.
      6. Leaf basal, linear-lanceolate .... 2. *Calopogon*
      6. Leaf above the base, elliptic to ovate to ovate-lanceolate.
        7. Leaf 2.5–3.0 cm wide .... 5. *Malaxis*
        7. Leaf 1.5–1.8 cm wide .... 7. *Pogonia*
    5. Leaves 2 or more at flowering time.
      8. Leaves at flowering time always 2 .... 4. *Liparis*
      8. Leaves usually 3 or more at flowering time .... 6. *Platanthera*

1. **Arethusa** L.—Dragon's Mouth Orchid

Only the following species comprises the genus. A second species from Japan is now considered to be in a different genus known as *Eleorchis.*

1. **Arethusa bulbosa** L. Sp. Pl. 2:950. 1753. Fig. 99.

Perennial from a small corm and fibrous roots; stems glabrous, unbranched, up to 20 cm tall; leaf one, appearing after the flowers are borne, lanceolate, subacute at the tip, tapering to the base, glabrous, 4–6 cm long, 6–10 mm wide; inflorescence with 1 (–2) flowers, subtended by a bract 2–4 mm long; lip petal pink with pinkish purple markings, 20–25 mm long, 12–15 mm wide, obovate, sinuate along the margins and with short lateral lobes, glabrous except for a fringe of yellow bristles at the central part; other 2 petals pink, 2.0–2.5 cm long, 3–4 mm wide; sepals similar in appearance to the petals, 3–4 cm long, 8–10 mm wide; fruit a capsule. May–June.

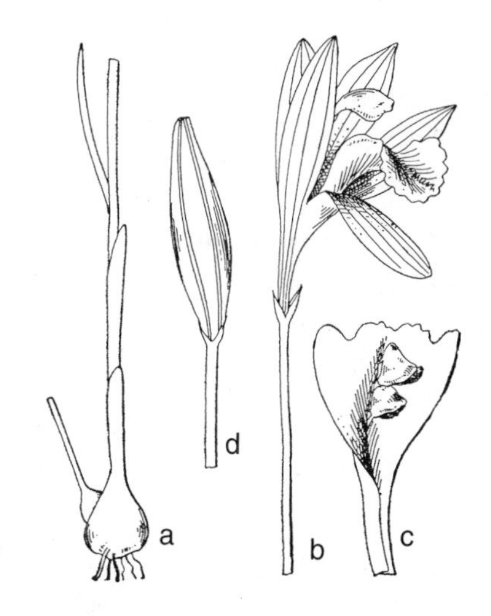

99. *Arethusa bulbosa* (Dragon's mouth orchid). a. Lower part of plant. b. Upper part of plant. c. Lip petal. d. Capsule.

Sphagnum bogs.

IL, IN, OH (OBL).

Dragon's mouth orchid.

When possessing its pretty pink flower, this plant has no leaves present. After flowering, however, a single leaf appears on the stem.

2. **Calopogon** L. C. Richard—Grass Pink Orchid

Perennial from a solid bulb; stems slender, glabrous, with a single nearly basal leaf and 1 or 2 sheathing scalelike leaves; flowers racemose, resupinate; sepals 3, free, spreading; petals free, spreading, the lip clawed, bearded on the inner face and papillose at the apex; column winged near the apex; anther 1, with 4 pollen masses; fruit a capsule.

There are five species in the genus, all North American.

1. **Calopogon tuberosus** (L.) BSP. Prel. Cat. N. Y. 1888. Fig. 100.
Limodorum tuberosum L. Sp. Pl. 1:950. 1753.
Limodorum pulchellum Salisb. Prod. Hort. Chap. Allerton 8. 1796.
Calopogon pulchellus (Salisb.) R. Br. ex Ait. Hort. Kew. 5:204. 1813.

Leaf usually 1, linear to linear-lanceolate, to 50 cm long, to 5 cm broad, glabrous, long-sheathing at the base; stem to 75 cm tall, glabrous; flowers (2–) 3–15, resupinate, 2.5–4.5 cm broad, subtended by a bract to 1 cm long; pedicels 3–9 mm long; sepals broadly elliptic, acute, concave, to 2.5 cm long, to 1 cm broad, pink or pinkish purple; petals broadly lanceolate, subacute or obtuse, to 2.5 cm long, to 1 cm broad, pink or pinkish purple; lip clawed below, expanded above, three-lobed, white-pubescent on the inner face, pink or pinkish purple tipped with yellow, to 2 cm long, to nearly 1 cm broad; capsules angular, to 2.5 cm long, strongly ribbed. May–August.

Sphagnum bogs, fens, wet prairies, moist sand flats, marly lake borders.

IA, IL, IN, KS, KY, MO, OH (OBL).

Grass pink orchid.

This showy species is distinguished by its single, basal, grasslike leaf that is present at flowering time.

3. **Cypripedium** L.—Lady's-slipper Orchid

Perennials with thick, fibrous roots; stems pubescent; leaves basal or cauline, large, pubescent; flowers 1–3, large; sepals 3, spreading, the lower 2 usually united; petals spreading; lip saclike, inflated; anthers 2; staminodium 1, petaloid; fruit a capsule.

The genus is readily distinguished by the large, slipper-shaped lip petal. There are about fifty species in the genus. The following may occur in shallow water, particularly in bogs, throughout some of the growing season.

1. Lip cleft down the middle, pink; leaves 2, basal .... 1. *C. acaule*
1. Lip not cleft, yellow, white, or pink or purple streaked; leaves more than 2, cauline.
    2. Sepals and petals, except for the lip petal, acuminate, yellow or greenish yellow or white and streaked with purple.
        3. Lip yellow .... 3. *C. calceolus*
        3. Lip white, occasionally marked with purple streaks.

100. *Calopogon tuberosus* (Grass pink orchid).
a. Habit.
b. Flower.
c. Habit (in fruit).

4. Lip white; staminodium narrowly triangular .... 2. *C. X andrewsii*
4. Lip white, streaked with purple; staminodium oblong .... 4. *C. candidum*
2. Sepals and petals, except for the lip petal, obtuse, white .... 5. *C. reginae*

1. **Cypripedium acaule** Ait. Hort. Kew. 3:303. 1789. Fig. 101.

Perennial with fibrous roots; leaves 2, basal, broadly elliptic, subacute to acute, tapering to the base, glandular-pubescent, ribbed, to 25 cm long, to 15 cm broad; scape to 50 cm tall, without bracts; flower 1, terminal; upper sepal narrowly lanceolate, greenish brown to greenish purple, to 5 cm long, to 1.5 cm broad; lateral sepals united, proportionately broader, to 4 cm long, to 2.5 cm broad, greenish brown to greenish purple; petals lanceolate, greenish brown, pubescent within, scarcely twisted, to 5 cm long, to 1.5 cm broad; lip pink or rose, 3–7 cm long, drooping, with a deep cleft down the middle, pubescent on the inner surface; staminodium ovate, purplish; capsule to 5 cm long. May–June.

Bogs, acid woodlands.

IL, IN (FACW), OH (FACU).

Pink lady's-slipper.

This is the only *Cypripedium* with two basal leaves and a pink flower on a leafless scape.

2. **Cypripedium X andrewsii** A.M. Fuller, Rhodora 34:100. 1932. Fig. 102.
Cypripedium candidum L. X *Cypripedium parviflorum* Salisb. A.M. Fuller, Bull. Publ. Mus. Milwaukee 14:72. 1933.

Perennial with fleshy roots; leaves several, cauline, pubescent, lance-ovate, acute, tapering to the base, ribbed; stem to 50 cm tall, pubescent; flowers 1–2, terminal, slightly fragrant, each subtended by a foliaceous bract; sepals purple, the upper ovate-lanceolate, 2.5–3.7 cm long, the lower two connate; lateral petals purple or greenish purple, lanceolate, twisted, 3–4 cm long; lip creamy white, marked with purple streaks, 2.0–2.5 cm long; staminodium narrowly triangular. June.

Springy areas.

IL, IN (FACW).

Hybrid lady's-slipper.

This handsome species is a hybrid between *C. calceolus* var. *parviflorum* and *C. candidum*. In most of its characters, it is intermediate between these two taxa. Its resemblance, however, is more with *C. candidum*.

3. **Cypripedium calceolus** L. var. **parviflorum** (Salisb.) Fern. Rhodora 48:4. 1946. Fig. 103.
Cypripedium parviflorum Salisb. Trans. Linn. Soc. 1:77. 1791.

Perennial with thickened roots; leaves 3–4, cauline, pubescent, ovate-lanceolate to ovate, acute, tapering to the base, ribbed, to 15 cm long, to 9 cm broad; stems to 50 cm tall, pubescent; flowers 1–2, terminal, each subtended by a foliaceous bract, to 10 cm long, to 5 cm broad; sepals purple, the upper broadly lanceolate, to 5 cm long, 2.0–2.5 cm broad, the lower two usually connate; lateral petals purplish, linear-lanceolate, twisted, 3.5–5.0 cm long, up to 8 mm broad; lip yellow, glossy, usually faintly purple-veined or -spotted, 2–4 cm long; staminodium ovoid, truncate or tapering to base; capsule to 4 cm long. May–June.

101. *Cypripedium acaule* (Lady's-slipper orchid).

a. Habit.
b. Capsule.
c. Habit (shaded).

102. *Cypripedium X andrewsii* (Lady's-slipper orchid). a. Habit. b. Habit (shaded).

103. *Cypripedium calceolus* (Small Yellow Lady's-slipper Orchid)
a. Habit.
b. Capsule.
c. Habit (shaded).

Bogs, marshes, fens, wet sandy prairies, dune swales.
IA, IL, IN, KY, MO, OH (FACW+), KS, NE (FACW).
Small yellow lady's-slipper.
The woodland variety, var. *pubescens,* has larger flowers that are not fragrant.

4. **Cypripedium candidum** Muhl. ex Willd. Sp. Pl. 4:142. 1805. Fig. 104.

Perennial from thickened roots; leaves 4–6, cauline, lanceolate to elliptic, acute, tapering to the base, pubescent, ribbed, to 15 cm long, to 14 cm broad, the lowest reduced to sheathing scales; stems to 40 cm tall, glandular-pubescent; flower 1, terminal, slightly fragrant, each subtended by an erect, foliaceous bract, to 6 cm long, to 2 cm broad; sepals greenish yellow striped with purple, the upper narrowly lanceolate, 2.0–3.5 cm long, up to 1 cm broad, the lower two connate; lateral petals greenish yellow striped with purple, linear-lanceolate, long-tapering at the apex, twisted, to 5 cm long, up to 0.4 cm broad; lip white, striped with purple, waxy-looking, to 2.5 cm long; staminodium oblongoid; capsule erect, to 3 cm long. May–June.
Fens, sedge meadows, marshes.
IA, IL, IN, KY, MO, NE, OH (OBL).
White lady's-slipper.
This species is readily distinguished by its waxy-looking, white lip.

5. **Cypripedium reginae** Walt. Fl. Carol. 222. 1788. Fig. 105.
Cypripedium spectabile Salisb. Trans. Linn. Soc. 1:78. 1791.

Perennial with thickened roots; leaves 4–6, cauline, pubescent on the margins and veins, broadly elliptic to ovate, short-acuminate, tapering to the base, ribbed, to 30 cm long, to 15 cm broad; stem to 90 cm tall, hirsute; flowers 1–3, terminal, subtended by a foliaceous bract to 10.5 cm long; sepals white, waxy-looking, obtuse, the upper suborbicular, to 5 cm long, to 3.5 cm broad, at length arching over the lip, the lower two connate; lateral petals white, obtuse, oblong, to 4 cm long, flat; lip white but strongly marked with pink or rose, 3.0–4.5 cm long; capsule to 4.5 cm long. May–June.
Fens, seep springs, bogs.
IA, IL, IN, MO (FACW), KY, OH (FACW+).
Showy lady's-slipper.
The lip of the flower is white with pinkish tints on the front and sides.

4. **Liparis** Rich.—Twayblade Orchid

Perennials from bulbs; foliage leaves 2, nearly basal, with several scale leaves present; inflorescence racemose, bracteate; flowers not fragrant; sepals spreading; petals twisted, threadlike, lip entire; anther 1, easily detached; fruit a capsule.
The genus consists of about 250 species found in most parts of the world.
Of the two species of *Liparis* in the central Midwest, only the following may be found sometimes in shallow water of fens.

1. **Liparis loeselii** (L.) Rich. Mem. Mus. Par. 4:60. 1817. Fig. 106.
Ophrys loeselii L. Sp. Pl. 1:947. 1753.

Bulb solid; leaves 2, nearly basal, lanceolate to lance-ovate, to 15 cm long, 2–3 cm wide, glabrous, shining; scape to 25 cm tall, somewhat angled, glabrous; raceme 3-

104. *Cypripedium candidum* (White Lady's-slipper orchid) a. Habit. b. Capsule.

105. *Cypripedium reginae* (Showy Lady's-slipper orchid). a. Habit. b. Capsule. c. Habit (shaded).

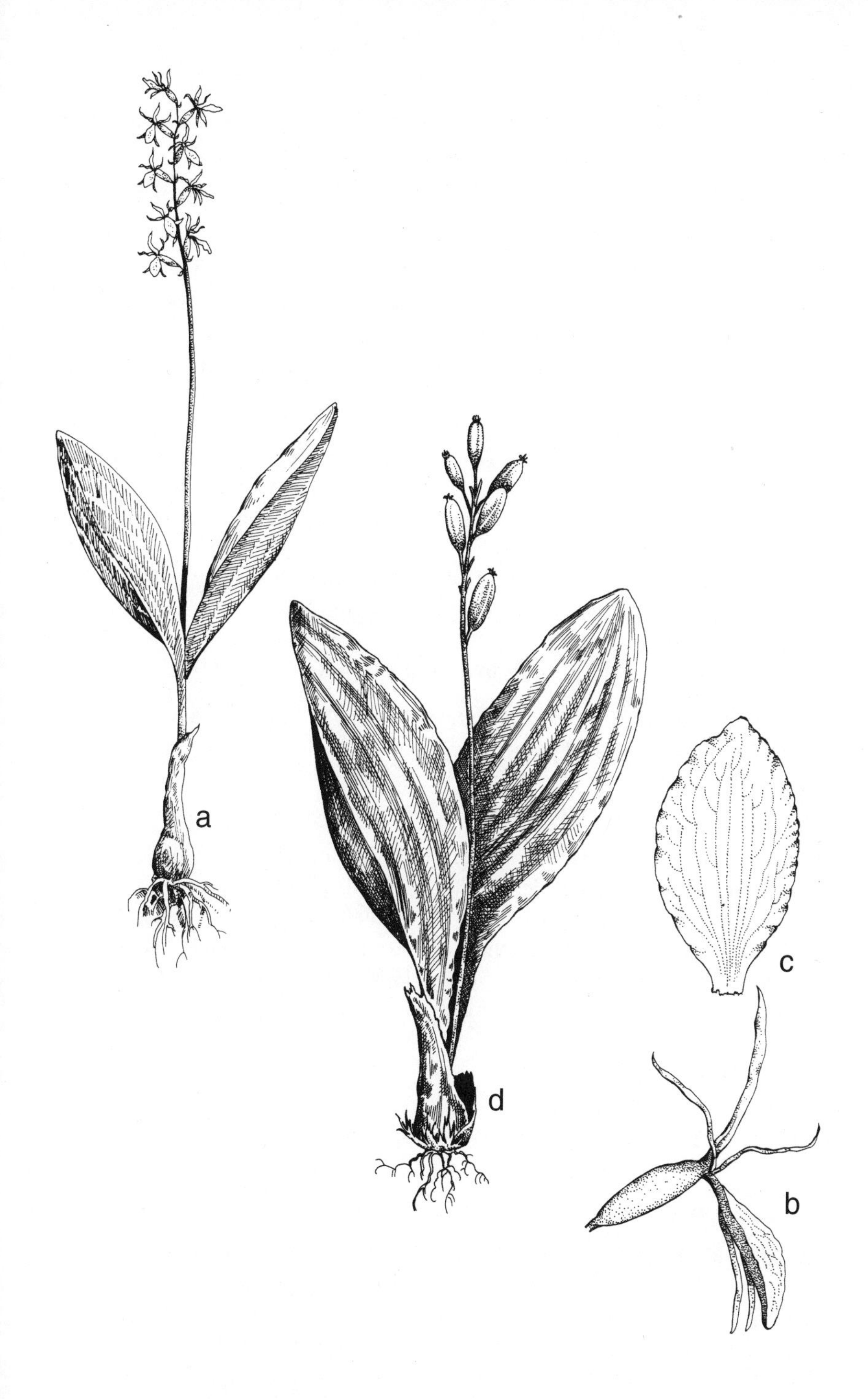

106. ***Liparis loeselii***
(Loesel's orchid).

a. Habit, in flower.
b. Flower.

c. Lip.
d. Habit, in fruit.

to 20-flowered; flowers pedicellate, the pedicels 4–5 mm long, usually a little shorter than the flowers; sepals narrowly oblong, 4–6 mm long, about 2 mm broad, yellow-green; lateral petals linear, twisted, 4–6 mm long, yellow-green; lip oblong, entire, apiculate, turned up slightly along the margin, yellow-green, 4–6 mm long; capsule up to 1 cm long. May–June.

Fens, sedge meadows, marshes, seep springs, floating mats, dune swales, rich woods.

IA, IL, IN, MO, OH (FACW+), KS, NE (FACW).

Loesel's twayblade.

The distinguishing characteristics of this species are the pair of basal leaves, the yellow-green lip of the flower, and the absence of a spur on the flower.

5. **Malaxis** Sw.—Adder's-mouth Orchid

Perennials from bulbs; leaf 1 (in the central Midwest), rarely 2, cauline; inflorescence racemose, bracteate; flowers not fragrant; sepals and the much smaller petals spreading; lip auriculate at base, entire or lobed; anther 1, easily detached; fruit a capsule.

Species of this genus in the central Midwest have a single, cauline leaf. All other orchids have either basal leaves or more than 2 cauline leaves. The genus consists of more than two hundred species.

Only the following species occurs in central Midwest wetlands.

1. **Malaxis monophylla** (L.) Sw. var. **brachypoda** (Gray) F. Morris. Morris & Eames, Our Wild Orchids 358. 1929. Fig. 107.
Microstylis brachypoda Gray, Ann. Lyc. N. Y. 3:228. 1828.
Malaxis brachypoda (Gray) Fern. Rhodora 28:176. 1926.

Bulb solid, ovoid; leaf 1 (rarely 2), borne below the middle of the stem, broadly elliptic, to 6 cm long, 3–4 cm wide; stem to 25 cm tall, glabrous; raceme slender, less than 1 cm thick, 4- to 25-flowered; flowers pedicellate, the pedicels 1–3 mm long; sepals broadly lanceolate, subacute, the upper 2.0–2.5 mm long; petals oblanceolate, about 1 mm long; lip broadly cordate at base, long-tapering and pointed at apex, unlobed, 2–3 mm long, deflexed, greenish white. July.

Bogs.

IL, OH (FACW).

Adder's-mouth orchid.

This tiny species has a single, elliptic leaf situated a little below the middle of the stem. The lip of the flower is unlobed.

6. **Platanthera** Rich.—Wood Orchid

Perennials with fleshy roots; leaves (1–) 2–several, cauline and alternate, or with one or more basal leaves; inflorescence spicate or racemose, bracteate; flowers usually fragrant; sepals and petals separate, similar; lip spurred; anther 1, persistent, difficult to detach; fruit a capsule. Although the genus name *Platanthera* dates back to 1818, all of our species were placed in *Habenaria* until 1972. *Habenaria* is considered to be a smaller, mostly tropical genus that differs from *Platanthera* by bearing tubers, by having 2-parted petals, and by having a 3-parted lip. There are about two hundred species of *Platanthera* worldwide.

107. ***Malaxis monophylla*** **(Adder's mouth orchid).** a. Habit, in flower. b. Flower. c. Habit, in fruit. d. Capsule.

The following species may occur in water in seep springs, bogs, or fens at some time during the year.

1. Lip entire, erose, or very shallowly 3-lobed, with the lobes entire; spur 3–12 mm long; flowers sessile.
   2. Leaves 1–3, with the lowest much larger; inflorescence 2–6 cm long; all bracts shorter than the flowers; spur 8–12 mm long; roots slender .... 3. *P. clavellata*
   2. Leaves 2–several; inflorescence 5–30 cm long; at least the lower bracts longer than the flowers; spur 3–8 mm long; roots thickened.
      3. Lip erose, with a tubercle borne near the summit .... 5. *P. flava*
      3. Lip entire, without a tubercle.
         4. Flowers greenish white, faintly odorous; lip gradually broadened at base .... 6. *P. hyperborea*

4. Flowers creamy white, with a spicy fragrance; lip abruptly broadened at base .... 4. *P. dilatata*

1. Lip fringed or deeply 3-lobed and toothed; spur 12 mm long or longer; flowers pedicellate.

5. Lip simple, fringed.

6. Flowers orange; leaves lanceolate, some or all over 2 cm broad .... 2. *P. ciliaris*

6. Flowers white; leaves linear to linear-lanceolate, less than 2 cm broad .... 1. *P. blephariglottis*

5. Lip deeply 3-lobed, the lobes fringed or long-toothed.

7. Flowers yellow-green or white; lobes of the lip fringed.

8. Flowers yellow-green or greenish white; sepals 4.5–7.0 mm long; spur 14–20 mm long .... 7. *P. lacera*

8. Flowers white; sepals 7–12 mm long; spur 20–48 mm long .... 8. *P. leucophaea*

7. Flowers reddish purple; lobes of the lip long-toothed .... 9. *P. psycodes*

1. **Platanthera blephariglottis** (Willd.) Lindl. Gen. & Sp. Orchid. Pl. 291. 1835. Fig. 108.
Orchis blephariglottis Willd. Sp. Pl. 4:9. 1805.
Habenaria blephariglottis (Willd.) Hook. Exot. Fl. 2:t. 87. 1824.
Platanthera holopetala Lindl. Gen. & Sp. Orch. 291. 1835.

Roots somewhat thickened; leaves 1–3, cauline, glabrous, linear-lanceolate, to 20 cm long, 1–2 cm broad; stem to 75 cm tall, glabrous; inflorescence 5–15 cm long, rather crowded; flowers white, appearing pedicellate; sepals somewhat longer than the petals, 5–9 mm long; lip broadly oblong, 8–11 mm long, fringed; spur 15–25 mm long. June–July.

Bogs.

IL (OBL).

White fringed orchid.

This pretty bog species is distinguished by its white flowers with an unlobed fringed lip.

2. **Platanthera ciliaris** (L.) Lindl. Gen. & Sp. Orchid. Pl. 292. 1835. Fig. 109.
Orchis ciliaris L. Sp. Pl. 1:939. 1753.
Habenaria ciliaris (L.) R. Br. Ait. Hort. Kew. 5:194. 1813.
Blephariglottis ciliaris (L.) Rydb. Britt. Man. Fl. N. U.S. 296. 1901.

Roots thickened; leaves (1–) 2–5, cauline, lanceolate, glabrous, to 25 cm long, 1.5–6.0 cm broad; stem to 90 cm tall, glabrous; inflorescence 5–15 cm long, rather lax, with 20–30 flowers; flowers orange, appearing pedicellate because of the pedicel-like ovary, each subtended by a bract to 2.5 cm long; sepals somewhat longer than the petals, ovate, 6–10 mm long, to 5 mm broad; petals orange, linear, fringed at tip, 4–5 mm long, 1.0–1.5 mm broad; lip broad, oblong, 10–13 mm long, 7–9 mm broad, long-fringed; spur 15–30 mm long. June–September.

Bogs, moist sand flats.

IL, IN, KY, MO, OH (FACW).

Orange fringed orchid.

This is the only orange-flowered orchid in the central Midwest. Its beauty cannot be overstated.

3. **Platanthera clavellata** (Michx.) Luer, Nat. Orch. Fl. 148. 1872. Fig. 110.
Orchis clavellata Michx. Fl. Bor. Am. 2:155. 1803.

108. *Platanthera blephariglottis* (White fringed orchid).

a. Inflorescence and leaves.
b. Flower.
c. Lip.
d. Fruiting branch.

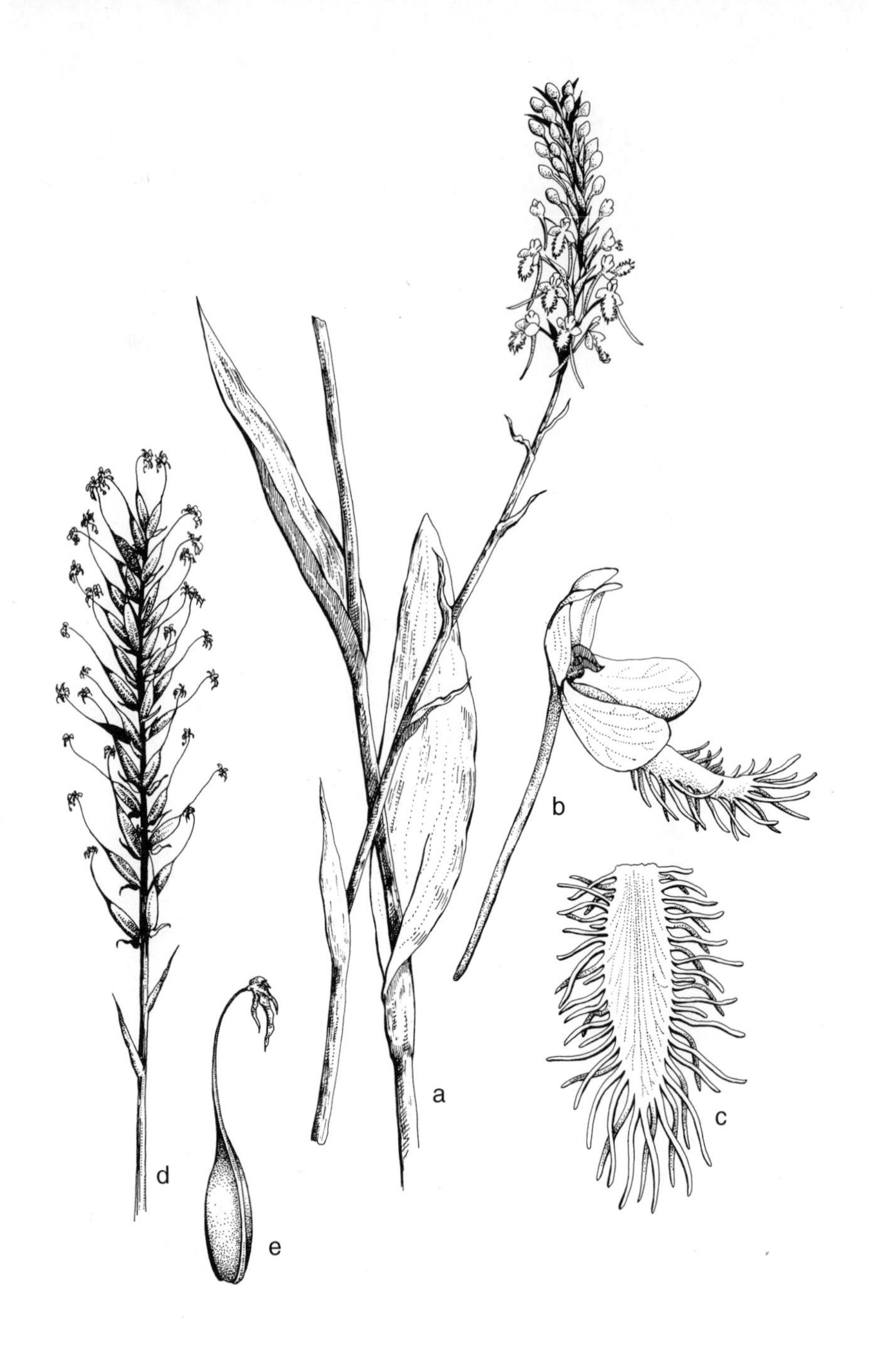

109. ***Platanthera ciliaris*** **(Yellow fringed orchid).**
a. Inflorescence and leaves.
b. Flower.
c. Lip.
d. Fruiting branch.
e. Capsule.

110. *Platanthera clavellata* (Wood orchid).

a. Habit, in flower.
b. Flower.
c. Lip.
d. Habit, in fruit.

Orchis tridentata Muhl. ex Willd. Sp. Pl. 4:41. 1805.
Habenaria tridentata (Muhl.) Hook. Exot. Fl. 2:pl. 81. 1825.
Habenaria clavellata (Michx.) Spreng. Syst. 3:689. 1826.
Gymnadeniopsis clavellata (Michx.) Rydb. Britt. Man. 293. 1901.

Roots slender, fleshy; leaves 1–3, near the base of the plant, glabrous, oblanceolate, to 20 cm long, 1.5–3.0 cm broad; stem to 45 cm tall, glabrous; inflorescence slender, rather sparse, 2–6 cm long, with 7–10 flowers; flowers greenish white or greenish yellow, often twisted, each subtended by a bract, to 12 mm long; sepals and petals about equal, oval, 2–4 mm long, nearly as broad; lip broadly oblong, shallowly 3-lobed, 4–5 mm long; spur curved, slender, 8–12 mm long. July–August.

Bogs, fens, seep springs.

IA, IL, IN, MO (OBL), KY, OH (FACW).

Small green wood orchid.

Because of its nonshowy appearance, this species is often difficult to locate. The slender, curved spur is 8–12 mm long and is distinctive for this species.

4. **Platanthera dilatata** (Pursh) Lindl. ex L. C. Beck, Bot. N. & Midl. States 347. 1833. Fig. 111.
Orchis dilatata Pursh, Fl. Am. Sept. 2:588. 1814.
Habenaria dilatata (Pursh) Hook. Exot. Fl. t. 95. 1824.
Limnorchis dilatata (Pursh) Rydb. Britt. Man. Fl. N. U.S. 294. 1901.
Limnorchis fragrans Rydb. Britt. Man. Fl. N. U.S. 294. 1901.
Habenaria fragrans (Rydb.) Niles, Bog-Trott. Orch. 253. 1904.

Roots fleshy, thickened; leaves 4–6, cauline, linear-lanceolate to lanceolate, glabrous, to 10 cm long, 7–10 mm broad; stem to 80 cm long, glabrous; inflorescence 10–30 cm long, crowded, with 25–35 flowers; flowers white, strongly aromatic, sessile; sepals slightly longer than the petals, 4–5 mm long, about 3 mm broad; petals white, 4–5 mm long, 1.0–1.5 mm broad; lip suborbicular, abruptly widened at the base, 6–8 mm long, 2–3 mm broad, entire; spur 6–8 mm long. May–June.

Fens, spring-fed areas.

IL, IN (FACW+).

Tall white bog orchid.

This is a northern species that barely reaches the central Midwest. The white flowers are very fragrant. The lip tapers to a slender point but is abruptly widened at the base.

5. **Platanthera flava** (L.) Lindl. Gen. & Sp. Orchid. Pl. 293. 1835.
Orchis flava L. Sp. Pl. 1:942. 1753.
Habenaria flava (L.) R. Br. ex Spreng. Syst. Veg. 3:691. 1826.
Perularia flava (L.) Farw. Ann. Rep. Parks Det. 11:54. 1900.

Roots fleshy, thickened; leaves 2, or with a greatly reduced third one, cauline, linear-lanceolate to lanceolate, to 15 cm long, to 3.5 cm broad, glabrous; stem to 60 cm tall, glabrous; inflorescence 5–20 cm long, crowded or lax, with 20–30 flowers; flowers green or greenish yellow, fragrant, subtended by longer or shorter bracts, sessile; sepals and petals about equal, 2–3 mm long, about half as wide; lip subor-

111. *Platanthera dilatata* (Tall white bog orchid).
a. Habit.
b. Flower.
c. Lip.
d. Fruiting branch.
e. Capsule.

bicular to oblong, 3–4 mm long, about 2 mm wide, green to yellow-green, erose, with a tubercle borne near the apex; spur slender, 5–6 mm long. May–September.

Two varieties occur in the central Midwest.

a. Bracts of lowest flowers equalling the flowers; inflorescence lax; lip suborbicular .... 5a. *P. flava* var. *flava*

a. Bracts of lowest flowers much surpassing the flowers; inflorescence crowded; lip oblong .... 5b. *P. flava* var. *herbiola*

5a. **Platanthera flava** (L.) Lindl. var. **flava.** Fig. 112.
Orchis scutellata Nutt. Trans. Am. Phil. Soc. n.s. 5:161. 1837.

Bracts of lowest flowers equalling the flowers; inflorescence lax; lip suborbicular, yellow-green. June–September.

Swamps, wet depressions in woods.

IA, IL, IN, KY, MO, OH (FACW).

Southern tubercled orchid.

5b. **Platanthera flava** (L.) Lindl. var. **herbiola** (R. Br.) Luer, Nat. Orch. U.S. & Can. 214. 1875. Fig. 113.
Orchis virescens Muhl. ex Willd. Sp. Pl. 4:37. 1805.
Habenaria herbiola R. Br. Ait. Hort. Kew. 5:193. 1813.
Habenaria virescens (Muhl.) Spreng. Syst. 3:688. 1826.
Habenaria flava (L.) R. Br. var. *virescens* (Muhl.) Fern. Rhodora 23:148. 1921.
Habenaria flava (L.) R. Br. var. *herbiola* (R. Br.) Ames & Correll, Bot. Mus. Leafl. Harv. Univ. 11:61. 1943.

Bracts of lowest flowers much surpassing the flowers; inflorescence crowded; lip oblong, green. May–August.

Swamps, depressions in woods, sand prairies.

IA, IL, IN, KY, MO, OH (FACW).

Northern tubercled orchid.

This variety blooms much earlier in the year than var. *flava.*

6. **Platanthera hyperborea** (L.) Lindl. var. **huronensis** (Nutt.) Luer, Fig. 114.
Orchis huronensis Nutt. Gen. 2:189. 1818.
Limnorchis huronensis (Nutt.) Rydb. Britt. Man. F. N. U.S. 294. 1901.
Habenaria hyperborea (L.) R. Br. var. *huronensis* (Nutt.) Farw. Papers Mich. Acad. Sci. 1:92. 1923.

Roots fleshy, thickened; leaves 3–4, cauline, lanceolate to oblanceolate, glabrous, to 12 cm long, to 3 cm broad; stem to 80 cm tall, glabrous; inflorescence 5–30 cm long, usually crowded, with 30–40 flowers; flowers greenish to greenish white, scarcely aromatic, sessile; sepals slightly longer than the petals, lanceolate to ovate, 2–4 mm long, 1–2 mm wide, green; petals lanceolate, 3–4 mm long, about 1 mm wide, green; lip lanceolate, subacute at the apex, 3–4 mm long, 1.5–2.0 mm wide, entire; spur 3–5 mm long. June–July.

Fens, marshes, seep springs.

IA, IL, IN (FACW+), NE, OH (FACW).

112. *Platanthera flava* var. *flava* (Tubercled orchid). a. Habit. b. Flower. c. Lip.

113. *Platanthera flava* var. *herbiola* (Tubercled orchid). a. Inflorescence. b. Flower. c. Capsule. d. Habit (shaded).

114. *Platanthera hyperborea* (Green orchid).
a. Habit.
b. Flower.
c. Inflorescence.

Northern green orchid.

This species is distinguished by its green lip that gradually widens toward its base.

7. **Platanthera lacera** (Michx.) G. Don ex Sweet, Hort. Brit. 650. 1839. Fig. 115.
Orchis lacera Michx. Fl. Bor. Am. 2:156. 1803.
Habenaria lacera (Michx.) Lodd. Bot. Cab. 3:pl. 229. 1818.
Blephariglottis lacera (Michx.) Rydb. Britt. Man. Fl. N. U.S. 296. 1901.

Roots fleshy, thickened; leaves 2–4, cauline, lanceolate to oval, glabrous, to 15 cm long, 1.5–2.0 cm broad; stem to 75 cm tall, glabrous; inflorescence 5–20 cm long, rather lax, with 20–30 flowers; flowers yellow-green, appearing short-pedicellate, subtended by bracts shorter than the flowers; sepals and petals about equal, 3–5 mm long, 2–3 mm wide, green, linear-oblong to ovate; lip 3-lobed 8–12 mm long, 8–10 mm wide, green, with each lobe fringed; spur 8–12 mm long. June–July.

Fens, bogs, seep springs, mesic woods, as well as dry fields, sand prairies, and sedge meadows.

IA, IL, IN, KS, KY, MO, OH (FACW).

Green fringed orchid.

This species is readily distinguished by its green, 3-lobed lip with each lobe long-fringed.

8. **Platanthera leucophaea** (Nutt.) Lindl. Gen. & Sp. Orch. 294. 18. Fig. 116.
Orchis leucophaea Nutt. Trans. Am. Phil. Soc. 5:161. 1834.
Habenaria leucophaea (Nutt.) Gray, Man. Bot. 502. 1867.
Blephariglottis leucophaea (Nutt.) Rydb. Britt. Man. Fl. N. U.S. 296. 1901.

Rootstocks tuberous; leaves 3–9, cauline, linear-lanceolate to lanceolate, glabrous, to 20 cm long, 1.5–3.5 cm broad; stems to 1 m tall, glabrous; inflorescence 8–20 cm long, lax; flowers white, appearing short-pedicellate, each with a bract to 5 cm long, to 0.5 cm broad; petals creamy white, three-parted, with each lobe cut deeply into narrow fringes, slightly longer than the sepals, 8–17 mm long, to 10 mm broad; lip 3-lobed, 15–30 mm long, each lobe fringed; spur 2–6 cm long. June–July.

Marshes, sedge meadows, wet prairies, bogs, fens.

IA, IL, IN, MO, OH (FACW+), NE (OBL).

Eastern prairie fringed orchid.

This species is distinguished by its creamy white flowers and petals with fringed lobes.

9. **Platanthera psycodes** (L.) Lindl. Gen. & Sp. Orch. 294. 1835. Fig. 117.
Orchis psycodes L. Sp. Pl. 943. 1753.
Habenaria psycodes (L.) Spreng. Syst. Veg. 3:693. 1826.
Blephariglottis psycodes (L.) Rydb. Britt. Man. Fl. N. U.S. 296. 1901.

Rootstocks tuberous; leaves (1–) 2–5 cauline, elliptic, glabrous, to 22 cm long, to 5 (–7) cm broad; stems to 75 cm tall, glabrous; inflorescence 5–20 cm long, rather crowded; flowers red-purple, each subtended by a bract to 5 cm long, to 0.5 cm broad; sepals and petals about equal, 4–8 mm long, to 5 mm wide; lip 3-lobed, 7–9 mm long, each lobe long-toothed or fringed, the terminal lobe without a notch, lavender to rose-purple; spur 16–25 mm long. June–August.

Bogs, seepage areas in forests.

IA, IL, IN, OH (FACW).

115. *Platanthera lacera* (Green fringed orchid).

a. Inflorescence and leaves.
b. Flower.
c. Fruiting branch.

116. *Platanthera leucophaea* (White fringed orchid). a. Upper part of plant. b. Flower.

117. *Platanthera psycodes* (Purple fringed orchid). a. Upper part of plant. b. Flower. c. Fruiting branch.

Small purple-fringed orchid.

This species is recognized by its fringed lip petal and purple flowers.

7. **Pogonia** Juss.—Pogonia

Perennials from a short rhizome with thickened roots; leaves basal plus one cauline leaf near the base of the stem; inflorescence terminal, 1- to 3-flowered, each flower subtended by a foliaceous bract; flowers fragrant; sepals and petals free, spreading; lip fringed on the margin, bearded on the upper face; anther 1, easily detached; fruit a capsule.

Pogonia consists of fewer than ten species, one of them in the central Midwest.

1. **Pogonia ophioglossoides** (L.) Ker, Bot. Reg. 2:pl.148. 1816. Fig. 118.
Arethusa ophioglossoides L. Sp. Pl. 951. 1753.

Rhizome short, with elongated thickened roots; leaves lanceolate to elliptic, the basal petiolate, the single cauline leaf sessile, to 10 cm long, to 2.5 cm broad; stems to 65 cm tall; flowers 1–3, fragrant; bracteal leaf lanceolate, foliaceous, smaller than the single cauline leaf; sepals and petals oblong, obtuse, pink, 15–25 mm long; lip elliptic, 15–20 mm long, pink, with a yellow beard; capsules to 3 cm long. June–August.

Peatlands, swamps, bogs, sedge meadows, floating mats.

IA, IL, IN, KY, MO, OH (OBL).

Rose pogonia.

This handsome species is distinguished by its single cauline leaf and its 1–3 pink flowers with the lip petal fringed on the margin.

8. **Spiranthes** Rich.—Ladies' Tresses

Perennials from tuberous thickened roots; leaves several, often all basal, the upper strongly reduced or seldom like the basal, present or absent at flowering time; inflorescence spicate, bracteate; flowers usually not fragrant, white or yellowish; lateral sepals free, the upper united with the petals; lip clawed, usually erose or crisped, with 2 callosities near the base; anther 1, easily detached; fruit a capsule.

There are approximately three hundred species in the genus, found in most parts of the world. Most of the species in the central Midwest do not occur in shallow water at any time.

1. Flowers up to 5 mm long, blooming during May and June .... 2. *S. lucida*
1. Flowers at least 7 mm long, blooming during September and October.
    2. Lip petal recurved, constricted at the middle; petals and sepals (excluding the lip) united to form a helmet-shaped hood .... 3. *S. romanzoffiana*
    2. Lip petal not recurved, not constricted at the middle; petals and sepals (excluding the lip) not forming a helmet-shaped hood .... 1. *S. cernua*

1. **Spiranthes cernua** (L.) Rich. Orch. Eur. Annot. 37. 1817. Fig. 119.
Ophrys cernua L. Sp. Pl. 2:946. 1753.
Ibidium cernuum (L.) House, Bull. Torrey Club 32:381. 1905.

Roots thickened, several; foliage leaves basal and cauline, oblanceolate, to 25 cm long, to 25 cm wide, present at flowering time; stems rarely over 50 cm tall, pubes-

118. *Pogonia ophioglossoides* (Rare pogonia).

a. Habit, in flower.
b. Flower.
c. Habit, in fruit.

119. *Spiranthes cernua* **(Nodding ladies'-tresses).** a. Inflorescence and bracts. b. Flower. c. Inflorescence and leaves.

cent, rather stout; spike 3–12 cm long, crowded, the flowers in 2–3 ranks, the rachis puberulent, the bracts to 1.5 cm long; sepals and petals more or less deltoid, 7–12 mm long; lip white to cream, oblong to ovate, 7–12 mm long, crisped, with two rather inconspicuous, rounded basal callosities. September–October.

Dry woods, old fields, marshes, fens, prairies.

IA, IL, IN, MO (FACW-), KS, NE, OH (FACW).

Nodding ladies' tresses.

This is one of the most common species of orchid in the central Midwest. It is distinguished by its flowers more than 5 mm long and its lip petal that is not constricted at the middle.

2. **Spiranthes lucida** (H. H. Eaton) Ames, Orchidac. 2:258. 1908. Fig. 120.
Neottia plantaginea Raf. Am. Monthly Mag. 2:206. 1818, non *S. plantaginea* Lindl. (1840).
Neottia lucida H. H. Eaton, Trans. Journ. Med. 5:107. 1832.
Spiranthes latifolia Torr. ex. Lindl. Gen. & Sp. Orch. pl. 467. 1840.
Ibidium plantagineum (Raf.) House, Bull. Torrey Club 32:381. 1905.

Roots thickened, several; foliage leaves all basal, present at flowering time, oblong to oblanceolate, glabrous, to 15 cm long, 1.5–2.5 cm broad, the cauline leaves all greatly reduced; stems to 15 cm tall, slender, pubescent; spike 2–8 cm long, crowded, the flowers borne in three ranks, the rachis puberulent; sepals and petals broadly linear, 5–7 mm long; lip yellowish, oblong, 4–5 mm long, crisped, with two inconspicuous, rounded callosities at the base. May–June

Fens, borders of lakes, stream banks, rarely in shallow water.

IA, IL, IN, MO (FACW+), KY, OH (FACW).

Shining ladies' tresses.

This species is distinguished by its small flowers up to 5 mm long with a yellowish lip.

3. **Spiranthes romanzoffiana** Cham. Linnaea 3:32. 1828. Fig. 121.

Roots thickened; leaves basal, usually persistent at flowering time, linear to elliptic, glabrous, to 25 cm long, to 3 cm broad; stems to 50 cm tall; spike with a tight spiral of 3–5 flowers per cycle, the rachis glabrous or sparsely pubescent; flowers white to ivory, tubular, the sepals and lateral petals connivent, forming a hood, 5–12 mm long, the lip petal 5–10 mm long, the apex broadly dilated. July–September.

Fens, marshes.

IA, IL, IN (OBL).

Hooded ladies' tresses.

The hood-shaped flowers are distinctive for this species.

26. POACEAE—GRASS FAMILY

Annual or perennial herbs (woody in the Bambuseae); culms cylindrical, with usually hollow internodes and closed nodes; leaves alternate, 2-ranked; sheaths usually free; ligule mostly present; inflorescence composed of (1–) several spikelets; spikelets 1- to several-flowered, each usually with a pair of sterile scales (glumes) at the base; flowers usually perfect, without a true perianth, the perianth reduced to rudiments (lodicules) or absent; each flower subtended by a lemma and a palea;

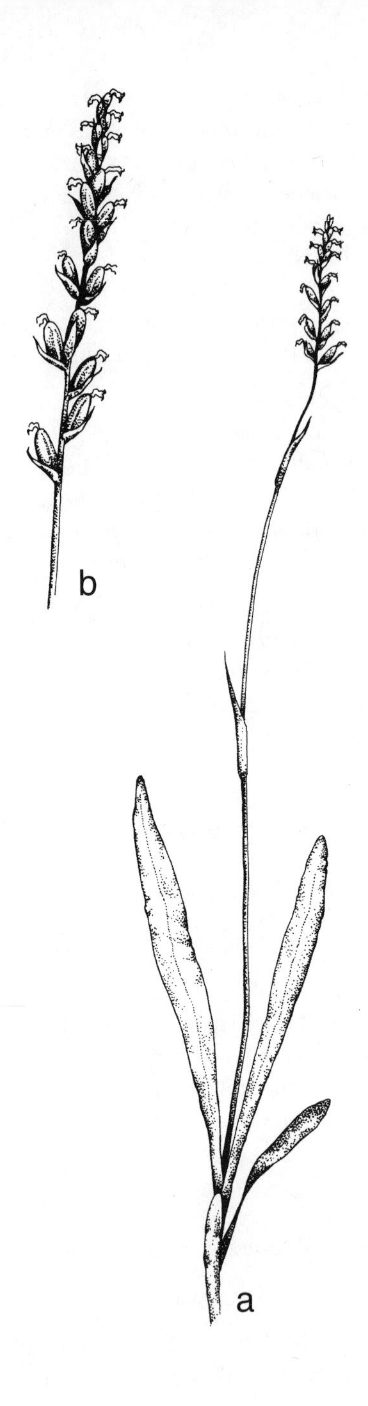

120. *Spiranthes lucida* (Shining ladies' tresses). a. Habit. b. Inflorescence.

121. *Spiranthes romanzoffiana* (Hooded ladies' tresses). a. Habit. b. Inflorescence. c. Flower part.

stamens (1–) 3 (–6); ovary 1-celled, with 1 ovule; stigmas 2 (–3); fruit usually a caryopsis (grain).

This family is often known as the Gramineae. It is one of the largest and economically most important families of flowering plants in the world. There are approximately six hundred genera and more than ten thousand species. The species enumerated below may occur in standing water at some time during the year.

1. Spikelets unisexual (i.e., either all staminate or all pistillate).
  2. Plants to 40 cm tall, dioecious; spikelets 10- to 75-flowered; glumes present .... 9. *Eragrostis*
  2. Plants 1–4 m tall, rarely shorter, monoecious; spikelets 1-flowered; glumes absent.
    3. Pistillate spikelets confined to the uppermost erect branches of the inflorescence, the staminate spikelets confined to the lower spreading branches; margin of leaf more or less smooth .... 22. *Zizania*
    3. Pistillate and staminate spikelets on the same branches of the inflorescence; margin of leaf harsh and cutting .... 23. *Zizaniopsis*
1. Spikelets with one or more perfect florets.
  4. Inflorescence appearing spicate, with one spike per culm.
    5. Spikelets disarticulating above the glumes; lemma with a bearded callus .... 4. *Calamagrostis*
    5. Spikelets disarticulating below the glumes; lemmas not bearded .... 2. *Alopecurus*
  4. Inflorescence racemose or paniculate.
    6. Each spikelet with two or more perfect florets.
      7. Some part of the spikelet awned.
        8. Spikelets disarticulating below the glumes; second glume dilated near tip; awn of lemma bent .... 20. *Sphenopholis*
        8. Spikelets disarticulating above the glumes; neither glume dilated near tip; awn of lemma straight .... 6. *Deschampsia*
      7. Spikelet without any awns.
        9. Plants at least 2 m tall; panicle at least 10 cm across .... 16. *Phragmites*
        9. Plants rarely more than 1 m tall; panicle less than 10 cm across.
          10. Lemmas 3-nerved.
            11. Lemmas keeled on back, with scarious margins, usually with cobwebby hairs at base; sheaths without conspicuous septations .... 17. *Poa*
            11. Lemmas rounded on back, without scarious margins and without cobwebby hairs at base.
              12. Sheaths with conspicuous septations .... 4. *Catabrosa*
              12. Sheaths lacking conspicuous septations .... 9. *Eragrostis*
          10. Lemmas with 5 or more nerves.
            13. Lemmas keeled on back, usually with cobwebby hairs at base .... 17. *Poa*
            13. Lemmas rounded on back, without cobwebby hairs at base.
              14. Lemmas toothed at tip .... 18. *Scolochloa*
              14. Lemmas not toothed at tip.
                15. Lemmas nerveless or with obscure nerves .... 21. *Torreyochloa*
                15. Lemmas conspicuously nerved .... 10. *Glyceria*
    6. Each spikelet with one perfect floret (1–2 sterile lemmas present in addition in *Phalaris*).
      16. Some part of the spikelets awned.
        17. Lemmas awnless; glumes awned .... 19. *Spartina*
        17. Lemmas awned; glumes usually awnless.
          18. Spikelets arranged in four or more crowded ranks, each spikelet composed of one fertile and one sterile floret .... 8. *Echinochloa*

18. Spikelets not arranged in four or more crowded ranks, each spikelet composed of one fertile floret .... 4. *Calamagrostis*

16. Spikelets without any awns.

19. Spikelets arranged in rows on one side of a slightly winged rachis .... 14. *Paspalum*

19. Spikelets not in rows on one side of a winged rachis.

20. Glumes absent or reduced to a cuplike structure.

21. Spikelets with all lemmas fertile; glumes absent; nodes bearded .... 11. *Leersia*

21. Spikelets with two sterile lemmas below a single perfect one; glumes reduced to a cuplike structure; nodes not bearded .... 12. *Oryza*

20. Glumes present.

22. Spikelets appearing to have one glume (second glume present but resembling the lemma).

23. Plants usually flowering during the spring and again during the autumn, forming overwintering basal rosettes of leaves .... 7. *Dichanthelium*

23. Plants usually flowering once, usually in the summer, not forming overwintering basal rosettes of leaves .... 13. *Panicum*

22. Spikelets appearing to have two similar glumes.

24. Glumes 1-nerved .... 1. *Agrostis*

24. Glumes 3-nerved.

25. All lemmas fertile .... 3. *Beckmannia*

25. Two lemmas above glumes sterile .... 15. *Phalaris*

1. **Agrostis** L.—Bent Grass

Tufted annuals or cespitose or rhizomatous perennials; blades flat or involute; inflorescence paniculate, spreading or contracted; spikelets 1-flowered, disarticulating above the glumes; glumes subequal, more or less keeled; lemmas smaller than the glumes, rounded on the back, obscurely nerved, awnless (in the wetland species); palea sometimes absent.

This genus is recognized by its 1-flowered spikelets that disarticulate above the glumes, by its lemmas that are awnless (in our species) and rounded on the back, and by its often capillary branches of the inflorescence. *Agrostis* consists of one hundred species found in temperate and subarctic regions of the world. Only the following sometimes occurs in shallow water in the central Midwest.

1. **Agrostis stolonifera** L. var. **palustris** (Huds.) Farw. Rep. Mich. Acad. Sci. 21:351. 1919. Fig. 122.

Agrostis palustris Huds. Fl. Angl. 27. 1762.
Agrostis polymorpha var. *palustris* (Huds.) Huds. Fl. Angl. ed. 2, 32. 1778.
Agrostis maritima Lam. Encycl. 1:61. 1783.
Agrostis alba L. var. *palustris* (Hud.) Pers. Syn. Pl. 1:76. 1805.
Agrostis stolonifera L. var. *compacta* Hartm. Skand. Fl. Handb., ed. 4, 24. 1843.

Matted perennial with extensively creeping stolons; culms decumbent, rarely erect, up to 1 m long; blades 1–5 mm broad; panicle usually straw-colored, with ascending scabrous branches, to 30 cm long; spikelets 2.0–3.5 mm long; glumes lanceolate, acute, 1-nerved, 2.0–3.5 mm long, scabrous on the keel; lemma acute,

122. *Agrostis stolonifera* (Creeping bent grass). a. Inflorescence. b. Base of plant. c. Sheath, with ligule. d. Spikelet.

1.5–3.0 mm long, awnless; palea one-half to two-thirds as long as the lemma. July–September.

Wet ground, marshes, sometimes in shallow water.

IA, IN, IL, KY, MO, OH (FACW), KS, NE (FAC+).

Creeping bent grass; marsh bent grass.

This species is distinguished by its very narrow leaves, its small, 1-flowered spikelets,and its very extensive creeping stolons.

2. **Alopecurus** L.—Meadow Foxtail

Annuals or perennials; blades flat; inflorescence a dense panicle, appearing spikelike; spikelets 1-flowered, disarticulating below the glumes; glumes equal, keeled; lemma 5-nerved, awned, the margins partly connate; palea none. There are twenty-five species in this genus, all in the north temperate region.

The following occasionally occur in standing water in the central Midwest.

1. Awn attached at the base of the lemma, straight, bent, or twisted, exserted 1.5–4.0 mm.
   2. Annual with fibrous roots; spikelets 2.0–2.5 (–2.7) mm long; awn straight .... 2. *A. carolinianus*
   2. Perennial with creeping off-shoots; spikelets 2.5–3.0 mm long; awn bent or twisted .... 3. *A. geniculatus*
1. Awn attached near middle of lemma, straight, exserted up to 1 mm .... 1. *A. aequalis*

1. **Alopecurus aequalis** Sobol. Fl. Petrop. 16. 1799. Fig. 123.
Alopecurus aristulatus Michx. Fl. Bor. Am. 1:43. 1803.
Alopecurus geniculatus L. var. *aristulatus* (Michx.) Torr. Fl. N. & Mid. U. S. 1:97. 1824.

Tufted perennial; culms erect or decumbent, glabrous, to 50 cm long, at least the uppermost internodes strongly glaucous; blades 1–4 mm broad; spikelike panicle to 7.5 cm long, 3.0–5.5 mm thick; spikelets 2.0–2.7 mm long; glumes obtuse, 2.0–2.7 mm long, villous at base, ciliate on keel; lemma 2.0–2.7 mm long, obtuse, 5-nerved, the awn straight or nearly so and exserted up to 1 mm. May–July.

Shallow water, muddy shores, and banks.

IA, IL, IN, KS, KY, MO, OH (OBL).

Glaucous meadow foxtail.

The uppermost internode is usually very glaucous. This perennial differs from *A. geniculatus*, another perennial, by its very shortly exserted awn that is attached near the middle of the lemma.

2. **Alopecurus carolinianus** Walt. Fl. Carol. 74. 1788. Fig. 124.
Alopecurus ramosus Poir. Lam. Encycl. 8:776. 1808.
Alopecurus geniculatus L. var. *ramosus* (Poir.) St. John, Rhodora 19:167. 1917.

Tufted annual with fibrous roots; culms usually erect, glabrous, to 60 cm tall; blades 1–4 mm broad; spikelike panicle to 5 cm long, 4–5 mm thick; spikelets 2.0–2.7 mm long, 1-flowered; glumes obtuse, 2.0–2.6 mm long, villous at base, ciliate on keel; lemma 2.0–2.7 mm long, obtuse, 5-nerved, the nearly straight awn exserted 1.5–4.0 mm. April–July.

Shallow water, low, moist soil.

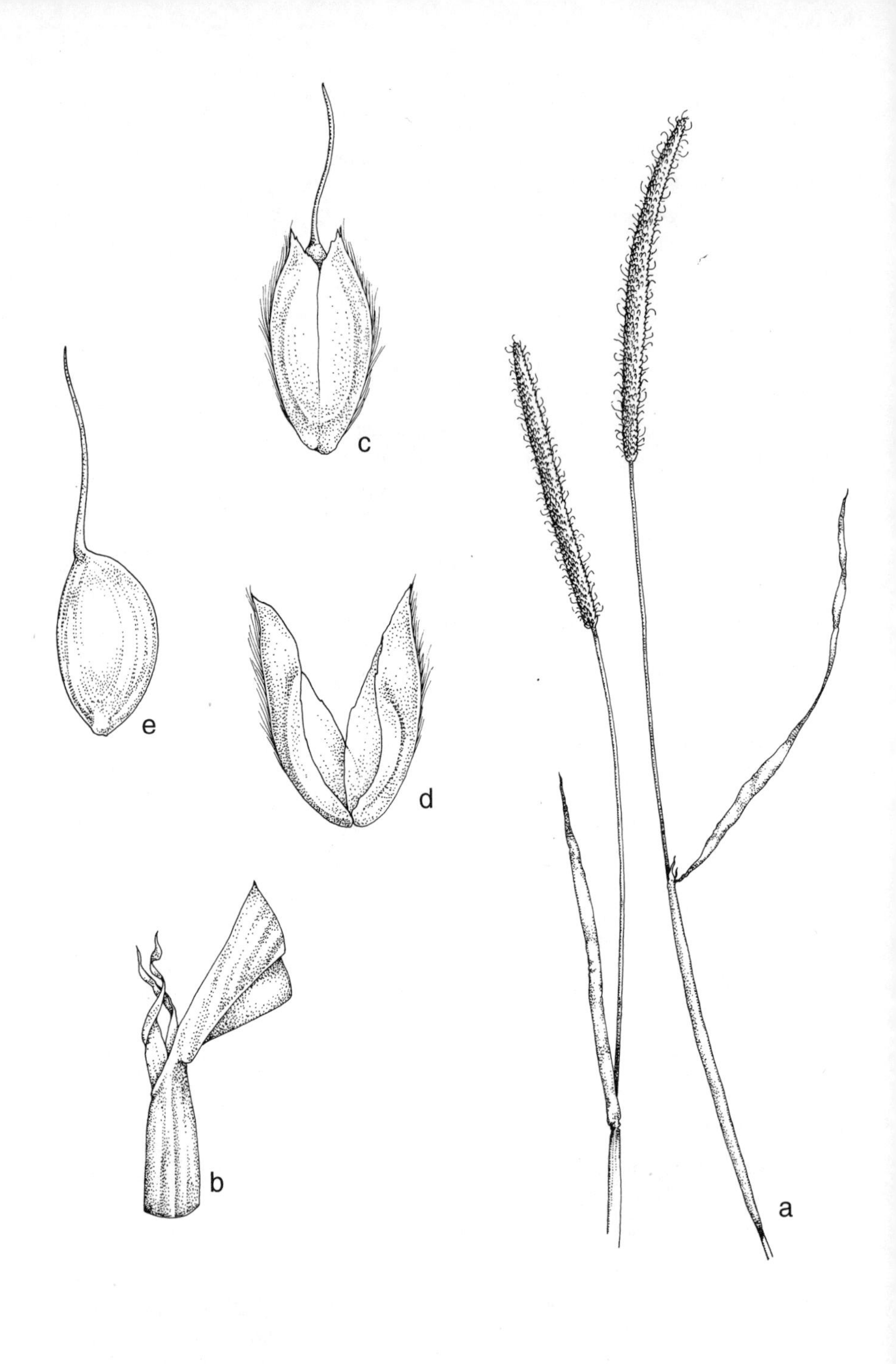

123. ***Alopecurus aequalis*** **(Glaucous meadow foxtail).**
a. Inflorescences.
b. Sheath, with ligule.
c. Spikelet.
d. Glumes.
e. Lemma.

124. *Alopecurus carolinianus* (Meadow foxtail). a. Habit. b. Spikelet.

IA, IL, IN, KS, KY, MO, NE, OH (FACW).

Meadow foxtail.

This species is similar to *A. aequalis* but differs by its annual growth form and its shorter spikes that are usually no longer than 5 centimeters.

3. **Alopecurus geniculatus** L. Sp. Pl. 1:60. 1753. Fig. 125.

Perennial; culms decumbent, rooting at the nodes, to 75 cm tall, glabrous; sheaths glabrous, with a membranaceous ligule; blades 4–5 mm broad, glabrous; inflorescence a spikelike panicle up to 5 cm long; spikelets 2.5–3.5 mm long, 1-flowered; glumes equal, 3-nerved, keeled, 2–3 mm long; lemma 2–3 mm long, 5-nerved, ciliate on the keel, with an awn bent or twisted, exserted 1.5–3.0 mm beyond the lemma. May–June.

Shallow water, muddy shores, often in disturbed areas.

IL, KY, OH (OBL).

Marsh foxtail.

This species is native to Europe and is not often seen in the central Midwest. It is a perennial like *A. aequalis*, but differs by its longer, twisted awn on the lemma.

125. *Alopecurus geniculatus* (Marsh foxtail). Habit with spikelet, in upper left.

3. **Beckmannia** Host—Slough Grass

Stout annuals; blades flat; inflorescence paniculate, composed of ascending spikes; spikelets 1-flowered, compressed, disarticulating below the glumes; glumes subequal, papery, inflated; lemma firmer but narrower than the glumes, 5-nerved, partly enclosing the slightly shorter, rigid palea.

There are two species in this genus, the other one occurring in Europe.

1. **Beckmannia syzigachne** (Steud.) Fern. Rhodora 30:27. 1928. Fig. 126.
Panicum syzigachne Steud. Flora 29:19. 1846.

Annuals; culms solitary or tufted, to nearly 1 m tall; blades flat, light green, scaberulous, to 8 mm broad; panicle slender, erect, to 25 cm long, composed of strongly ascending spikes to 1 cm long; spikelets 2–3 mm long, equally as broad, with one perfect flower and sometimes one imperfect flower; glumes broadly triangular, wrinkled, keeled, cuspidate, 3-nerved, 2–3 mm long, the margins sometimes or nearly meeting; lemmas lanceolate, acuminate, 2–3 mm long, 5-nerved. June–August.

Shallow water, marshes, edges of ponds and lakes.

IA, IL, MO, NE, OH (OBL).

American slough grass.

The panicle of strongly ascending spikes each about 1 cm long is distinctive. The spikelets often consist of one perfect and one imperfect flower. The glumes are broadly triangular.

4. **Calamagrostis** Adans.—Reed Grass

Perennials from creeping rhizomes; blades flat or involute; inflorescence paniculate, sometime spikelike; spikelets 1-flowered, disarticulating above the glumes; glumes subequal; lemmas shorter than the glumes, 3- or 5-nerved, awned from the back, with a bearded callus.

This genus is recognized by its 1-flowered spikelets with a lemma subtended by white hairs on the callus and a slender awn. There are about 150 species in the genus in the temperate and cool regions of the world.

1. Panicle open, more or less nodding .... 1. *C. canadensis*
1. Panicle contracted, spikelike.
  2. Awn attached well above the middle of the lemma .... 2. *C. cinnoides*
  2. Awn attached at the middle of the lemma .... 3. *C. stricta*

1. **Calamagrostis canadensis** (Michx.) Beauv. Ess. Agrost. 15:122. 1812. Fig. 127.
Arundo canadensis Michx. Fl. Bor. Am. 1:73. 1803.

Perennial from creeping rhizomes; culms to 1.5 m tall, glabrous or nearly so; sheaths glabrous; blades flat, 4–8 mm broad, more or less glaucous, becoming involute on drying; panicle more or less nodding, open, 5–20 cm long, purplish or greenish; spikelets 2.2–3.8 mm long; glumes subequal, lanceolate to narrowly ovate, obtuse to acute to acuminate, rounded or weakly keeled on the back, glabrous to puberulent, 1.7–3.5 mm long, the tips spreading in fruit; lemmas translucent at the erose

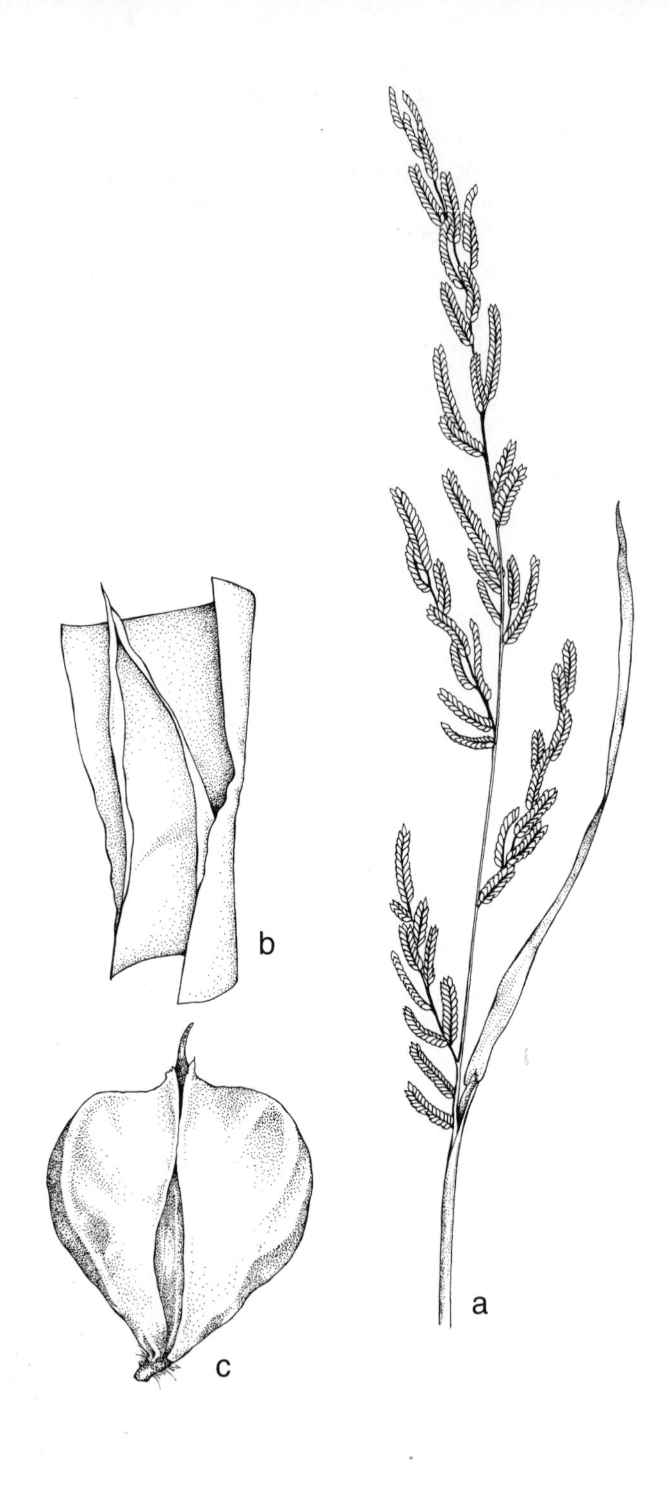

126. ***Beckmannia syzigachne*** **(American slough grass).** a. Inflorescence. b. Sheath, with ligule. c. Spikelet.

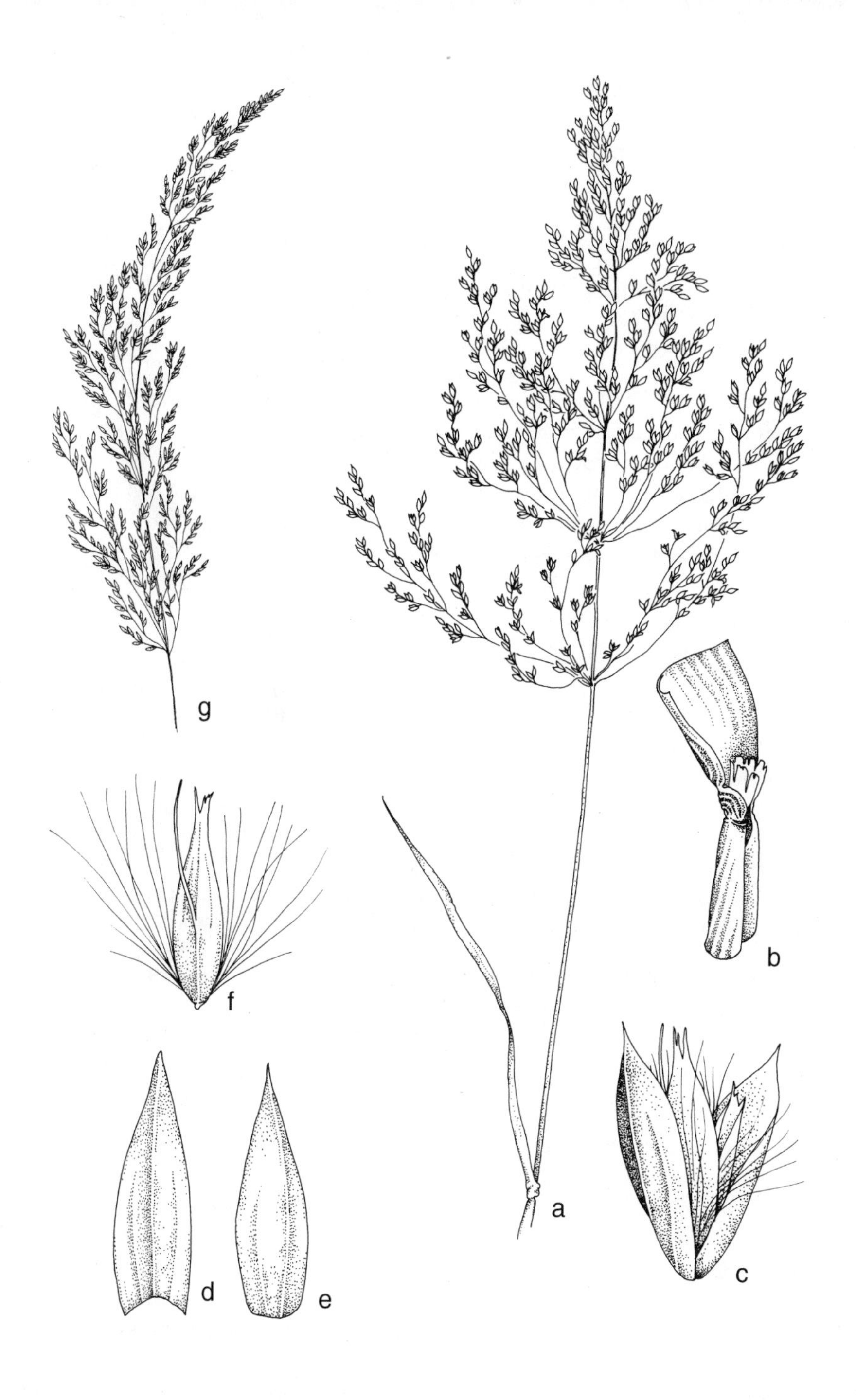

127. ***Calamagrostis canadensis*** **(Bluejoint grass).**
a. Inflorescence.
b. Sheath, with ligule.
c. Spikelet.
d. First glume.
e. Second glume.
f. Lemma.
g. Inflorescence.

tip, 1.5–3.0 mm long, mostly glabrous, with a straight, included awn inserted near the middle; callus of lemma usually as long as the lemma. June–July.

Shallow water, marshes, swamps, wet meadows, wet prairies.

IA, IL, IN, KS, MO, NE (OBL), KY, OH (FACW+).

Bluejoint grass.

This is the only wetland *Calamagrostis* in the central Midwest with an open panicle.

2. **Calamagrostis cinnoides** (Muhl.) W.P.C. Barton, Comp. Fl. Philadelph. 1:45. 1818. Fig. 128.

Arundo cinnoides Muhl. Gram. 187. 1817.

Tufted perennial with creeping rhizomes; culms to 1.2 m tall, usually smooth beneath the panicle; sheaths smooth or slightly scabrous; blades flat, 5–10 mm broad; panicle spikelike, erect, 10–20 cm long; spikelets 6–7 mm long; glumes lance-acuminate, the tips more or less outcurved, 4.5–8.0 mm long, the second slightly shorter than the fruit; lemma 3.5–6.0 mm long, scabrous, the callus hairs 1/2 as long as the lemma, the awn straight, inserted well above the middle of the lemma. June–July.

Swamps, wet woods, sometimes in standing water.

KY, OH (OBL).

Eastern reed grass.

128. *Calamagrostis cinnoides* (Eastern reed grass). Inflorescence. Spikelet (right).

The awn is attached well above the middle of the lemma, distinguishing this species from *C. stricta.*

3. **Calamagrostis stricta** (Timm) Koeler, Saccardoa 105. 1802. Fig. 129.

Arundo stricta Timm, Meklenb. Mag. 2:235. 1795.

Calamagrostis stricta (Timm) Koeler var. *brevior* Vasey. Rothr. Wheeler, Rep. U.S. Survey W. 100th Merid. 6:285. 1878.

Calamagrostis inexpansa Gray var. *brevior* (Vasey) Stebbins, Rhodora 32:50. 1930.

Tufted perennial from creeping rhizomes; culms to 1 m tall, scabrous beneath the panicle, otherwise glabrous; sheaths glabrous or scabrous; blades involute, 2–4 mm broad, scabrous, glaucous; panicle spikelike, erect, 5–20 cm long; spikelets 3.0–4.5 mm long; glumes subequal, lanceolate, acute to acuminate, rounded or weakly keeled on the back, scabrous, the tips connivent in fruit, the first 2.5–5.0 mm long, the second 2.5–4.8 mm long, purplish; lemma firm, toothed a the tip, 2.5–3.5 mm long, with a straight, included awn inserted near the middle; callus of lemma shorter than the lemma. June–July.

Wet meadows, marshes, occasionally in standing water.

IA, IL, IN, KS, MO, NE (FACW+), OH (FACW).

Northern reed grass.

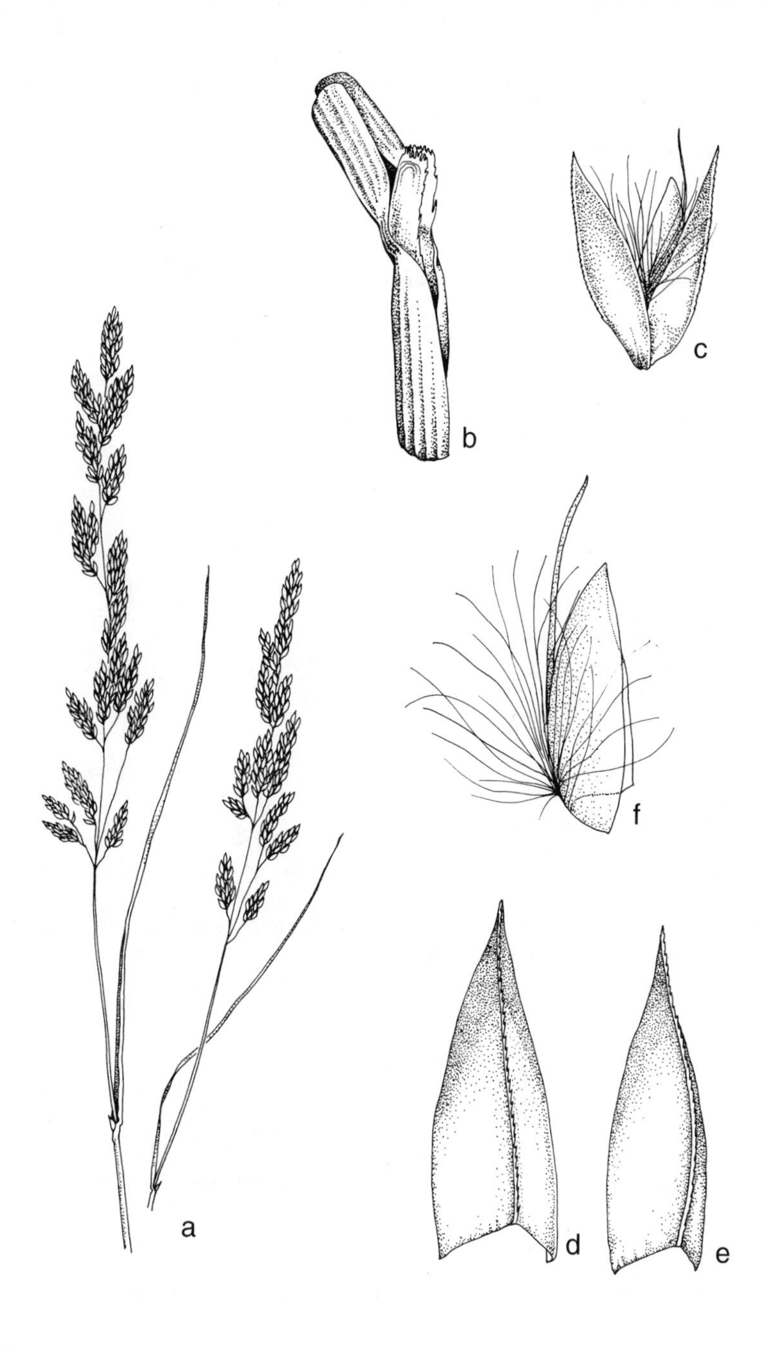

129. ***Calamagrostis stricta*** **(Northern reed grass).**
a. Inflorescences.
b. Sheath, with ligule.
c. Spikelet.
d. First glume.
e. Second glume.
f. Lemma.

This species differs from all other wetland species of *Calamagrostis* by its involute leaf blades. It also differs from the similar appearing *C. cinnoides* by its smaller spikelets and its awn of the lemma attached at the middle of the lemma.

The U.S. Fish and Wildlife Service calls this plant *C. inexpansa.*

5. **Catabrosa** Beauv.—Brook Grass

Only the following species comprises this genus.

1. **Catabrosa aquatica** (L.) Beauv. Agrost. 157. 1812. Fig. 130.
Aira aquatica L. Sp. Pl. 1:64. 1753.

Perennial with stolons; culms erect or decumbent, to 75 cm tall, glabrous, hollow; blades flat or plicate, up to 15 (–18) cm long, (3–) 6–10 (–12) cm broad, glabrous; sheaths glabrous, conspicuously cross-septate; ligules membranous, up to 6 mm long; panicle open, to 25 cm long, the lowest part usually included in the sheath; spikelets 2-flowered, 2.0–3.3 mm long; glumes lanceolate, to 2.2 mm long, scarious, 1- to 3-nerved; lemmas to 3 mm long, truncate, strongly 3-nerved. July–September.

In streams, springs, lakes, ponds.

NE (OBL).

Brook grass.

This totally glabrous aquatic grass is recognized by its 2-flowered spikelets and its strongly 3-nerved lemmas.

130. *Catabrosa aquatica* (Brook grass). Habit. Spikelet (lower right).

6. **Deschampsia** Beauv.—Hairgrass

Perennials; blades flat or plicate; inflorescence paniculate; spikelets 2-flowered, disarticulating above the glumes; glumes nearly equal, usually about as long as the spikelet; lemmas obscurely nerved, rounded on the back, awned, the callus bearded.

There are approximately thirty species of *Deschampsia* found in temperate and cooler regions of the world. Only the following may sometimes occur in shallow standing water in the central Midwest.

1. **Deschampsia cespitosa** (L.) Beauv. var. **glauca** (Hartm.) Lindm. f. Svensk. Fan. 81. 1918. Fig. 131.
Deschampsia glauca Hartm. Handb. Skand. Fl. 448. 1820.

Cespitose perennial to about 1 m tall; sheaths glabrous; blades flat or plicate, scabrous above, more or less glabrous below, 1–5 mm broad; panicles 3–20 cm long, more or less open; spikelets 3–5 mm long; glumes glabrous, acute, the first 2.5–4.5 mm

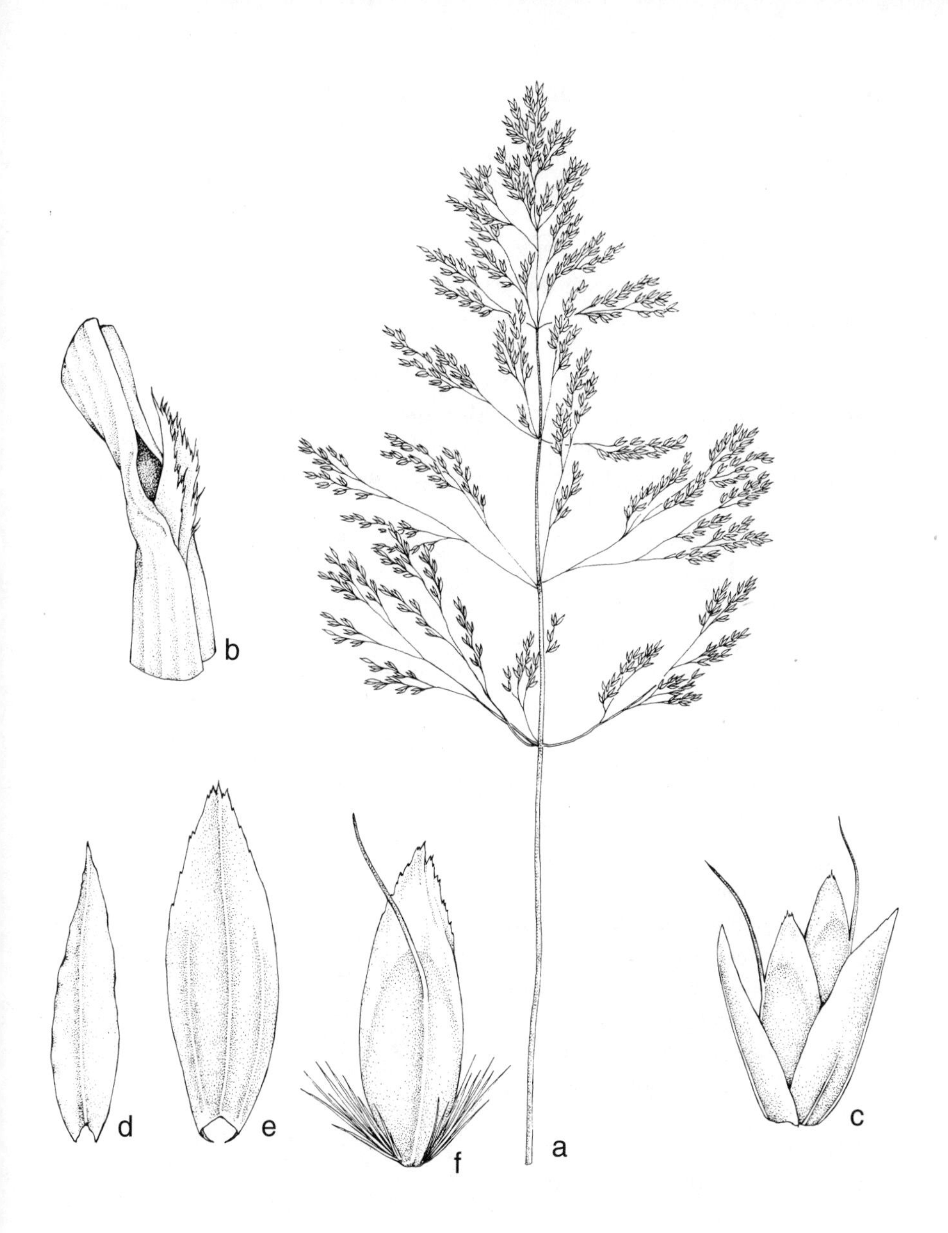

131. ***Deschampsia cespitosa* var. *glauca* (Tufted hairgrass).**
a. Inflorescence.
b. Sheath, with ligule.
c. Spikelet.
d. First glume.
e. Second glume.
f. Lemma.

long, the second 3–5 mm long; lemmas glabrous, except for the bearded callus, obscurely 5-nerved, 3–5 mm long, awned; awn more or less straight, 3–6 mm long, arising from near the middle or base of the lemma. June–July.

Swamps, bogs, along streams, sometimes in shallow water.

IL, IN (FACW+), OH (FACW).

Tufted hairgrass.

The distinguishing features of this grass are its two-flowered spikelets, with each lemma awned from the back.

7. **Dichanthelium** (Hitchc. & Chase) Gould

Spikelets 1-flowered, borne in panicles, with autumnal panicles different in appearance from vernal panicles; glumes very unequal, the first one small, the second one appearing similar in size and shape to the sterile lemma; winter rosettes of leaves produced.

I choose to follow some botanists in segregating *Dichanthelium* from *Panicum*. In *Dichanthelium*, spikelets are found in both the spring and the autumn and a basal rosette of leaves overwinters. In *Panicum*, spikelets are formed in the summer and there is no rosette of overwintering leaves.

Most species of *Dichanthelium* are upland, but the following sometimes may be found in shallow standing water.

1. **Dichanthelium spretum** (Schult.) Freckmann, Phytologia 48:102. 1981. Fig. 132.
Panicum spretum Schult, Mantissa 2:288. 1824.

Tufted perennial; culms erect, sometimes slightly decumbent, to 1 m tall, glabrous; sheaths glabrous or sparsely pilose; ligule a group of cilia up to 3 mm long; leaves ascending, lanceolate, 3–9 mm broad, glabrous or puberulent beneath and occasionally ciliate; panicles narrow, up to 12 cm long, up to nearly half as broad, the branches strongly ascending and glabrous; spikelets ellipsoid to obovoid, obtuse, pubescent, 1.3–1.8 mm long, less than 1 mm wide, frequently purplish. June–August.

132. *Dichanthelium spretum* **(Panic grass).** Habit. Spikelet (above).

Wet shores, occasionally in standing water.

IN (FACU).

Panic grass.

This is the only species of *Dichanthelium* with a fringe of cilia up to 3 mm long for its ligule and with a narrow panicle.

The U.S. Fish and Wildlife Service equates this species with *Dichanthelium sphaerocarpon* (Ell.) Gould.

8. **Echinochloa** Beauv.—Wild Millet

Annuals; blades flat; inflorescence paniculate, often contracted, composed of dense racemes; spikelets 1-flowered, arranged in 4 or more ranks; first glume up to half as long as the spikelet, 3-nerved; second glume and both lemmas equal in length, the second glume and sterile lemma usually awned, the fertile lemma papery, not awned.

Echinochloa consists of about thirty species in the warmer regions of the world.

Much variation exists in the treatment of North American species of *Echinochloa* by various authors.

1. Second glume with an awn 2–10 mm long; sheaths papillose-hirsute (glabrous in f. *laevigata*); grain about three times longer than broad .... 3. *E. walteri*
1. Second glume awnless or with an awn less than 2 mm long; sheaths glabrous or scabrous; grain at most about twice as long as broad, usually shorter.
   2. Fertile lemma with a weak, easily broken tip, with a ring of minute setae just below the tip .... 1. *E. crus-galli*
   2. Fertile lemma firm-tipped, without a setulose ring below the tip .... 2. *E. muricata*

1. **Echinochloa crus-galli** (L.) Beauv. Ess. Agrost. 53. 1812. Fig. 133.
Panicum crus-galli L. Sp. Pl. 1:56. 1753.
Echinochloa muricata (Micx.) Fern. var. *occidentalis* Wieg. Rhodora 23:58. 1921.
Echinochloa occidentalis (Wieg.) Rydb. Brittonia 1:82. 1931.
Echinochloa pungens (Poir.) Rydb. var. *occidentalis* (Wieg.) Fern. & Grisc. Rhodora 37:137. 1935.

Annual; culms erect or decumbent from the base, usually less than 1.5 m tall; sheaths glabrous; blades 5–20 mm broad, glabrous or scabrous above; panicle to 25 cm long, green or purple, the mostly ascending racemes to 4 cm long; spikelets ovoid; second glume hispidulous on the nerves, acute; sterile lemma hispidulous on the nerves, rarely with marginal papillose hairs, short-awned; fertile lemma with a soft, easily broken tip, with a ring of small setae where the tip adjoins the body; grain 2.5–3.5 mm long, ovoid, shining. June–November.

Disturbed soil, rarely in standing water.

IA, IL, IN, KS, MO, NE (FACW), KY, OH (FACU).

Barnyard grass.

Although this European grass is common in barnyards, it also occurs in disturbed wet soil. It very closely resembles *E. muricata,* but the tip of each lemma is soft and wilts readily and is surrounded by a ring of small setae.

2. **Echinochloa muricata** (Michx.) Fern. Rhodora 17:106. 1915. Figs. 134, 135, 136.
Panicum muricatum Michx. Fl. Bor. Am. 1:47. 1803.
Setaria muricata (Michx.) Beauv. Ess. Agrost. 51. 1812.
Panicum pungens (Poir.) Lam. Encycl. Sup. 4:273. 1816.
Echinochloa crus-galli Beauv. var. *muricata* (Michx.) Farw. Rep. Mich. Acad. Sci. 21:350. 1920.
Echinochloa muricata (Michx.) Fern. var. *microstachya* Wieg. Rhodora 23:58. 1921.
Echinochloa pungens (Poir.) Rydb. Brittonia 1:81. 1931.
Echinochloa microstachya (Wieg.) Rybd. Brittonia 1:82. 1931.

133. *Echinochloa crus-galli* (Barnyard grass).

a. Upper part of plant.
b. Sheath, with ligule.
c. Spikelet, front view.
d. Spikelet, back view.

134. *Echinochloa muricata* var. *muricata* (Wild millet). a. Inflorescence. b. Sheath, with ligule. c. Spikelet, front view. d. Spikelet, back view.

135. *Echinochloa muricata* var. *microstachya* **(Wild millet).** a. Inflorescence. b. Sheath, with ligule. c. Spikelet, front view. d. Spikelet, back view.

136. *Echinochloa muricata* var. *wiegandii* (Wild millet).

a. Habit.
b. Sheath, with ligule.
c. Spikelet, front view.
d. Spikelet, back view.

Echinochloa pungens (Poir.) Rydb. var. *microstachya* (Wieg.) Fern. & Grisc. Rhodora 37:137. 1935.
Echinochloa pungens (Poir.) Rydb. var. *wiegandii* Fassett, Rhodora 51:2. 1949
Echinochloa muricata (Michx.) Fern. var. *wiegandii* (Fassett) Mohlenbr. Ill. Fl. Il. Grasses: Panicum to Danthonia, ed. 2, 396. 2001.

Annual; culms erect or decumbent, branching from the base; sheaths scabrous or glabrous; blades to 20 mm broad, scabrous above; panicle to 20 cm long, more or less open to contracted, purplish or greenish, the spreading to ascending racemes to 6 cm long; spikelets ovoid; second glume and sterile lemma papillose-hispid to minutely pubescent, the second glume acuminate to awn-tipped, the sterile lemma awnless or with an awn up to 10 mm long; tip of fertile lemma firm, without a ring of small setae at its base. June–October.

Moist soil, occasionally in standing water.

IA, IL, IN, KS, MO, NE (OBL), KY, OH (FACW+).

Wild millet.

This common grass in wet areas resembles the non-native *E. crus-galli* but differs in its firm-tipped fertile lemma that lacks a small ring of setae at its base. This species in the past has been known as *E. pungens* (Poir.) Rydb. It is a variable species with several of the varieties occurring in the central Midwest. The most common variety in the central Midwest is var. *microstachya.* The three common varieties may be distinguished by the following key:

a. Spikelets and grain 3.5–4.5 mm long; sterile lemma with an awn over 3 mm long; anthers 0.7–0.9 mm long; panicle more or less open, the branches spreading .... 2a. *E. muricata* var. *muricata*
a. Spikelets and grain 2.5–3.5 mm long; sterile lemma awnless or with an awn less than 3 mm long; anthers 0.3–0.7 mm long; panicle contracted, the branches ascending.
  b. Panicle purple; second glume and sterile lemma papillose-hispid .... 2b. *E. muricata* var. *microstachya*
  b. Panicle green; second glume and sterile lemma appressed-pubescent, not papillose .... 2c. *E. muricata* var. *wiegandii*

3. **Echinochloa walteri** (Pursh) Heller, Cat. N. Am. Pl. 2:21. 1900. Fig. 137.
Panicum hirtellum Walt. Fl. Carol. 72. 1788, non L. (1759).
Panicum walteri Pursh, Fl. Am. Sept. 66. 1814.
Panicum hispidum Muhl. Descr. Gram. 107. 1817, non Forst. (1786).
Echinochloa walteri (Pursh) Heller f. *laevigata* Wieg. Rhodora 23:62. 1921.

Robust annual; culms erect, to 2 m tall; sheaths papillose-hirsute, rarely glabrous; blades harshly scabrous, 5–30 mm broad; panicle dense, 10–30 cm long, greenish, the racemes appressed-ascending; spikelets ellipsoid, 3.5–4.5 mm long; second glume strigose, with an awn 2–10 mm long; sterile lemma with an awn 10–25 mm long; fruit about three times longer than broad; anthers 0.6–1.2 mm long. August–October.

Marshes, swamps, often in standing water.

IL, IN, MO (OBL), KY, OH (FACW+).

Southern wild millet.

137. *Echinochloa walteri* forma *walteri* (Southern wild millet). a. Inflorescence. b. Sheath, with ligule. c. Spikelet, forma *laevigata*. d. Sheath, with ligule, forma *laevigata*.

Both glumes of the spikelets are long-awned and the lemmas lack bristles with swollen bases. The sheaths of the leaves are often hispid, except in f. *laevigata.*

9. **Eragrostis** Beauv.—Lovegrass

Annuals or perennials; blades usually flat or folded, rarely involute; inflorescence paniculate; spikelets compressed, 2- to many-flowered, disarticulating above the glumes; glumes 2, somewhat unequal, shorter than the spikelets; lemmas 3-nerved, keeled, without a tuft of cobwebby hairs at the base.

This genus consists of about 250 species distributed throughout the world. Only the following two species may sometimes be found in shallow water.

1. Sheaths more or less glabrous; inflorescence 2–8 cm long, the peduncle glabrous; lemmas 1.5–2.0 mm long, acute, glabrous; anthers 0.2–0.5 mm long; plants monoecious .... 1. *E. hypnoides*
1. Sheaths pubescent; inflorescence 10–25 cm long, the peduncle villous; lemmas 2–4 long, acuminate, sparsely villous along the nerves; anthers 1.5–2.0 mm long; plants dioecious .. .. 2. *E. reptans*

1. **Eragrostis hypnoides** (Lam.) BSP. Prel. Cat. N. Y. 69. 1888. Fig. 138.
Poa hypnoides Lam. Tabl. Encyl. 1:185. 1791.

Creeping, mat-forming, monoecious annual, rooting at the nodes; sheaths more or less glabrous; blades scabrous or puberulent above, 1–3 mm broad; inflorescence 2–8 cm long, the peduncles glabrous; spikelets 5–15 mm long, perfect, 1.0–1.5 mm broad, 10- to 35-flowered; glumes subacute, the first usually a little less than 1 mm long, the second 1.0–1.8 mm long; lemmas 1.5–2.0 mm long, acute, glabrous; anthers 0.2–0.5 mm long. July–October.

Shores, muddy banks, occasionally in standing water.

IA, IL, IN, KY, MO, OH (OBL), KS, NE (FAC).

Pony grass.

This mat-forming species is distinguished by its 10- to 35-flowered perfect spikelets and its very narrow and short leaves.

2. **Eragrostis reptans** (Michx.) Nees, Agrost. Bras. 514. 1829. Fig. 139.
Poa reptans Michx. Fl. Bor. Am. 1:69. 1803.
Poa dioica Michx. ex Poir. Lam. Encycl. 5:87. 1804.
Poa weigeltiana Reichenb. ex Trin. Mem. Acad. St. Petersb. VI Math. Phys. Nat. 1:410. 1830.
Eragrostis weigeltiana (Reichenb.) Bush, Trans. Acad. Sci. St. Louis 13:180. 1903.
Neeragrostis reptans (Michx.) Nicora, Rev. Argent. Agron. 29:5. 1962.

Creeping, mat-forming, dioecious annual, rooting at the nodes; sheaths pubescent; blades pubescent, 1–3 mm broad; inflorescence 10–25 cm long, the peduncles villous; spikelets unisexual, 5–18 mm long, 2–4 mm broad, 10- to 75-flowered; glumes acute, the first 1.0–1.5 mm long, the second 1.5–2.2 mm long; lemmas 2–4 mm long, acuminate, sparsely villous along the nerves; anthers 1.5–2.0 mm long. July–October.

Around ponds and lakes, along streams, occasionally in standing water.

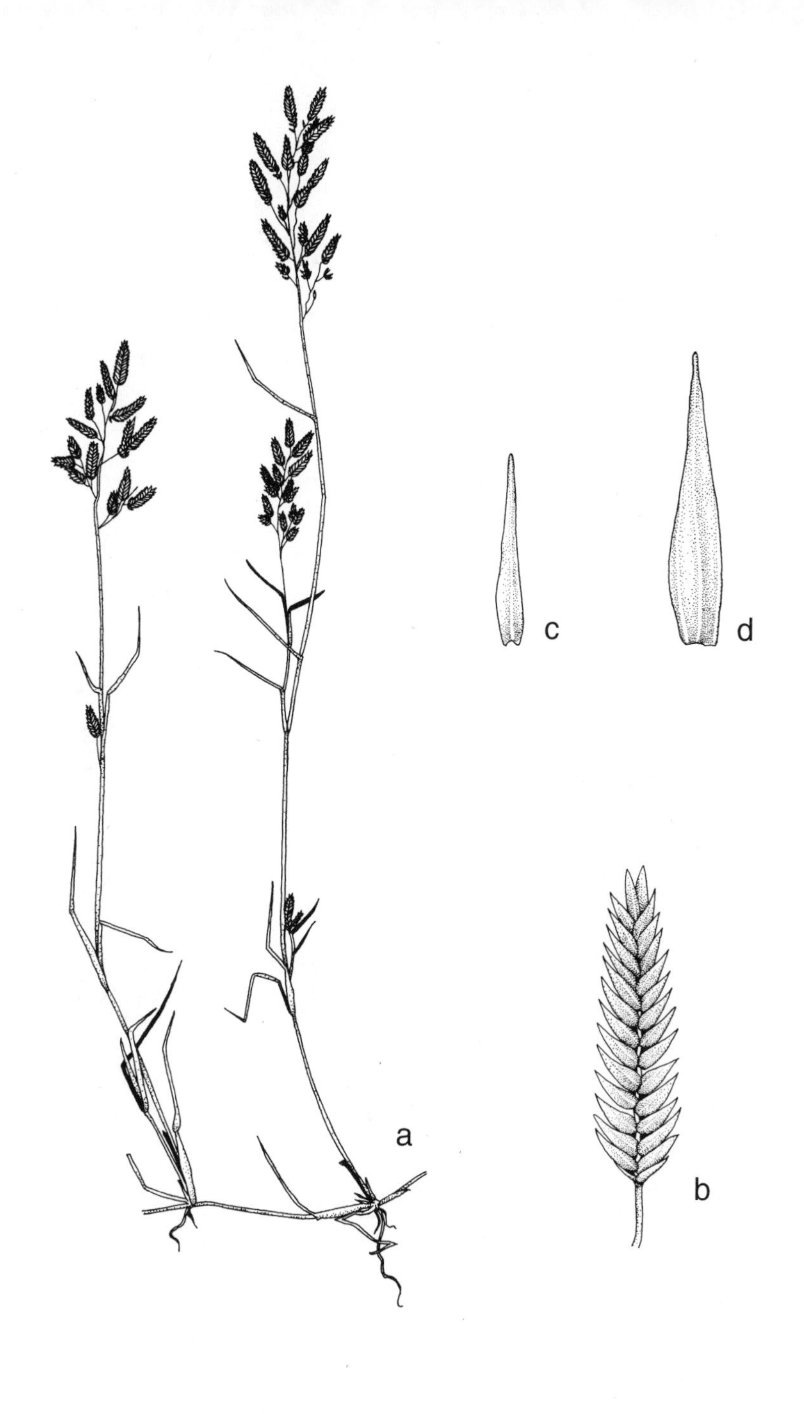

138. *Eragrostis hypnoides* (Pony grass).

a. Habit.
b. Spikelet.
c. First glume.
d. First lemma.

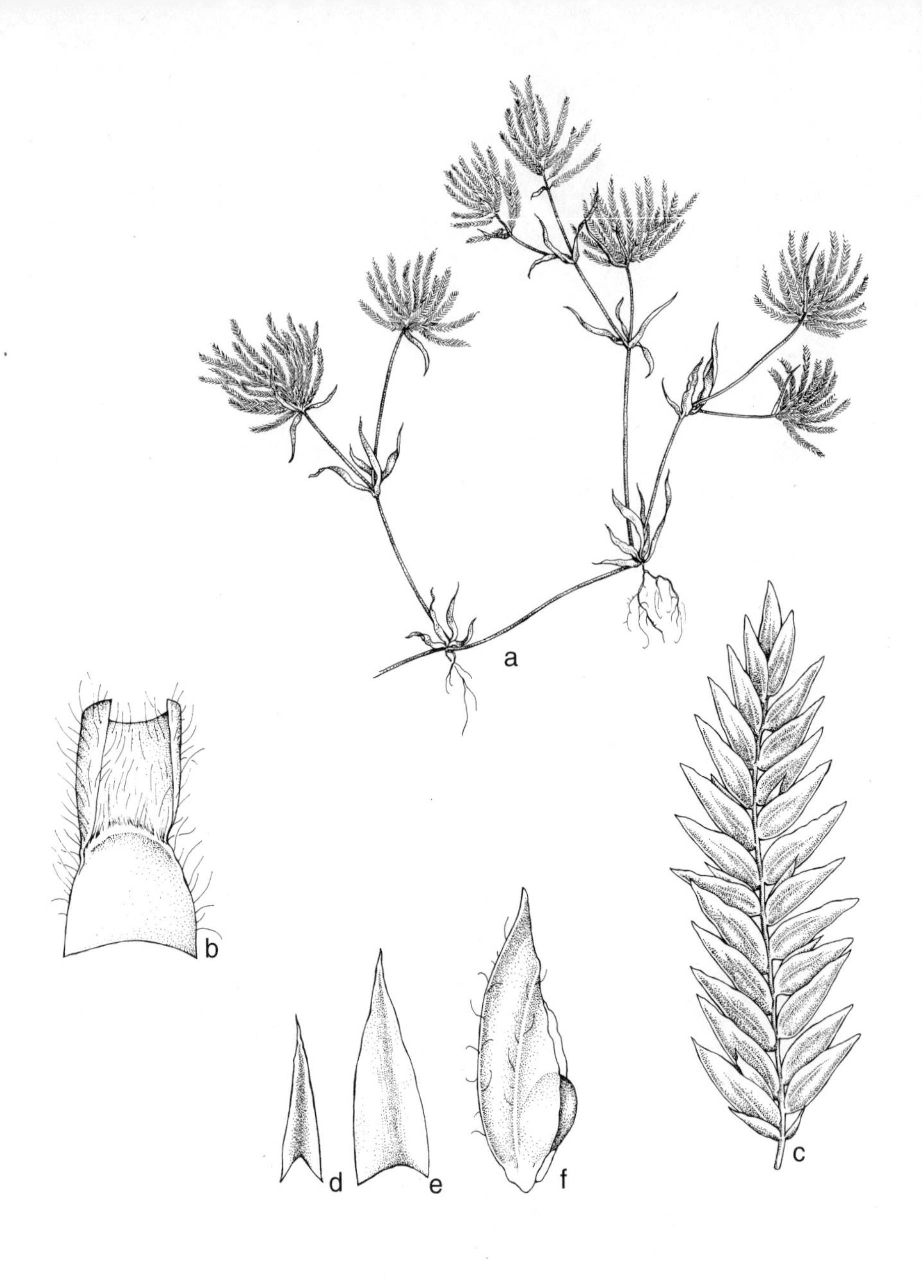

139. ***Eragrostis reptans*** **(Creeping lovegrass).**
a. Habit.
b. Sheath, with ligule.
c. Pistillate spikelet.
d. First glume.
e. Second glume.
f. Lemma.

IA, IL, IN, KS, MO, NE (OBL), OH (FACW).

Creeping lovegrass.

This is another mat-forming species with usually 5- to 20-flowered spikelets, but it is distinctive because of the dioecious nature of the plants. Because of this difference, this species is sometimes placed in the genus *Neeragrostis*.

10. **Glyceria** R. Br.—Manna Grass

Perennials; sheaths usually closed; blades flat; inflorescence paniculate; spikelets several-flowered, disarticulating above the glumes; glumes 2, unequal, shorter than the spikelets; lemmas rounded on the back, distinctly 5- to 9-nerved, awnless; lodicules united; style present.

Chief apparent differences separating this genus from *Torreyochloa* are the usually closed sheaths. The plant usually known as *G. pallida* now is considered to be a species of *Torreyochloa*.

There are about thirty-five species of *Glyceria* found in the north temperate regions of the world.

1. Spikelets at least 10 mm long; sheaths compressed.
  2. Lemmas acute .... 1. *G. acutiflora*
  2. Lemmas obtuse.
    3. Principal leaves 2–5 mm broad; lemmas shining, scabrous only on the nerves; pedicels very slender, all one-fourth to two-thirds the length of the spikelets .... 3. *G. borealis*
    3. Principal leaves 6–18 mm broad; lemmas dull, scabrous or hirtellous between the nerves; pedicels thickened upward, less than one-fourth the length of the spikelets (except for the terminal ones).
      4. Principal blades 6–12 mm broad; lemmas obscurely nerved, scabrous, 3.5–5.5 mm long; anthers over 1 mm long .... 7. *G. septentrionalis*
      4. Principal blades 10–18 mm broad; lemmas sharply nerved, hirtellous, 2.5–3.0 mm long; anthers less than 1 mm long .... 2. *G. arkansana*
1. Spikelets 2–8 mm long; sheaths terete or subterete.
  5. Panicle strict and not open .... 6. *G. melicaria*
  5. Panicle open.
    6. Lemmas 3–4 mm long, obscurely nerved; spikelets 3–4 mm broad .... 4. *G. canadensis*
    6. Lemmas 1.5–2.7 mm long, sharply nerved; spikelets 2.0–2.5 mm broad.
      7. Inflorescence 5–20 cm long; spikelets 2.0–4.5 mm long; first glume 0.5–1.0 mm long; second glume 0.8–1.3 mm long; lemmas 1.5–2.0 mm long .... 8. *G. stricta*
      7. Inflorescence 20–40 cm long; spikelets 5–6 mm long first glume 1.2–2.0 mm long; second glume 1.5–2.5 mm long; lemmas 2.0–2.7 mm long .... 5. *G. grandis*

1. **Glyceria acutiflora** Torr. Fl. N. Middle U.S. 1:104. 1823. Fig. 140.

Perennial, rooting at the nodes, with culms to 1 m tall; sheaths overlapping, glabrous; ligule 5–7 mm long; blades glabrous, or scabrous on the upper surface, 3–6 mm broad; inflorescence narrow, to 30 cm long, the branches erect and flattened, the base of the inflorescence enclosed in the uppermost sheath; spikelets 15–40 mm long, 5- to 13-flowered; glumes narrowly ovate, acute, the first up to 3.5 mm long, the second 4.0–5.7 mm long; lemmas lanceolate, acute, 6–10 mm long, scabrous. May–July.

Shallow water, edge of ponds.

KY, MO (OBL).

Sharp-scaled manna grass.

This species differs from the other species of *Glyceria* whose spikelets are longer than 10 mm by its acute lemmas.

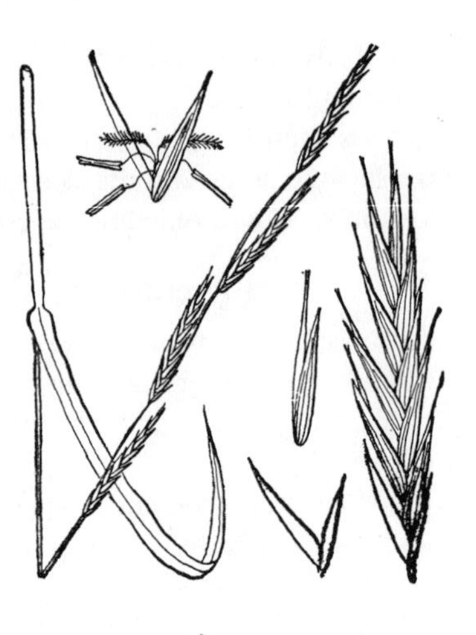

140. *Glyceria acutiflora* (Sharp-scaled manna grass). Floret (center). Spikelet (left). Spikelet, close up (right).

2. **Glyceria arkansana** Fern. Rhodora 31:49. 1929. Fig. 141.
Glyceria septentrionalis Hitchc. var. *arkansana* (Fern.) Steyerm. & Kucera, Rhodora 63:24. 1961.

Perennial, rooting at the lower nodes, with culms to nearly 2 m tall; sheaths compressed, glabrous; blades glabrous, 10–18 mm broad; inflorescence 40–70 cm long, ascending; spikelets 15–20 mm long, 10- to 15-flowered; glumes elliptic to obovate, scarious throughout, rather obscurely nerved, the first 1.5–3.0 mm long, the second 2.5–3.5 mm long; lemmas elliptic, obtuse and erose at the apex, sharply 7-nerved, hirtellous throughout on the back, 2.5–3.0 mm long, slightly shorter than the palea; anthers less than 1 mm long. May–June.

Shallow water, swamps.

IL, KY, MO (OBL).

Southern manna grass.

This manna grass differs from the closely related *G. septentrionalis* by its broader blades, its shorter, more strongly nerved, hirtellous lemmas, and its tiny anthers.

The U.S. Fish and Wildlife Service considers this plant to be the same as *G. septentrionalis.*

3. **Glyceria borealis** (Nash) Batchelder, Proc. Manchester Inst. 1:74. 1900. Fig. 142.
Panicularia borealis Nash, Bull. Torrey Club 24:348. 1897.

Perennial, rooting at the lower nodes, with culms to 1.2 m tall; sheaths compressed, glabrous; blades glabrous, 2–6 mm broad; inflorescence 15–45 cm long, ascending; spikelets 10–18 mm long, 6- to 12-flowered, on slender pedicels one-fourth to two-thirds as long; glumes elliptic, obtuse to subacute, obscurely nerved, with a scarious margin, the first 1–2 mm long, the second 2–3 mm long; lemmas obtuse, erose and scarious at the apex, scabrous only on the nerves, 7-nerved, 3–4 mm long, longer than the palea; anthers less than 1 mm long. June–July.

Shallow water, wet shores.

IA, IL, IN, NE (OBL).

Northern manna grass.

The membranaceous lemmas are glabrous except for the hirtellous nerves, distinguishing it from *G. arkansana.* The similar *G. septentrionalis* has coriaceous lemmas.

141. *Glyceria arkansana* (Southern manna grass).

a. Inflorescence.
b. Sheath, with ligule.
c. Spikelets.
d. Lemma.

142. ***Glyceria borealis*** **(Northern manna grass).**
a. Inflorescence.
b. Sheath, with ligule.
c. Spikelets.
d. First glume.
e. Second glume.
f. Lemma.

143. *Glyceria canadensis* (Rattlesnake manna grass).

a. Inflorescence.
b. Sheath, with ligule.
c. Spikelet.
d. Glumes.

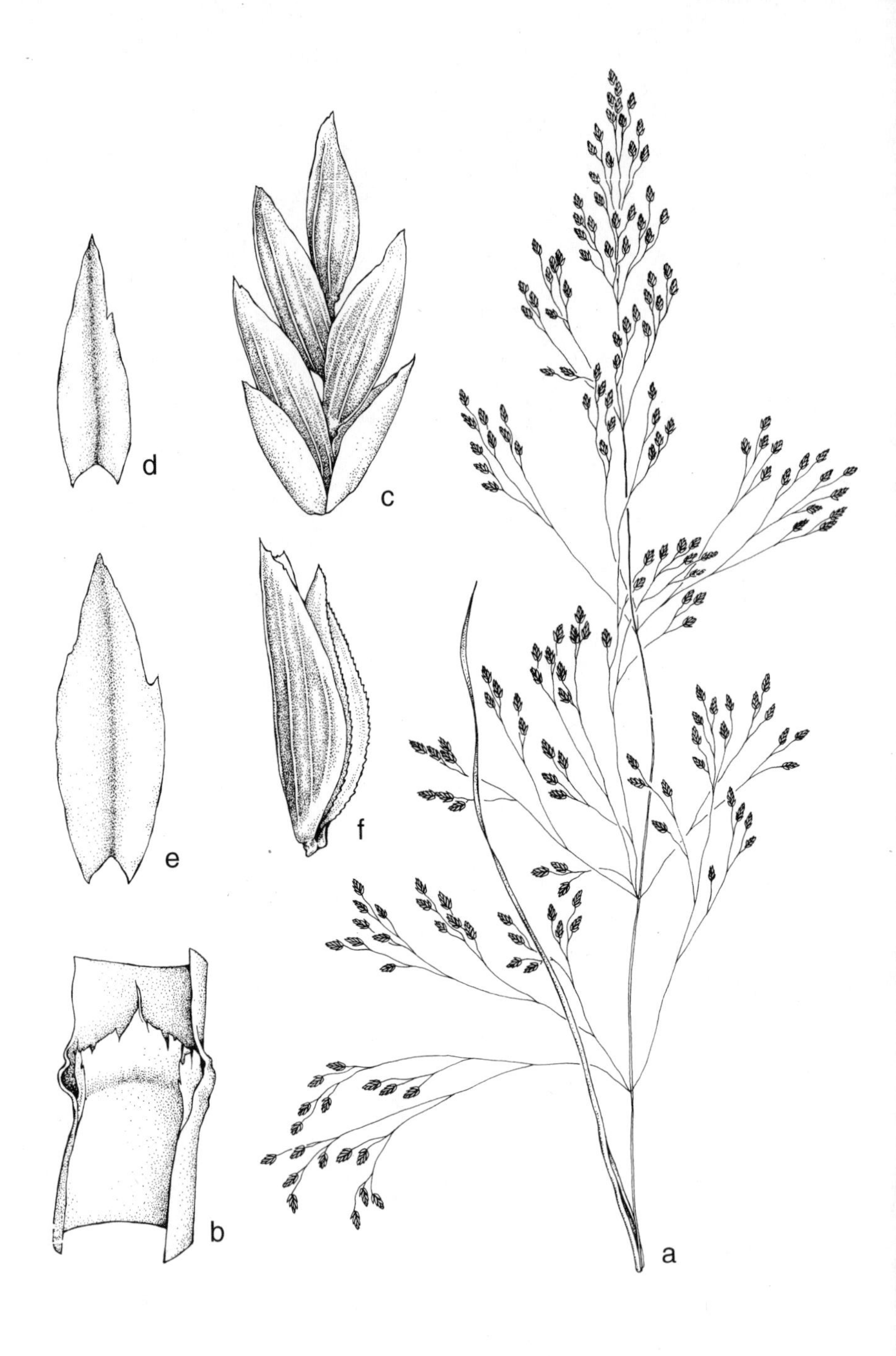

144. *Glyceria grandis* (American manna grass).
a. Inflorescence.
b. Sheath, with ligule.
c. Spikelet.
d. First glume.
e. Second glume.
f. Lemma and palea.

4. **Glyceria canadensis** (Michx.) Trin. Mem. Acad. St. Petersb. VI Math. Phys. Nat. 1:366. 1830. Fig. 143.
Briza canadensis Michx Fl. Bor. Am. 1:71. 1803.
Panicularia canadensis (Michx.) Kuntze, Rev. Gen. Pl. 1:783. 1891.

Solitary or tufted perennial to 1.5 m tall; sheaths terete or subterete, glabrous; blades scabrous, 2.5–8.5 mm broad; inflorescence open, drooping at the tip, 5–25 cm long; spikelets 4–8 mm long, 3–5 mm broad, 4- to 10-flowered; glumes obscurely nerved, with a scarious margin, the first lanceolate, 1.5–2.5 mm long, the second ovate, 2–3 mm long; lemmas broadly ovate, with a scarious margin, obscurely 7-nerved, 3–4 mm long, longer than the palea. June–September.

Bogs, wet meadows, along shores, sometimes in standing water.

IL, IN, OH (OBL).

Rattlesnake manna grass.

This is the only *Glyceria* with spikelets less than 10 mm long but 3–5 mm broad. The branches of the inflorescence are capillary.

5. **Glyceria grandis** S. Wats. ex Gray, Man., ed. 6, 667. 1890. Fig. 144.
Poa aquatica L. var. *americana* Torr. Fl. N. & Mid. U.S. 1:108. 1824.
Glyceria americana (Torr.) Pammel, Rep. Iowa Geol. Surv. 1903:271. 1905.
Panicularia grandis (Gray) Nash. Britt. & Brown, Ill. Fl. 1:265. 1913.

Tufted perennial to 1.5 m tall; sheaths terete or subterete, glabrous; blades glabrous or scabrous, 6–14 mm broad; inflorescence 20–40 cm long, open, nodding at the tip; spikelets 5–6 mm long, 2.0–2.5 mm broad, 2- to 9-flowered, purplish; glumes scarious, acute, the first 1.2–2.0 mm long, the second 1.5–2.5 mm long; lemmas narrowly ovate, obtuse, 7-nerved, 1.0–1.7 mm long. June–July.

Wet meadows, ditches, shores of ponds, occasionally in standing water.

IA, IL, IN, OH, NE (OBL).

American manna grass; reed manna grass.

The purplish spikelets of this species recall *G. striata*, but *G. grandis* has a more open panicle and longer glumes and lemmas.

6. **Glyceria melicaria** (Michx.) F. T. Hubbard, Rhodora 14:186. 1912. Fig. 145.
Panicum melicaria Michx. Fl. Bor. Am. 1:50. 1803.

Perennial with a creeping base; culms solitary or few, slender, to 1 m tall; ligule up to 1 mm long; blades 2–8 mm wide, scabrous; inflorescence narrow, strict, the branches ascending, nodding at the tip, to 12 cm long; spikelets 3- to 4-flowered, 3.5–4.0 mm long; first glume narrowly lan-

145. *Glyceria melicaria* (Northeastern manna grass). Inflorescence. Spikelets (right).

ceolate, 1.3–2.5 mm long; second glume lanceolate, 1.7–3.0 mm long; lemmas broadly elliptic, abruptly acute, 2.0–2.8 mm long, glabrous. June–August.

Swamps, wet woods, occasionally in standing water.

KY, OH (OBL).

Northeastern manna grass.

This is the only *Glyceria* with narrow, strict inflorescences.

7. **Glyceria septentrionalis** Hitchc. Rhodora 8:211. 1906. Fig. 146.
Panicuilaria septentrionalis (Hitchc.) Bickn. Bull. Torrey Club 35:196. 1908.

Perennial, rooting at the lower nodes, with culms to 1.5 m tall; sheaths compressed, glabrous; blades glabrous, the principle ones 4–12 mm broad; inflorescence 20–45 cm long, ascending; spikelets 10–20 mm long, 6- to 15-flowered, on upwardly thickened pedicels less than one-fourth as long (except in the terminal spikelets); glumes elliptic to obovate, scarious throughout, obscurely nerved, the first 2–4 mm long, the second 3–5 mm long; lemmas elliptic, obtuse and erose at the apex, obscurely 7-nerved, scabrous between the nerves, 3.5–5.5 mm long, slightly shorter than the palea; anthers over 1 mm long. May–August.

Swampy woods, wet meadows, marshes, often in standing water.

IA, IL, IN, KY, MO, OH (OBL).

Manna grass.

This species is recognized by its spikelets at least 10 mm long and its obtuse, scabrous lemmas that are 3.5–5.5 mm long.

8. **Glyceria striata** (Lam.) Hitchc. Proc. Biol. Soc. Wash. 41:157. 1928.
Poa striata Lam. Tabl. Encycl. 1:183. 1791.

Tufted perennial to 1.2 m tall; sheaths terete or subterete, glabrous; blades flat or conduplicate, scabrous above, 2–8 mm broad; inflorescence 5–20 cm long, open, usually drooping at the tip; spikelets 2.0–4.5 mm long, 2.0–2.5 mm broad, 3- to 7-flowered, green or purple; glumes obovate, obscurely nerved, the first 0.5–1.0 mm long, the second 0.8–1.3 mm long; lemmas elliptic to obovate, obtuse, more or less scarious at the apex, 7-nerved, 1.5–2.0 mm long. May–August.

Moist soil, occasionally in standing water.

IA, IL, IN, KS, KY, MO, NE, OH (OBL).

Fowl manna grass.

Two varieties may be distinguished in the central Midwest.

1. Spikelets green; uppermost branches of the panicle more or less nodding; lemmas with a minutely scarious apex .... 8a. *G. striata* var. *striata*
1. Spikelets purple; uppermost branches of the panicle ascending; lemmas with a broadly scarious apex .... 8b. *Glyceria striata* var. *stricta*

8a. **Glyceria striata** (Lam.) Hitchc. var. **striata.** Fig. 147.
Poa nervata Willd. Sp. Pl. 1:389. 1797.
Poa lineata Pers. Syn. Pl. 1:89. 1805.
Glyceria nervata (Willd.) Trin. Mem. Acad. St. Petersb. VI Math. Phys. Nat. 1:365. 1830.
Panicularia nervata (Willd.) Kuntze, Rev. Gen. Pl. 2:783. 1891.

146. ***Glyceria septentrionalis*** **(Manna grass).**
a. Inflorescence.
b. Sheath, with ligule.
c. Spikelet.
d. First glume.
e. Second glume.
f. Lemma and palea.

147. ***Glyceria striata*** **var.** ***striata*** **(Fowl manna grass).** a. Inflorescence. b. Sheath, with leaf. c. Spikelet.

Leaves flat; inflorescence 10–20 cm long, the uppermost branches more or less nodding; spikelets green, 2–4 mm long; lemmas with a minutely scarious apex.

This typical variety differs from var. *stricta* primarily by its green spikelets and more open inflorescence.

8b. **Glyceria striata** (Lam.) Hitchc. var. **stricta** (Scribn.) Fern. Rhodora 31:47. 1929. Fig. 148.
Panicularia nervata (Willd.) Kuntze var. *stricta* Scribn. Bull. U.S.D.A. Div. Agrost. 13:44. 1898.
Glyceria nervata (Willd.) Trin. var. *stricta* Scribn. ex Hitchc. Gray, Man., ed. 7, 159. 1908.

Leaves flat or plicate; inflorescence 5–15 cm long, the uppermost branches ascending; spikelets purple, 3.0–4.5 mm long; lemmas with a broadly scarious apex.

There is considerable overlapping of characters between this variety and var. *striata* so that it is questionable whether the varieties should be distinguished. Spikelet color seems to be the most reliable character.

11. **Leersia** Sw.—Cut-grass

Perennials from scaly rhizomes; blades flat; inflorescence paniculate, composed of short racemes; spikelets 1-flowered, sometimes sterile, disarticulating at the base; glumes none; lemma papery, broad, keeled, 5-nerved; palea papery, nearly as long as the lemma, keeled, usually 3-nerved.

The nodes of the culms of this genus are always puberulent or ciliate. The oblong to ovate, 5-nerved lemmas and the absence of glumes are distinctive.

Seven species comprise this genus worldwide, but only the following are in the central Midwest.

1. Spikelets broadly rounded, 3–4 mm broad, over half as wide as long. .... 1. *L. lenticularis*
1. Spikelets oblongoid, 1–2 mm broad, less than half as wide as long.
    2. Sheaths conspicuously retrorse-scabrous; blades spinulose on the margins; lowest panicle branches whorled; stamens 3 .... 2. *L. oryzoides*
    2. Sheaths glabrous or scaberulous; blades scaberulous; lowest panicle branches solitary; stamens 2 .... 3. *L. virginica*

1. **Leersia lenticularis** Michx. Fl. Bor. Am. 1:39. 1803. Fig. 149.
Asprella lenticularis (Michx.) Beauv. Ess. Agrost. 2:153. 1812.
Homalocenchrus lenticularis (Michx.) Kuntze, Rev. Gen. Pl. 2:777. 1891.

Perennial from scaly rhizomes; culms erect, to 1 m tall; sheaths more or less glabrous; blades 5–20 mm broad, glabrous or softly villous; panicle to 25 cm long, freely branched, spreading; spikelets broadly rounded, 4.0–5.5 mm long, 3–4 mm broad, over half as wide as long, arranged in short racemes to 2 cm long; lemma 3.8–5.5 mm long, ciliate on the keel and the margins; stamens 2. August–October.

Swamps and low woods, occasionally in standing water.

IL, IN, KS, KY, MO, OH (OBL).

Catchfly grass.

The broadly oval to suborbicular spikelets that are 3–4 mm broad distinguish this species from the other species of *Leersia*. The blades and sheaths are usually scabrous, but not harshly so as in *L. oryzoides*.

148. ***Glyceria striata***
var. *stricta* (Fowl manna grass).
a. Inflorescence.
b. Sheath, with ligule.
c. Spikelet.
d. First glume.
e. Second glume.

149. *Leersia lenticularis* (Catchfly grass).

a. Upper part of plant.
b. Sheath, with ligule.
c. Spikelet.

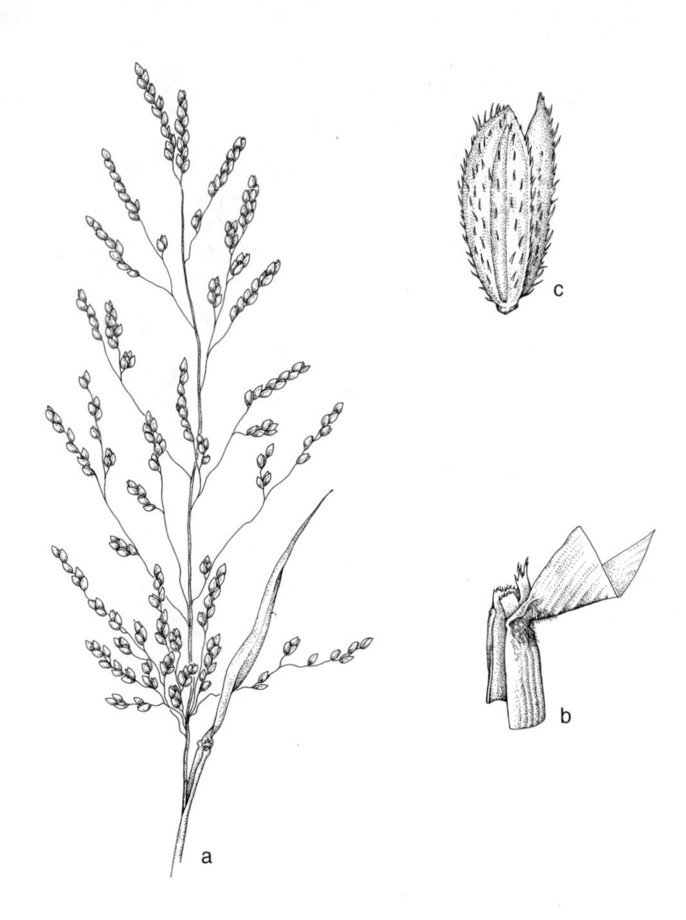

150. *Leersia oryzoides* (Rice cut-grass).

a. Inflorescence.
b. Sheath, with ligule.
c. Spikelet.

2. **Leersia oryzoides** (L.) Swartz, Prodr. Veg. Ind. Occ. 21. 1788. Fig. 150.
Phalaris oryzoides L. Sp. Pl. 55. 1753.
Homalocenchrus oryzoides (L.) Poll. Hist. Pl. Palat. 1:52. 1776.
Oryza clandestinus L. f. *inclusa* Wiesb. Baenitz. Deut. Bot. Monat. 15:19. 1897.
Leersia oryzoides (L.) Sw. f. *inclusa* (Wiesb.) Dorfl. Herb. Norm. Sched. Cent. 55–56, 164. 1915.

Perennial from slender rhizomes; culms erect or decumbent, to 1.5 m tall; sheaths conspicuously retrorse-scabrous; blades 6–12 mm broad, spinulose on the margins; panicle to 20 cm long, spreading to ascending, the lowest branches whorled; spikelets oblongoid, 3.8–6.0 mm long, 1–2 mm broad, less than half as wide as long, arranged in short racemes to 1 cm long; lemma 3.8–6.0 mm long, pilose, ciliate on the keel; stamens 3. August–October.

Swampy woods, marshes, ditches, sometimes in standing water.

IA, IL, IN, KS, KY, MO, NE, OH (OBL).

Rice cut-grass.

The extremely harshly scabrous culms, blades, and sheaths are the most severe of any grass and make the grass dangerous to handle. Care should be taken to avoid letting the plant come in contact with bare skin.

Specimens in which the panicle scarcely becomes exserted from the sheath have been called f. *inclusa*.

3. **Leersia virginica** Willd. Sp. Pl. 1:325. 1797. Fig. 151.
Leersia ovata Poir. Lam. Encycl. Sup. 3:329. 1813.
Homalocenchrus virginicus (Willd.) Britt. Trans. N. Y. Acad. Sci. 9:14. 1889.
Leersia virginica Willd. var. *ovata* (Poir.) Fern. Rhodora 38:386. 1936.

Perennial from short, thick rhizomes; culms erect or decumbent, rooting at the base, to 1.0 (–1.5) m tall; sheaths smooth or scaberulous; blades 3–15 mm broad, scaberulous; panicle to 20 cm long, spreading, the lowest branches solitary; spikelets oblongoid, 2.5–4.0 mm long, 1–2 mm broad, less than half as wide as long, arranged in a short raceme to 2 cm long; lemma 3–4 mm long, glabrous to ciliate on the keel and margins; stamens 2. July–September.

Moist or wet woods, rarely in standing water.

IA, IL, IN, KS, KY, MO, NE, OH (FACW).

White grass.

There is variation in width of the blade and in pubescence of the lemma. Some specimens with densely ciliate lemmas have been referred to as var. *ovata*.

12. **Oryza** L.—Rice

Annuals; spikelets numerous in open panicles, flattened; glumes reduced to cuplike structures; fertile lemma 1, keeled, awned or awnless; sterile lemmas 2, glumelike in appearance; grains enclosed by the lemmas.

There are eighteen species in the genus, most of them in the tropics and subtropics.

1. **Oryza sativa** L. Sp. Pl. 1:333. 1753. Fig. 152.

Tufted annual; culms to 1.5 m tall, erect, glabrous; sheaths glabrous or puberulent, the ligule membranaceous; blades up to 18 mm broad, glabrous but scabrous on the margins, sometimes auriculate at the base; panicles open, the branches spreading or sometimes pendulous, bearing numerous spikelets; spikelets flattened, consisting of 2 sterile lemmas below a single perfect floret; glumes reduced to a cuplike structure; sterile lemmas glumelike in appearance, 1.5–3.0 mm long, narrowly lanceolate, acute, 1-nerved, glabrous; fertile lemma 7–10 mm long, keeled, oblong to obovate, acute, awnless or with an awn up to 5 mm long, glabrous or puberulent, rugose, 5-nerved; grains 6–7 mm long, ellipsoid, white, enclosed by the lemma. July–September.

Wet depressions.

IL, MO (OBL).

Rice.

This plant occasionally escapes into wet areas.

13. **Panicum** L.—Panic Grass

Annuals or perennials; blades flat; inflorescence paniculate; spikelets disarticulating below the glumes; glumes 2, the lower much smaller, the upper more or less

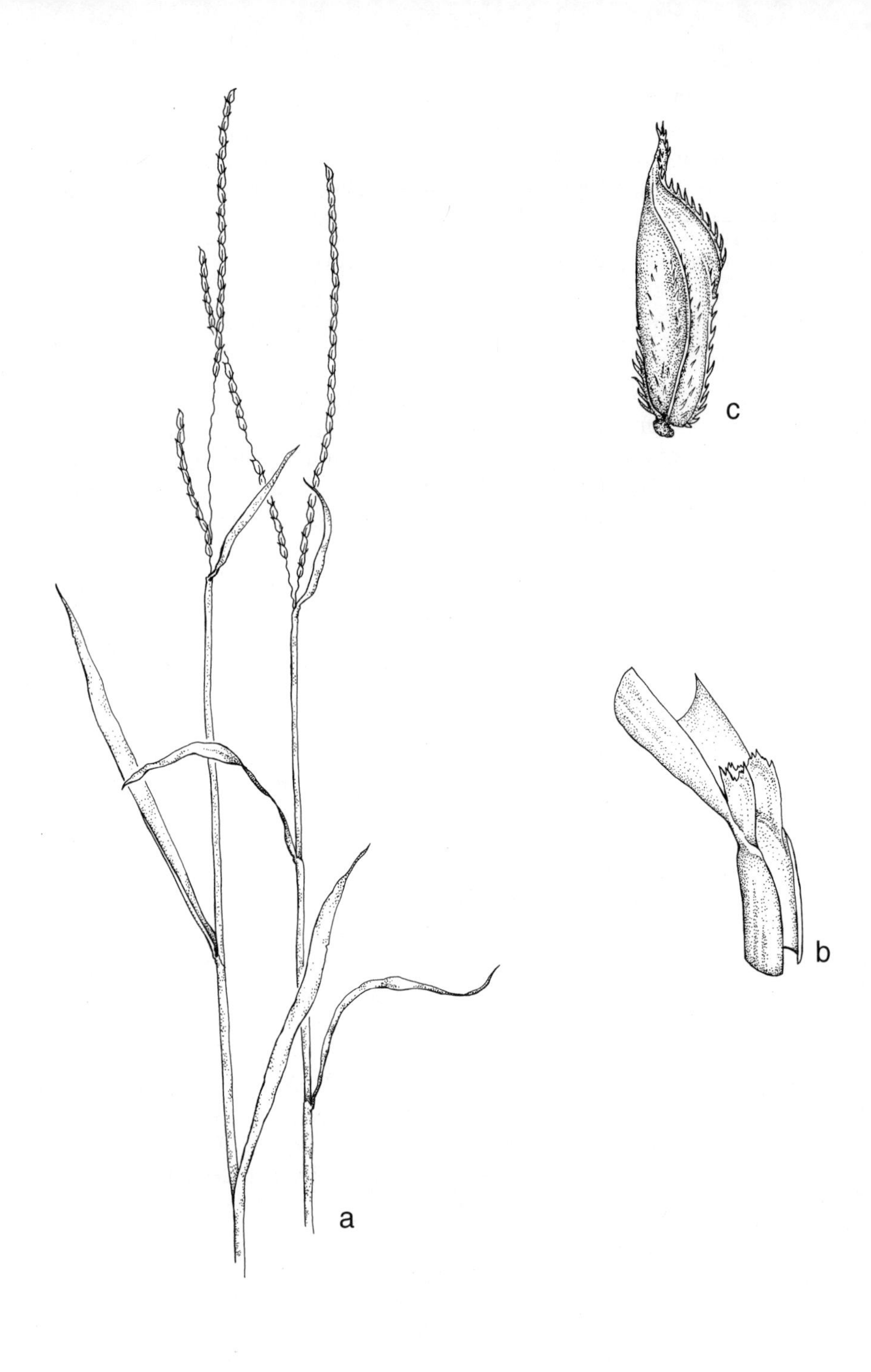

151. *Leersia virginica* **(White grass).**

a. Upper part of plants.
b. Sheath, with ligule.
c. Lemma.

152. ***Oryza sativa*** **(Rice).** Habit with spikelet in upper left corner.

equaling and of similar texture to the sterile lemma; sterile lemma with a hyaline palea; fertile lemma indurate, the margins inrolled over the enclosed palea; grain free from the lemma and palea.

Some botanists, including me, believe that *Panicum* should be divided into at least two genera, *Panicum* and *Dichanthelium*. In *Panicum*, plants bloom a single time during the summer and then die back completely in the late autumn. In *Dichanthelium*, plants bloom in the spring and then again a second time in the autumn. Rosettes of leaves form in the autumn and overwinter; their withered remains may usually be found attached to the base of the next year's plants.

Many species of *Panicum* are upland species. The following three species sometimes live in standing water.

1. Second glume and sterile lemma papillose .... 3. *P. verrucosum*
1. Second glume and sterile lemma smooth, at least not papillose.
   2. Palea of sterile floret hardened, longer than the glume; blades 3–6 mm broad; culms wiry, not compressed .... 1. *P. hians*
   2. Palea of sterile floret not hardened, shorter than the glume; blades to 12 mm broad; culms not wiry, compressed .... 2. *P. rigidulum*

1. **Panicum hians** Ell. Bot. S.C. & Ga. 1:118. 1816. Fig. 153.
Steinchisma hians (Ell.) Nash, Fl. S. E. U.S. 165. 1903.

Cespitose perennial; culms sparsely branched, erect or geniculate at the base, to 60 cm tall; sheaths glabrous, keeled; ligules 0.5 mm long; blades 1–5 mm broad, flat or occasionally folded, pilose on the upper surface near the base, glabrous below; panicle to 20 cm long, loose and open, the branches ascending, spreading, or drooping; spikelets 1.5–2.4 mm long, 1.2–2.0 mm broad, glabrous, more or less secund; first glume about half as long as the spikelet, acute; second glume and sterile lemma subequal; palea indurate and enlarged at maturity, forcing the spikelet open; grain 1.4–1.9 mm long, 0.7 mm broad, ellipsoid, acute. June–October.

Banks of ponds, ditches.

MO (FACW). The U.S. Fish and Wildlife Service calls this plant *Steinchisma hians*.

Panic grass.

The hardened palea enlarges to force the lemma and sterile glume apart at maturity. Because of this unusual feature, some botanists have placed this species in a different genus, calling it *Steinchisma hians*.

2. **Panicum rigidulum** Bosc ex Nees. Mart. Fl. Bras. 2 (1):163. 1829.
Panicum agrostoides Spreng. Pl. Pugill. 2:4. 1815, *nomen illeg.*

Densely clumped perennial from a short caudex; culms erect, compressed, often geniculate at the base, glabrous, to 1 m tall; sheaths pilose on the sides, or appressed-pubescent, or glabrous; ligule erose, about 1 mm long; blades flat near the tip, to 12 mm broad, becoming folded near the base, glabrous or sparsely pilose above near the base; panicles terminal and axillary, to 30 cm long, 1/2–2/3 as broad, the axis glabrous to scabrous, the branches spreading, ascending, or strongly erect; spikelets 1.8–2.5 mm long, 0.7–0.8 mm broad, narrowly ellipsoid to lanceoloid, acute to short-acuminate, glabrous or with the nerves somewhat scabrous; first

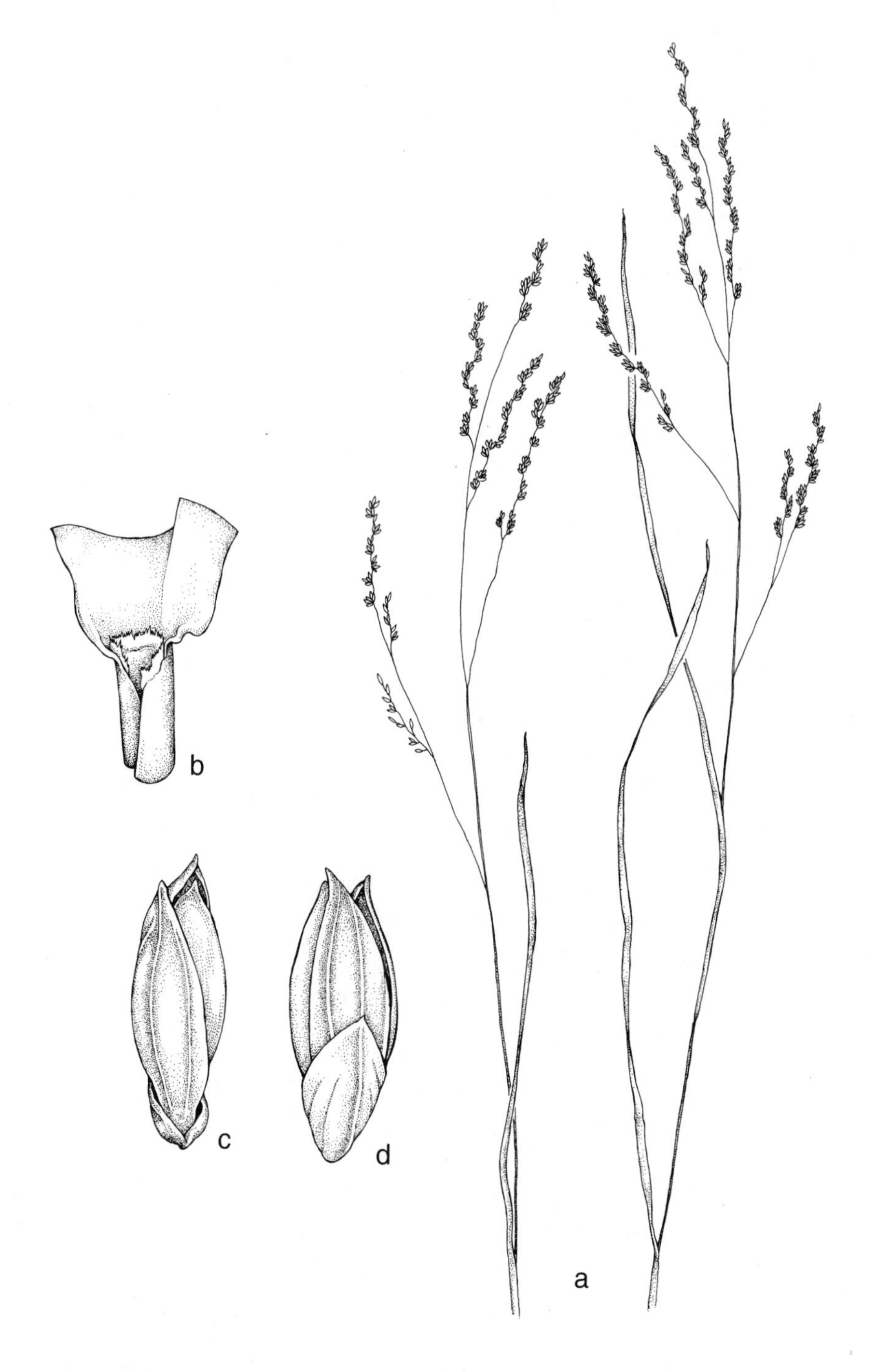

153. *Panicum hians* (Panic grass).

a. Upper part of plants.
b. Sheath, with ligule.
c. Spikelet, front view.
d. Spikelet, back view.

glume about half as long as the spikelet, acute or acuminate, glabrous; second glume and sterile lemma subequal, with scabrous nerves near the tips; grain 1.3–1.5 mm long, 0.6–0.7 mm broad, ellipsoid, apiculate.

Panicum rigidulum is distinguished by its compressed culms, its spikelets mostly arranged on one side of the axes of the inflorescence, and by its shiny spikelets.

Two varieties occur in the central Midwest.

a. Panicle branches spreading to ascending; spikelets 1.8–2.2 mm long .... 2a. *P. rigidulum* var. *rigidulum*

a. Panicle branches strongly erect; spikelets 2.2–2.5 mm long. .... 2b. *P. rigidulum* var. *condensum*

2a. **Panicum rigidulum** Nees var. **rigidulum.** Fig. 154.
Panicum agrostoides Spring. Pl. Pugill. 2:4. 1815, *nomen illeg.*
Panicum elongatum Pursh var. *ramosior* Mohr, Contr. U.S. Nat. Herb. 6:357. 1901.
Panicum agrostoides Spreng. var. *ramosius* (Mohr) Fern. Rhodora 38:390. 1936.

Culms to 1 m tall; sheaths pilose on the sides, otherwise glabrous; blades to 12 mm broad, glabrous; panicle to 30 cm long, the axis often glabrous, spreading to ascending; spikelets 1.8–2.2 mm long, 0.7–0.8 mm broad, narrowly ellipsoid; grain 1.3 mm long, 0.6 mm broad. July–October.

Moist soil, along ponds or creek, low woodlands.

IA, IL, IN, KS, MO, NE (FACW), KY, OH (FACW+).

Munro grass.

This variety differs from var. *condensum* in its more spreading panicle branches and in its smaller spikelets.

2b. **Panicum rigidulum** Nees var. **condensum** (Nash) Mohlenbr. Ill. Fl. Illinois Grasses 71. 1973. Fig. 155.
Panicum condensum Nash. Small, Fl. Southeast. U.S. 93. 1903.
Panicum agrostoides Spreng. var. *condensum* (Nash) Fern. Rhodora 36:74. 1934.

Culms to 75 cm tall; sheaths appressed-pubescent or glabrous; blades to 10 mm broad, sparsely pilose at the base of the upper surface, glabrous below; panicle to 25 cm long, very narrow, the axis scabrous, the branches strongly erect; spikelets 2.2–2.5 mm long, 0.8 mm broad, lanceoloid, acuminate, glabrous except for the scabrous nerves; grain 1.4–1.5 mm long, 0.7 mm broad. June–October.

Moist soil, rarely in standing water.

IA, IL, IN, KS, MO (FACW), KY, OH (FACW+).

Narrow munro grass.

3. **Panicum verrucosum** Muhl. Descr. Gram. 113. 1817. Fig. 156.

Annual from slender roots; culms erect to spreading, to 1.2 m long, glabrous; sheaths glabrous except for the puberulent margins, the ligule up to 0.3 mm long; blades to 10 mm broad, glabrous, pale green; panicle to 20 cm long, with slender, glabrous branches bearing spikelets only near the tip; spikelets 1.7–2.2 mm long, obovoid, glabrous; first glume 0.5–0.8 mm long, up to one-fourth as long as the spikelet, obtuse, faintly 1-nerved, usually with a few small tubercles; second glume

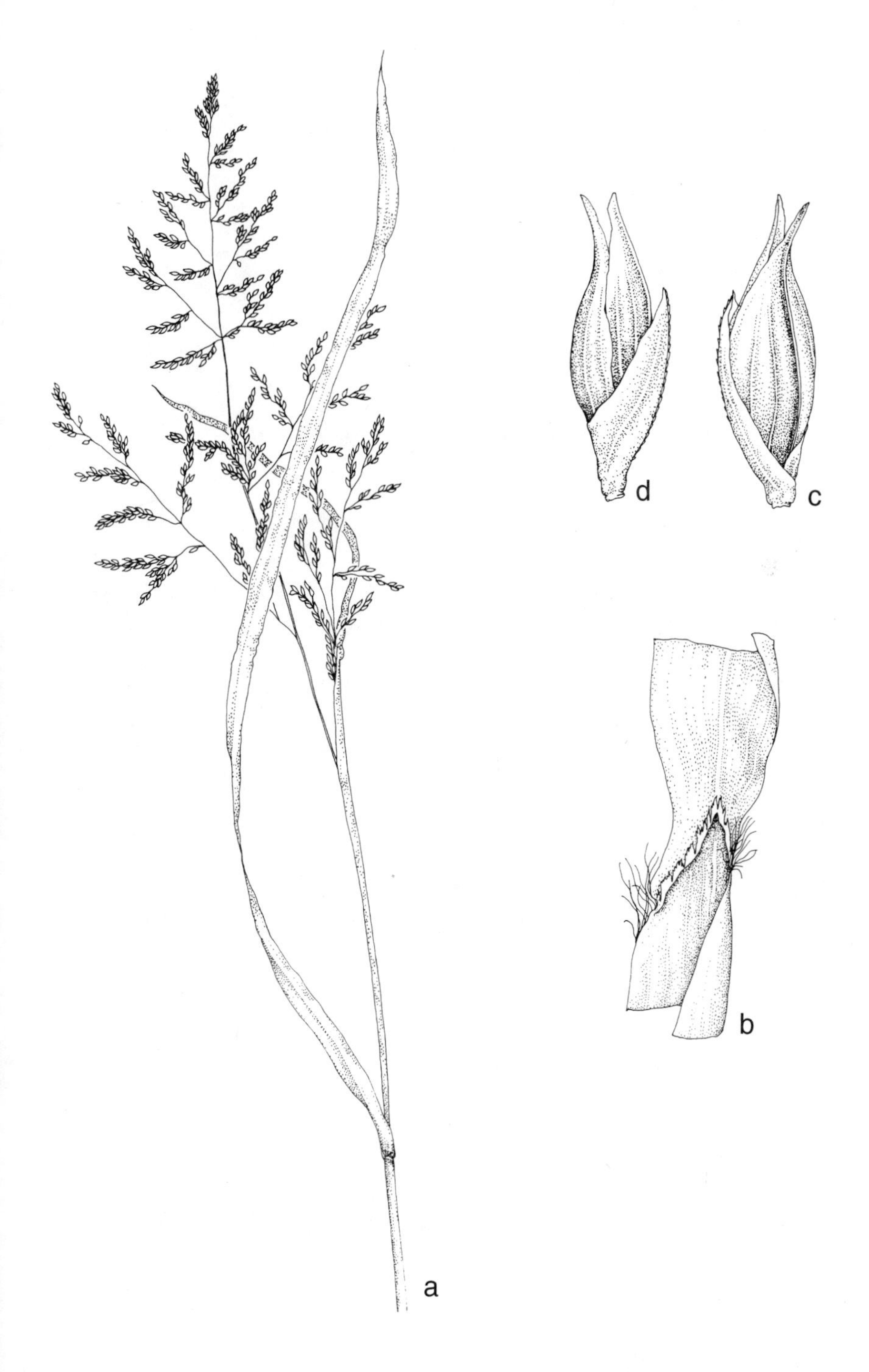

154. *Panicum rigidulum* var. *rigidulum* (Munro grass).

a. Upper part of plant.
b. Sheath, with ligule.
c. Spikelet, front view.
d. Spikelet, back view.

155. *Panicum rigidulum* var. *condensum* (Narrow munro grass).

a. Inflorescence.
b. Sheath, with ligule.
c. Spikelet, front view.
d. Spikelet, back view.

1.7–2.0 mm long, elliptic, acute, with 3 or 5 obscure nerves and numerous tubercles; lemma narrowly elliptic, acute, with numerous tubercles; grain apiculate. June–September.

Wet roadside ditches.

IL, KY, MO, OH (FACW).

Warty panicum.

This is the only species of *Panicum* in the central Midwest with small tubercles on the glumes and lemmas.

156. ***Panicum verrucosum*** **(Warty panic grass).** Upper part of plant with spikelet in upper left corner.

14. **Paspalum** L.—Bead-grass

Annuals or perennials; inflorescence of many 1-flowered, usually plano-convex, subsessile, solitary or paired spikelets arranged along a central axis in 2 or 4 rows with the convex sides toward the rachis, forming simple spikelike racemes; racemes 1–many, digitate or racemose, terminal; first glume usually wanting; second glume similar to sterile lemma; fertile lemma and pale chartaceous-indurate, the margins of the lemma inrolled at maturity; stamens 3; styles 2; stigmas plumose.

The spikelets are arranged in two or more rows on a dilated rachis. There are approximately 250 species of *Paspalum* in the tropics and warm temperate regions of the world. The following may sometimes be found in standing water.

1. Rachis of racemes broadly winged, always wider than the spikelets.
   2. Racemes usually up to 5 in number; spikelets 1.7–2.0 mm long .... 1. *P. dissectum*
   2. Racemes more than 10 in number; spikelets up to 1.5 mm long .... 3. *P. fluitans*
1. Rachis of the racemes narrowly winged or wingless, always narrower than the spikelets ... . 2. *P. distichum*

1. **Paspalum dissectum** (L.) L. Sp. Pl. ed. 2:81. 1762. Fig. 157.
Panicum dissectum L. Sp. Pl. 1:57. 1753.
Paspalum dimidiatum L. Syst. Nat. ed. 10, 2:855. 1759.
Paspalum walterianum Schult. Mant. 2:166. 1824.

Creeping, branching, glabrous, subaquatic perennial; culms repent, 20–60 cm long, often forming mats; leaves 3–6 cm long, 4–5 mm wide; racemes 1–5, 2–3 cm long, terminal or axillary, falling entire; rachis membranous, 2–3 mm wide, narrower and shorter than the rows of spikelets but folded over and clasping them; spikelets glabrous, ovoid to obovoid, 1.5–2.5 mm long, 1.4 mm wide; glume and sterile lemma 3- to 5-nerved, slightly longer than the grain. July–October.

Shallow water, muddy banks.

IL, KY, MO (OBL).

Matted bead-grass.

This bead-grass is readily recognized by its broad rachises and mat-forming habit. It differs from the similar *P. fluitans,* which has winged rachises, by its few racemes and smaller spikelets.

2. **Paspalum distichum** L. Syst. Nat. ed. 10, 855. 1759. Fig. 158.

Creeping perennial rooting at the nodes; culms to 50 cm tall, glabrous; blades short, crowded, 2–5 mm broad, glabrous except for marginal cilia; racemes two at the tip of each peduncle, up to 7 cm long, the rachis 1.0–1.5 mm wide; spikelets borne singly, 2.5–4.0 mm long, puberulent, ovoid, plano-convex, 5-nerved. July–October.

Marshy soil, occasionally in standing water.

KY (OBL).

Knot-grass.

This southern *Paspalum* almost always has two racemes at the tip of each peduncle.

3. **Paspalum fluitans** (Ell.) Kunth, Rev. Gram. 1:24. 1829. Fig. 159.
Ceresia fluitans Ell. Bot. S. C. & Ga. 1:109. 1816.
Paspalum mucronatum Muhl. Descr. Gram. 96. 1816.

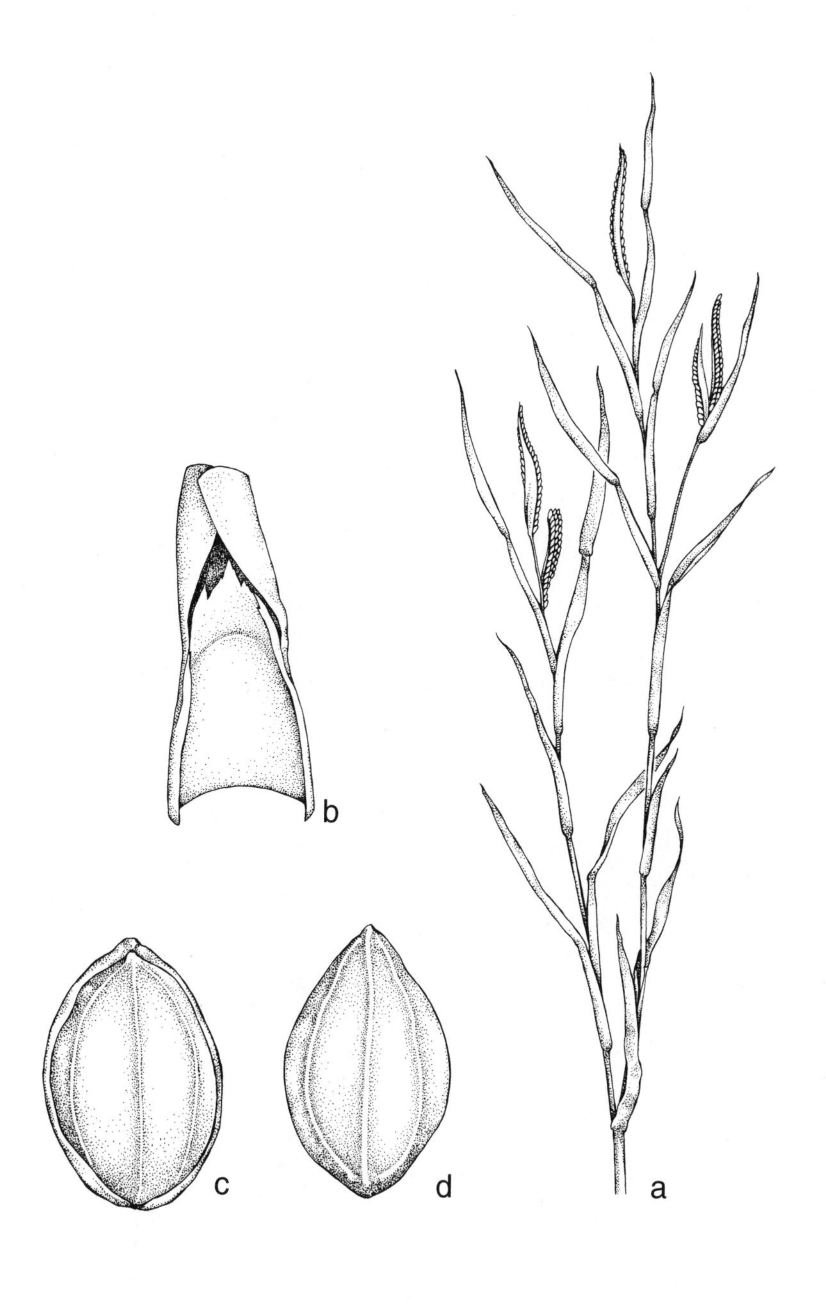

157. *Paspalum dissectum* (Matted bead-grass).

a. Upper part of plant.
b. Sheath, with ligule.
c. Spikelet, front view.
d. Spikelet, back view.

158. ***Paspalum distichum*** **(Knot-grass).** Habit with spikelet on left.

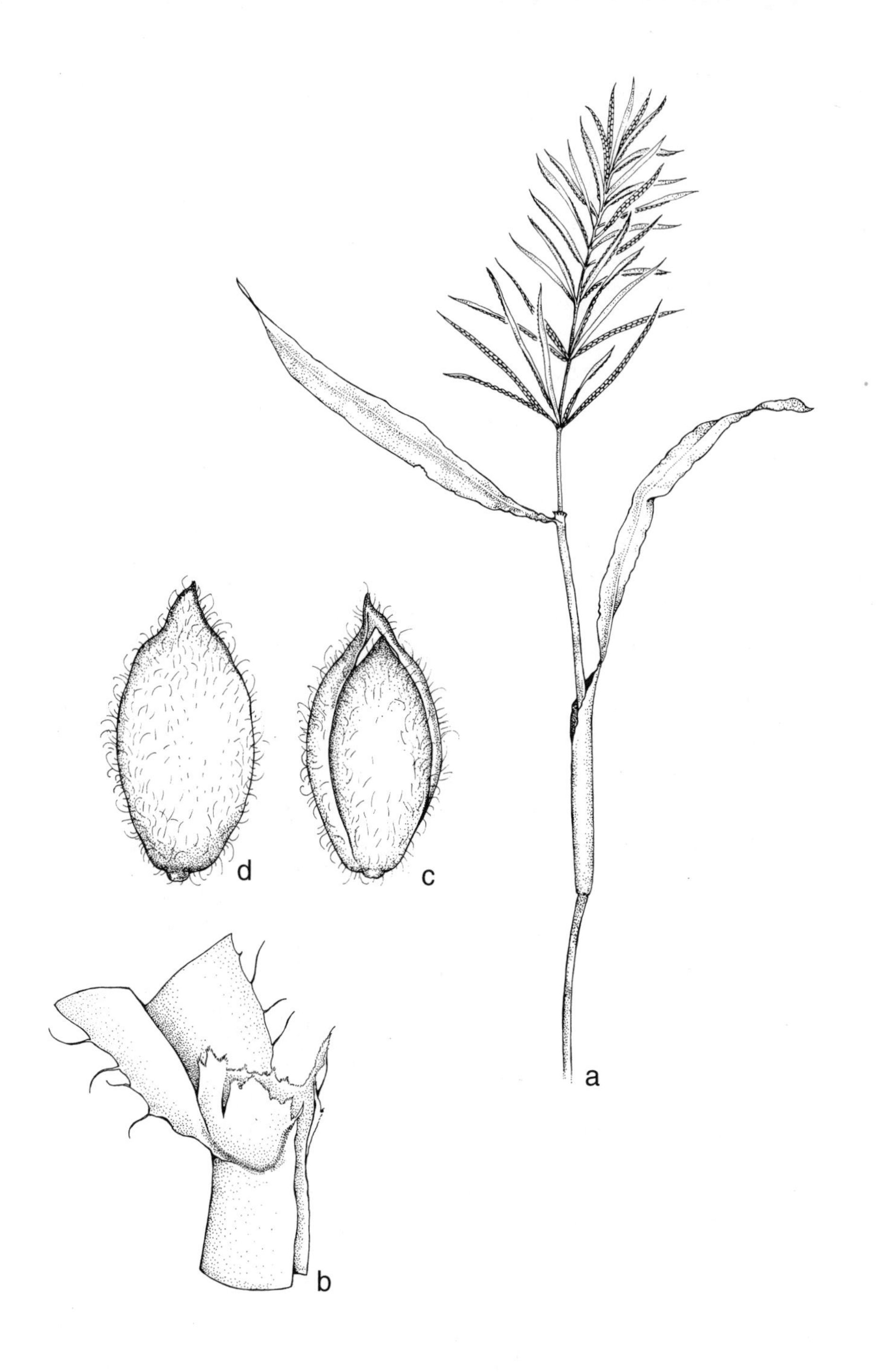

159. *Paspalum fluitans* (Swamp bead-grass).

a. Upper part of plant.
b. Sheath, with ligule.
c. Spikelet, front view.
d. Spikelet, back view.

Paspalum natans LeConte, Journ. Phys Chem. 91:285. 1820.
Paspalum frankii Steud. Syn. Pl. Glum. 1:119. 1854.

Sprawling or repent, branching, glabrous, aquatic annual; culms soft and spongy, to 1 m long; leaves 10–20 cm long, 10–15 mm wide; racemes 5–50, usually more than 10, 3–8 cm long, spreading or recurved; rachis herbaceous, 1.3–2.0 mm wide, wider and longer than the rows of spikelets but folded over and clasping them; spikelets minutely glandular-pubescent, ellipsoid, 1.3–2.0 mm long, 0.8 mm wide; glume and sterile lemma 2-nerved, the mid-nerve suppressed, slightly longer than the grain. August–October.

Floating in shallow standing water.

IL, IN, KS, KY, MO (OBL).

Swamp bead-grass.

Paspalum fluitans differs from the similar *P. dissectum* by its more numerous racemes and larger spikelets.

15. **Phalaris** L.—Canary Grass

Annuals or perennials; blades flat; panicles contracted, sometimes spikelike; spikelets with 1 terminal perfect flower and 1–2 empty lemmas below, disarticulating above the glumes; glumes subequal, keeled; sterile lemmas minute, awnless; fertile lemma indurate (at least in fruit), awnless; palea smaller than the lemma, 2-nerved.

The 1-flowered spikelets have 1–2 empty lemmas below the fertile one.

Phalaris consists of fifteen species in temperate regions of the world. Only the following sometimes occurs in standing water.

1. **Phalaris arundinacea** L. Sp. Pl. 1:55. 1753. Fig. 160.
Phalaris arundinacea L. var. *picta* L. Sp. Pl. 1:55. 1753.
Phalaris arundinacea L. var. *variegata* Parnell, Grasses Brit. 188. 1845.
Phalaris arundinacea L. f. *variegata* (Parness) Druce, Fl. Berks. 556. 1897.
Phalaris arundinacea L. f. *picta* (L.) Asch. & Graebn. Syn. Mitt. Eur. Fl. 2:24. 1898.

Perennial from scaly, creeping rhizomes; culms to 1.5 m tall; blades 10–20 mm broad, green or occasionally white-striped (f. *picta*); ligule membranous, 3–5 mm long, erose; panicle contracted, to 30 cm long, lobed at base; glumes 4.5–6.5 mm long, acute, wingless, the keel scabrous; fertile floret 2.7–4.5 mm long, lanceolate, glabrous or appressed-pubescent, conspicuously nerved; sterile florets 2, subulate, pubescent, 1–2 mm long. May–July.

Meadows, marshes, sometimes in standing water, becoming very aggressive.

IA, IL, IN, KS, KY, MO, NE, OH (FACW+).

Reed canary grass.

In the northern part of the central Midwest, *P. arundinacea* is a very invasive species of wetlands.

16. **Phragmites** Trin.—Common Reed

Very tall perennials from stout, creeping rhizomes; blades broad, flat; inflorescence paniculate, large, dense, much-branched; spikelets several-flowered, disarticulating

160. ***Phalaris arundinacea*** **(Reed canary grass).**

a. Inflorescence.
b. Sheath, with ligule and leaves.
c. Spikelet.

above the glumes, the rachilla silky-villous; glumes 2, unequal; lemmas glabrous, 3-nerved, the lowest sterile or bearing a staminate flower, the others perfect.

Phragmites consists of three species found worldwide.

1. **Phragmites australis** (Cav.) Trin. ex Steud. Nom. Bot. ed. 2, 2:324. 1841. Fig. 161.
Arundo phragmites L. Sp. Pl. 1:81. 1753.
Arundo australis Cav. Ann. Hist. Nat. 1:100. 1799.
Phragmites communis Trin. Fund. Agrost. 134. 1820.
Phragmites berlandieri Fourn. Bull. Bot. Soc. France 24:178. 1877.
Phragmites communis Trin. var. *berlandieri* (Fourn.) Fern. Rhodora 34:211. 1932.

Tall perennial with culms to 4 m high, usually forming extensive colonies; blades flat, glabrous, 10–50 mm broad; inflorescence large, 15–40 cm long, mostly ascending, yellowish to purplish; spikelets 3- to 7-flowered, 10–17 mm long, surpassed by the silky hairs of the rachilla; first glume narrowly elliptic, obtuse to subacute, 4–6 mm long; second glume linear, acute, 6.0–8.5 mm long; lemmas linear-lanceolate, glabrous, 3-nerved, the lowest 8–12 mm long, the upper progressively smaller, with all lemmas attaining the same height. July–September.

Moist soil, frequent around abandoned strip mines.

IA, IL, IN, MO (FACW+), KS, KY, NE, OH (FACW).

Common reed.

This is an aggressive invader of wetlands. It is often planted in areas where strip-mined lands are being restored. It has been suggested that a native, nonaggressive strain of this species occurs in the United States. This native strain usually does not form extensive colonies. Its internodes are often reddish purple.

17. **Poa** L.—Bluegrass

Annuals or tufted or rhizomatous perennials; blades flat, mostly boat-shaped at the tips; inflorescence paniculate; spikelets several-flowered, disarticulating above the glumes; glumes 2, more or less unequal, shorter than the spikelets; lemmas distinctly keeled and nerved, awnless, usually with a tuft of cobwebby hairs at the base (except *P. annua).*

The tuft of cobwebby hairs at the base of the lemmas (except for a few species) sets this genus apart from all other closely related genera. The lemmas are keeled.

Poa consists of approximately 150 species mostly in temperate parts of the world. Three species may sometimes be found in shallow water.

1. Annuals; lemmas without a tuft of cobwebby hairs at base .... 1. *P. annua*
1. Perennials; lemmas with a tuft of cobwebby hairs at base.
    2. Weak perennial; sheaths scabrous .... 2. *P. paludigena*
    2. Firm perennial; sheaths not scabrous .... 3. *P. palustris*

1. **Poa annua** L. Sp. Pl. 1:68. 1753. Fig. 162.

Tufted annual, sometimes rooting at the nodes, with terete culms to about 30 cm tall; sheaths loose, glabrous; blades soft, 1–3 mm broad; inflorescence 2–10 cm long, ascending; spikelets 3–6 mm long, 3- to 6-flowered; glumes narrowly ovate, obscurely nerved, with a scarious margin, the first 1.5–2.5 mm long, the second 2–

161. *Phragmites australis* (Common reed). a. Inflorescence. b. Leaf. c. Spikelet.

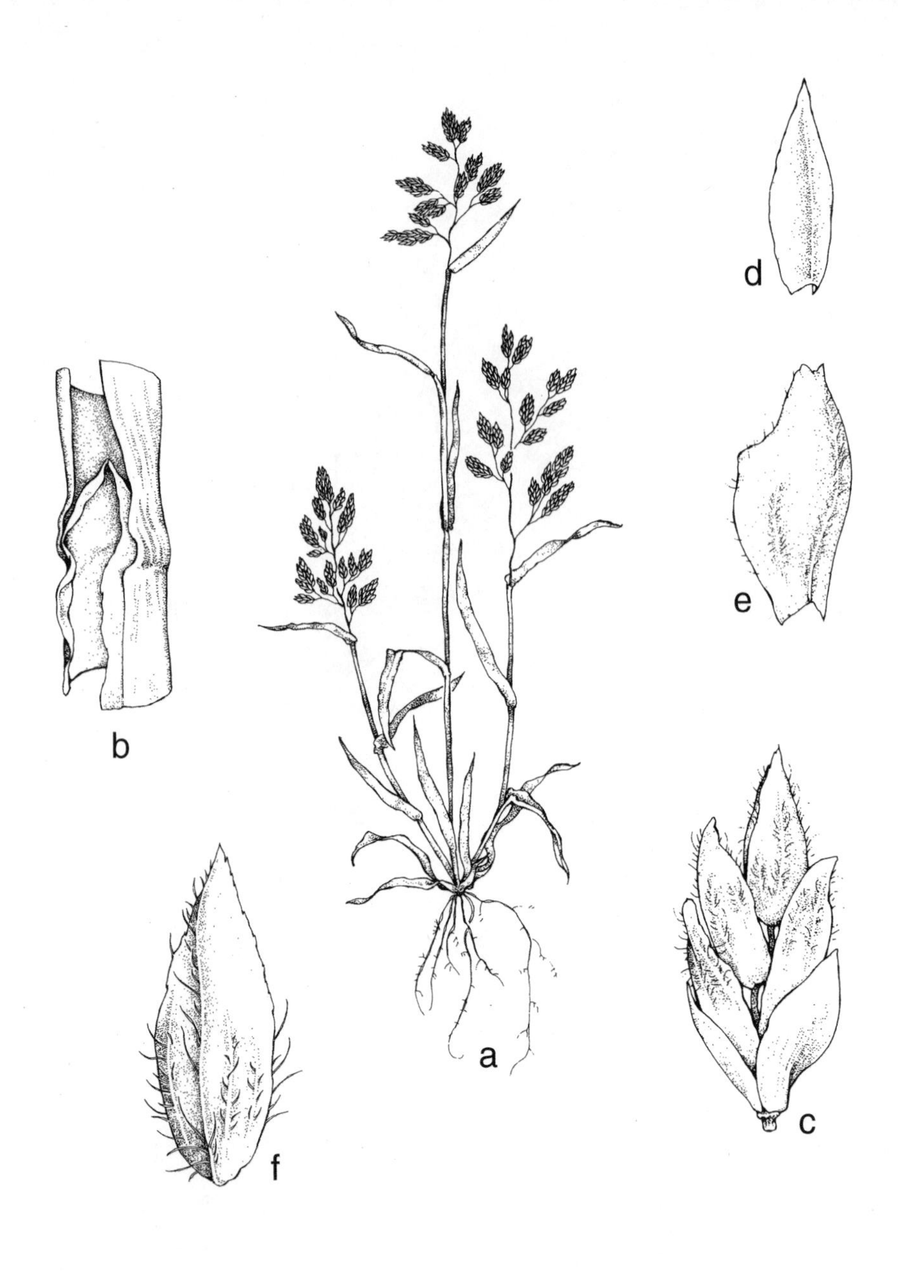

162. *Poa annua* (Annual bluegrass).
a. Habit.
b. Sheath, with ligule.
c. Spikelet.
d. First glume.
e. Second glume.
f. Lemma.

3 mm long; lemmas elliptic to ovate, obtuse, thin, 5-nerved, more or less pubescent throughout on the nerves, 2.5–3.5 mm long, without a web at the base; anthers 0.8–1.0 mm long. April–November.

Disturbed soil, rarely in standing water.

IA, IL, IN, MO (FAC-), KS, NE (FAC), KY, OH (FACU).

Annual bluegrass.

This plant is found in disturbed areas, but rarely in shallow standing water. It is the only annual *Poa* that lacks the tuft of cobwebby hairs at the base of the lemma.

2. **Poa paludigena** Fern. & Wieg. Rhodora 20:126. 1918. Fig. 163.

Perennial with culms solitary or in small tufts, weak, compressed, to 60 cm tall; sheaths scabrous; ligules 0.5–1.5 mm long; blades soft, thin, 1–2 mm broad; inflorescence 3–15 cm long, widely spreading; spikelets 3–6 mm long, 2- to 5-flowered; glumes lanceolate, with a scarious margin, the first 1.7–2.2 mm long, the second 2–3 mm long; lemmas lanceolate to narrowly ovate, acute, 2.5–3.5 mm long, with 3 distinct, pubescent nerves and 2 obscure, glabrous nerves, webbed at the base. June–July.

Bogs.

IL (OBL).

Bog bluegrass.

This rarely observed grass has very long, solitary culms, scabrous sheaths, and paired panicles branches near the base of the inflorescence.

3. **Poa palustris** L. Syst. Nat. ed. 10, 2:874. 1759. Fig. 164.
Poa serotina Ehrh. Beitr. Naturk. 6:83. 1791.
Poa triflora Gilib. Exerc. Phyt. 2:531. 1792.

Stout, tufted perennial, sometimes rooting at the lower nodes, to over 1 m tall; sheaths more or less loose; ligules 2–5 mm long; blades 1–3 mm broad; inflorescence 10–30 cm long, nodding; spikelets 3.0–4.5 mm long, 2- to 4-flowered; glumes ovate-lanceolate, acute, the first 2.0–2.5 mm long, the second 2–3 mm long; lemmas broadly lanceolate, acute, 2.0–3.5 mm long, with 3 distinct, pubescent nerves and 2 obscure, glabrous nerves, webbed at the base. June–September.

Marshes, fens, bogs, often in shallow water.

IA, IL, IN (FACW+), NE (not indicated for this state).

Marsh bluegrass.

This species resembles the very common, nonaquatic *Poa pratensis*, but *P. palustris* has longer ligules and lacks rhizomes.

18. **Scolochloa** Link—Marsh Sprangletop

Spikelets 3- to 4-flowered, disarticulating above the glumes; glumes unequal, the second about as long as the lemma; lemmas 2-nerved, densely pubescent in 2 rows; panicle wide-spreading.

There are two species of *Scolochloa*, the other in Siberia.

1. **Scolochloa festucacea** (Willd.) Link, Enum. Pl. 1:137. 1827. Fig. 165.
Arundo festucacea Willd. Enum. Pl. 1:126. 1809.

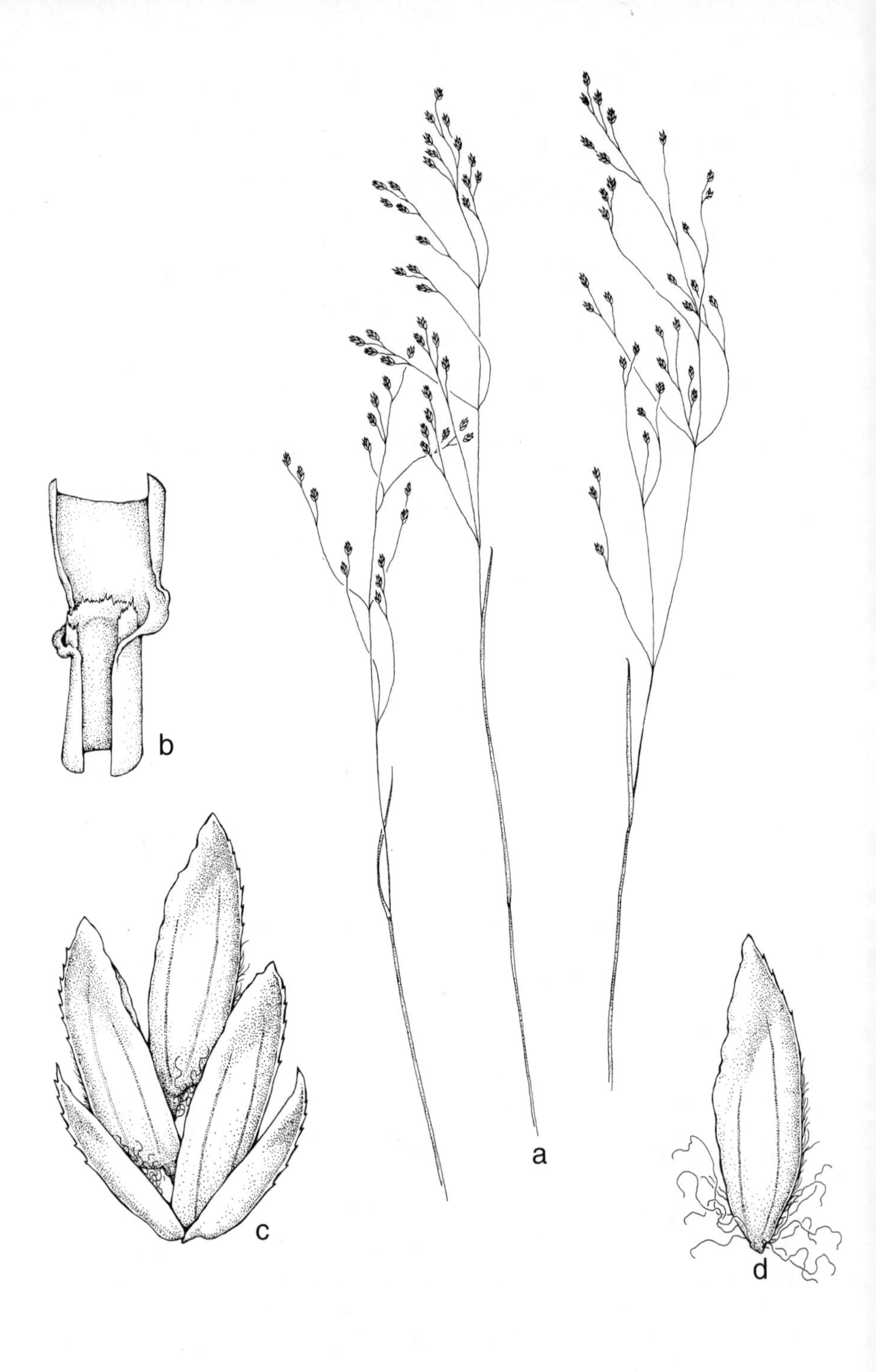

163. *Poa paludigena* (Bog bluegrass).

a. Upper part of plants.
b. Sheath, with ligule.
c. Spikelet.
d. Lemma.

164. ***Poa palustris***
(Marsh bluegrass).
a. Habit.
b. Sheath, with ligule.
c. Spikelet.
d. First glume.
e. Second glume.
f. Lemma.

Perennial from thick rhizomes; culms stout, to 1.5 m tall, hollow, glabrous; blades flat with an inrolled tip, 6–12 mm broad, scabrous, strongly veined; ligules membranaceous, up to 10 mm long; panicle open, up to 35 cm long; spikelets 3- to 4-flowered, (5–) 7–10 mm long; glumes unequal, acute, scarious, the lower 5.0–7.5 mm long, 3-nerved, the upper 5.5–9.0 mm long, 5-nerved; lemmas obtuse, erose, 5.0–7.5 mm long, bearded on the callus. June–August.

Marshes, potholes, sloughs, wet meadows, ponds.

IA, NE (OBL).

Marsh sprangletop; marsh grass.

This handsome grass is distinguished by its open panicles with spikelets up to 10 mm long. The lemmas are densely bearded on the callus.

165. *Scolochloa festucacea* (Marsh sprangletop). a. Inflorescence. b. Leaf with sheath. c. Lemma. d. Spikelet.

19. **Spartina** Schreb.—Cord Grass

Perennials from scaly rhizomes; blades flat (in species of the Midwest); inflorescence paniculate, composed of numerous 1-sided spikes; spikelets 1-flowered, disarticulating below the glumes; glumes unequal, keeled, awned, 1- to 3-nerved; lemma firm, keeled, 1- to 3-nerved; palea papery, 2-nerved, obscurely keeled, longer than the lemma.

There are fifteen species in this genus. Only the following occurs in the central Midwest.

1. **Spartina pectinata** Link, Jahrb. Gewachs. I 3:92. 1820. Fig. 166.
Trachynotia cynosuroides Michx. Fl. Bor. Am. 1:64. 1803, *non Dactylis cynosuroides* L. (1753).
Spartina michauxiana Hitchc. Contr. U.S. Nat. Herb. 12:153. 1908.

Perennial from rhizomes; culms to 2.5 m tall, erect, often solitary; sheaths glabrous or pilose at the summit, the margins more or less scabrous; blades flat, becoming involute when dry, to 15 mm broad, scabrous on the margin, glabrous on both surfaces; panicle to 50 cm long, composed of up to 50 mostly ascending spikes to 15 cm long; spikelets 10–25 mm long, appressed; glumes awned, scabrous on the keel, the first 5–10 mm long, the second (including the awn) 10–25 mm long; lemma bilobed, apiculate, glabrous except for the hispidulous margins and the more or less scabrous keel, 7–10 mm long. June–September.

Marshes, sloughs, wet prairies.

IA, IL, IN, KS, MO, NE (FACW+), KY, OH (OBL).

Prairie cord grass; slough grass.

The flattened, 1-flowered spikelets with awned lemmas are distinctive.

166. *Spartina pectinata* (Prairie cord grass).

a. Upper part of plant.
b. Sheath, with ligule.
c. Spikelet.

20. **Sphenopholis** Scribn.—Wedge Grass

Perennials; blades flat; inflorescence paniculate, narrow; spikelets 2- to 3-flowered, disarticulating below the glumes; glumes strongly unequal, rather obscurely nerved, keeled; lemmas obscurely nerved, more or less rounded on the back, usually awnless.

I am following recent botanical opinion by including *Trisetum* in this genus. As a result, *Sphenopholis* consists of eight species, all North American. Only the following is sometimes found in standing water.

1. **Sphenopholis pensylvanica** (L.) A. Hitchc. Am. Journ. Bot. 2:304. 1915. Fig. 167.
Avena pensylvanica L. Sp. Pl. 1:79. 1753.
Trisetum pensylvanicum (L.) Beauv. ex Roem. & Schult. Syst. Veg. 2:658. 1817.

Weak perennial to 1 m tall; blades 2–5 mm wide, slightly scabrous; sheaths glabrous, except for the lowermost; panicle lax, to 25 cm long; spikelets 2- to 3-flowered, 4–8 mm long; glumes subequal, acute, scabrous on the back, the first narrowly oblong, 3.5–5.5 mm long, 3-nerved, the second obovate to oblong, 4–6 mm long, 3- to 5-nerved; first lemma 4–7 mm long, awned or awnless; second lemma scabrous, 3.5–6.5 mm long, with a bent awn 3.5–7.0 mm long. May–July.

Wet meadows, swampy woods, occasionally in standing water.

KY, MO, OH (OBL).

Swamp oats.

This species is often placed in the genus *Trisetum*, and its spikelets are larger in all respects than in any species of *Sphenopholis*. The lemmas, or at least the second lemma, in *S. pensylvanica* have an awn up to 7 mm long.

167. *Sphenopholis pensylvanica* **(Swamp oats).** Upper part of plant. Spikelet (lower right).

21. **Torreyochloa** Church

Only the following species comprises this genus. The only species in the genus was originally described in the genus *Windsoria*. In 1836, it was transferred to *Glyceria* and remained in that genus until Church created *Torreyochloa* for it in 1949. Three years later, Clausen transferred this species to *Puccinellia*. I am retaining it as *Torreyochloa*. *Torreyochloa* differs from *Glyceria* species by its open sheaths. It differs from *Puccinellia* species by its conspicuously nerved lemmas.

1. **Torreyochloa pallida** (Torr.) Church, Am. Journ. Bot. 36:164. 1949. Fig. 168.
Windsoria pallida Torr. Cat. Pl. N. Y. 91. 1819.
Glyceria pallida (Torr.) Trin. Acad. St. Petersb. Mem. VI Sci. Nat. 2:57. 1836.
Panicularia pallida (Torr.) Kuntze, Rev. Gen. Pl. 2:783. 1891.
Puccinellia pallida (Torr.) Clausen, Rhodora 54:44. 1952.

Semiaquatic perennial with decumbent bases and culms to 1 m tall; leaves soft, 3–8 mm broad; inflorescence paniculate, 5–25 cm long; spikelets 4–7 mm long, 4- to 8-flowered, pale green; glumes glabrous, ovate, the first 1–2 mm long, the second 1.2–2.5 mm long; lemmas narrowly ovate, obtuse, puberulent near the apex, 5- to 7-nerved, 2.0–3.5 mm long. May–August.

Swamps, often in standing water.

IL, IN, KY, MO (OBL).

Pale manna grass.

This species resembles *Glyceria arkansana* but differs by its open sheaths and its smaller and fewer-flowered spikelets.

22. **Zizania** L.—Wild Rice

Annuals (in the central Midwest); blades flat; inflorescence paniculate, large, terminal; spikelets 1-flowered, unisexual, disarticulating at the base, the pistillate appressed in concave depressions at the summit of the pedicels; glumes none; staminate lemma acuminate or short-awned, 5-nerved; pistillate lemma awned, 3-nerved.

Zizania consists of four North American species and one Asian species.

1. Lemmas of pistillate flowers scabrous throughout; abortive lemma less than 1.5 mm wide .... 1. *Z. aquatica*
1. Lemmas of pistillate flowers scabrous only on the nerves; abortive lemma 1.5–2.0 mm wide.
    2. Leaves 10–30 mm broad; ligules 10–15 mm long; awns of pistillate lemmas 4 cm long or longer .... 2. *Z. interior*
    2. Leaves to 10 mm broad; ligules up to 10 mm long; awns of pistillate lemmas up to 4 cm long .... 3. *Z. palustris*

1. **Zizania aquatica** L. Sp. Pl. 991. 1753. Fig. 169.

Tall aquatic annual; culms slender, erect or decumbent at base, simple or branched, to nearly 3 m tall; blades 10–50 mm broad; panicle to 60 cm long, erect, the branches bearing staminate spikelets spreading to drooping, the branches bearing pistillate spikelets erect to ascending; staminate spikelets pendulous, 6–11 mm long, glabrous or nearly so, the lemma awnless or with an awn to 3 mm long; pistillate spikelets erect to ascending, linear, or the abortive ones subulate, glabrous or scabrous, the lemmas densely scabrous throughout, with an awn 1–6 cm long. June–September.

Standing water.

IL, IN, OH (OBL).

Wild rice.

This species differs from the other species of wild rice by its pistillate lemmas that are densely scabrous throughout.

2. **Zizania interior** (Fassett) Rydb. Brittonia 1:82. 1931. Fig. 170.
Zizania aquatica L. var. *interior* Fassett, Rhodora 26:158. 1924.

Robust annual; culms up to 3 m tall, glabrous; blades flat, 10–30 mm broad, glabrous; ligules 10–15 mm long; spikelets unisexual, the pistillate on appressed upper branchlets, the staminate on lower branchlets of the same panicle; glumes of pistillate spikelets absent; glumes of staminate spikelets 3- or 5-nerved; fertile

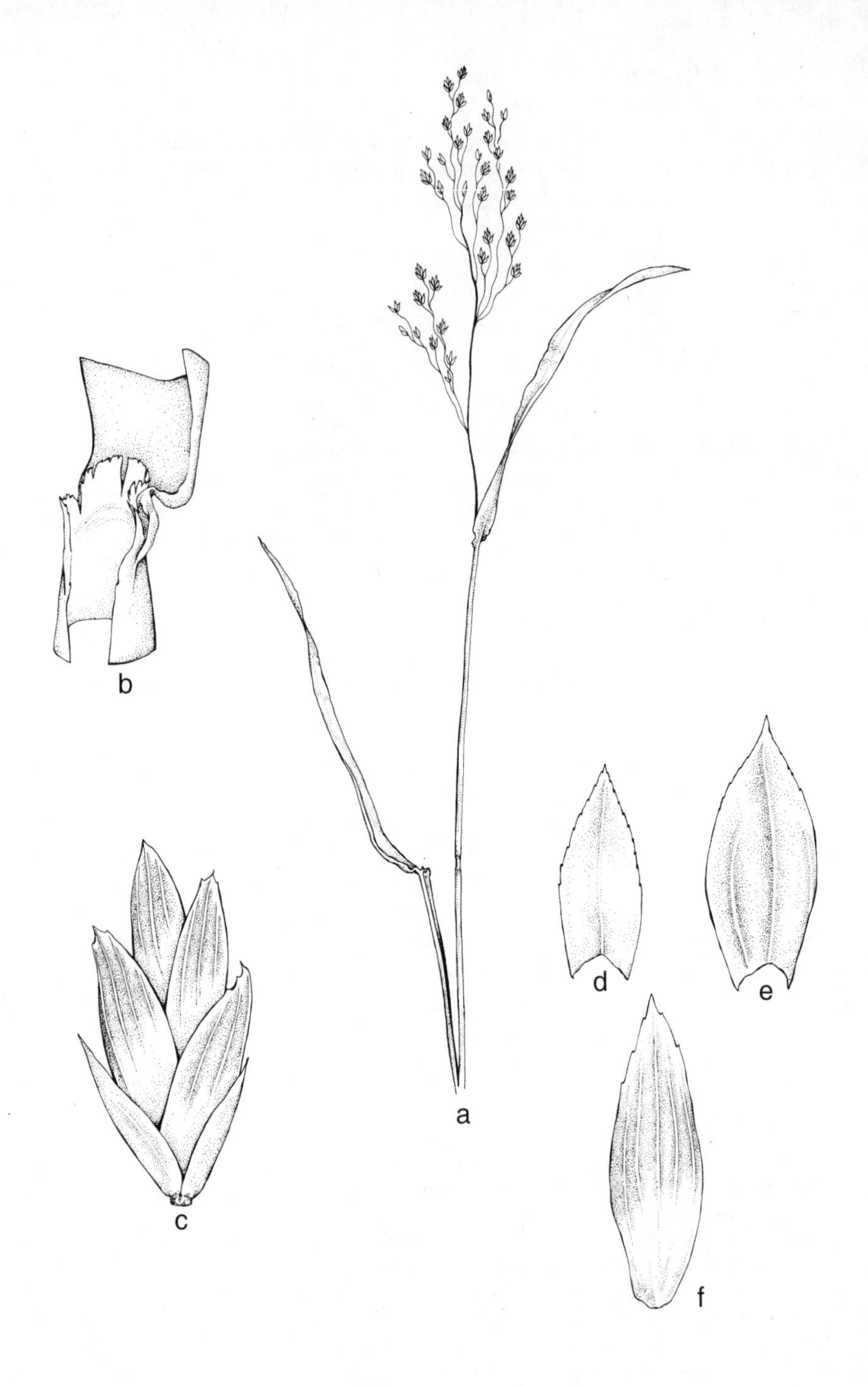

168. ***Torreyochloa pallida*** **(Pale manna grass).**
a. Inflorescence.
b. Sheath, with ligule.
c. Spikelet.
d. First glume.
e. Second glume.
f. Lemma.

169. *Zizania aquatica* (Wild rice).

a. Inflorescence.
b. Staminate lemma.

c., d. Pistillate lemma.

170. *Zizania interior* (Interior wild rice).

a. Habit.

b. Pistillate flower.

pistillate lemma coriaceous, shiny, 1–2 cm long, scabrous only on the nerves, the awn 4–7 cm long; abortive lemma 1.5–2.0 mm wide. June–September.

Standing water.

IA, IL, IN, KS, KY, MO, NE (OBL).

Interior wild rice.

This species, sometimes considered a variety of *Z. palustris* or of *Z. aquatica*, has scabrous nerves of the pistillate lemmas, ligules at least 10 cm long, and blades at least 10 mm broad.

3. **Zizania palustris** L. Mant. Pl. 295. 1771. Fig. 171.

More slender annual than the species above, up to 1.5 m tall; blades flat, 8–10 mm broad, glabrous; ligules 3–10 mm long; spikelets unisexual, the pistillate on appressed upper branchlets, the staminate on lower branchlets of the same panicle; glumes of pistillate spikelets absent; glumes of staminate spikelets 3- or 5-nerved; fertile pistillate lemmas coriaceous, shiny, 1–2 cm long, scabrous only on the nerves, the awn 1.5–4.0 cm long. June–September.

Standing water.

IA (OBL).

Northern wild rice.

This is the smallest of the three species of wild rice in the central Midwest. It differs from *Z. interior* by its shorter ligules, shorter awns, and narrower blades, and from *Z. aquatica* by the pistillate lemmas scabrous only on the nerves.

171. ***Zizania palustris*** **(Northern wild rice).** a. Inflorescence. b. Spikelet.

23. **Zizaniopsis** Doell & Aschers—Southern Wild Rice

Large perennials from stout rhizomes; inflorescence paniculate; spikelets 1-flowered, unisexual, with staminate flowers below and pistillate flowers above on same branches; disarticulation below the spikelet; glumes absent; lemma 5-nerved and awnless in the staminate spikelet, 7-nerved and short-awned in the pistillate spikelet; stamens 6.

Zizaniopsis differs from *Zizania* by having the staminate and pistillate spikelets intermingled on the same branches. The awns of the pistillate lemmas are slightly shorter than in *Zizania*.

Four species, all in the western hemisphere, comprises this genus.

1. **Zizaniopsis miliacea** (Michx.) Doell & Aschers in Doell in Mart. Fl. Bras. 2(2):13. 1871. Fig. 172.

Zizania miliacea Michx. Fl. Bor. Am. 1:74. 1803.

Robust perennial; culms to 3 m tall, glabrous; leaves to 2 cm broad, with very rough margins; panicle nodding, to 60 cm long, with numerous, whorled branches; lemmas 6–9 mm long, with scabrous nerves, the pistillate ones with an awn to 5 mm long. June–September.

Standing water, marshes.

IL, KY, MO (OBL).

Southern wild rice.

The very broad leaves and the extremely rough margins of the leaves are distinctive.

27. PONTEDERIACEAE—PICKERELWEED FAMILY

Perennial herbs of wet situations from rhizomes; leaves basal and cauline; flowers perfect, borne singly or in spikes or spikelike panicles; perianth parts 6, united below, of uniform color; stamens 3 or 6; ovary 1- or 3-celled, superior; fruit a capsule or utricle.

There are nine genera and thirty species, found worldwide, in this family.

The following genera occur in the central Midwest.

1. Inflorescence with 50 or more flowers; fruit a 1-seeded utricle .... 3. *Pontederia*
1. Inflorescence with 1–30 flowers; fruit a capsule with 10–many seeds.
    2. Leaves linear; inflorescence 1-flowered .... 4. *Zosterella*
    2. Leaves lanceolate to ovate to reniform; inflorescence 1- to several-flowered.
        3. Petiole swollen; perianth lobes more than 2 cm long; stamens 6 .... 1. *Eichhornia*
        3. Petiole not swollen; perianth lobes less than 2 cm long; stamens 3 .... 2. *Heteranthera*

1. **Eichhornia** Kunth—Water Hyacinth

Perennials or annuals; leaves with parallel veins, petiolate, usually with a swollen area on the petioles; stipules absent; inflorescence spicate or a spikelike panicle; perianth 6-parted, united below the middle, more or less 2-lipped; stamens 6; fruits capsules enclosed by the persistent perianth; seeds numerous, ribbed.

There are eight tropical and subtropical species in the genus.

1. **Eichhornia crassipes** (Mart.) Solms, Monogr. Phan. 4:527. 1883. Fig. 173.
Pontederia crassipes Mart. Nova Gen. & Sp. Pl. 1:9, pl. 4. 1823 (1824).

Annuals (in our area); stems stout; leaves reniform to orbicular, shiny, 3–10 cm wide, on petioles up to 20 (–35) cm long, bulbous-inflated at the middle; inflorescence a spicate panicle, glandular-pubescent; perianth more or less 2-lipped, the lobes 3–4 cm long, usually lilac, the upper lobe yellow at base; stamens 6, 3 long and 3 short; capsules to 2 cm long; seeds 1 mm long, longitudinally ribbed. June–July.

Shallow water.

IL, MO (OBL).

Water hyacinth.

This wetland adventive from South America is a nuisance plant in the southeastern United States, but it barely survives in the central Midwest.

2. **Heteranthera** Ruiz & Pav.—Mud Plantain

Perennials from rhizomes; leaves cauline, lanceolate to orbicular; flower solitary, or 2–8 in a spike, borne from a bladeless spathe; perianth salverform, regular; stamens

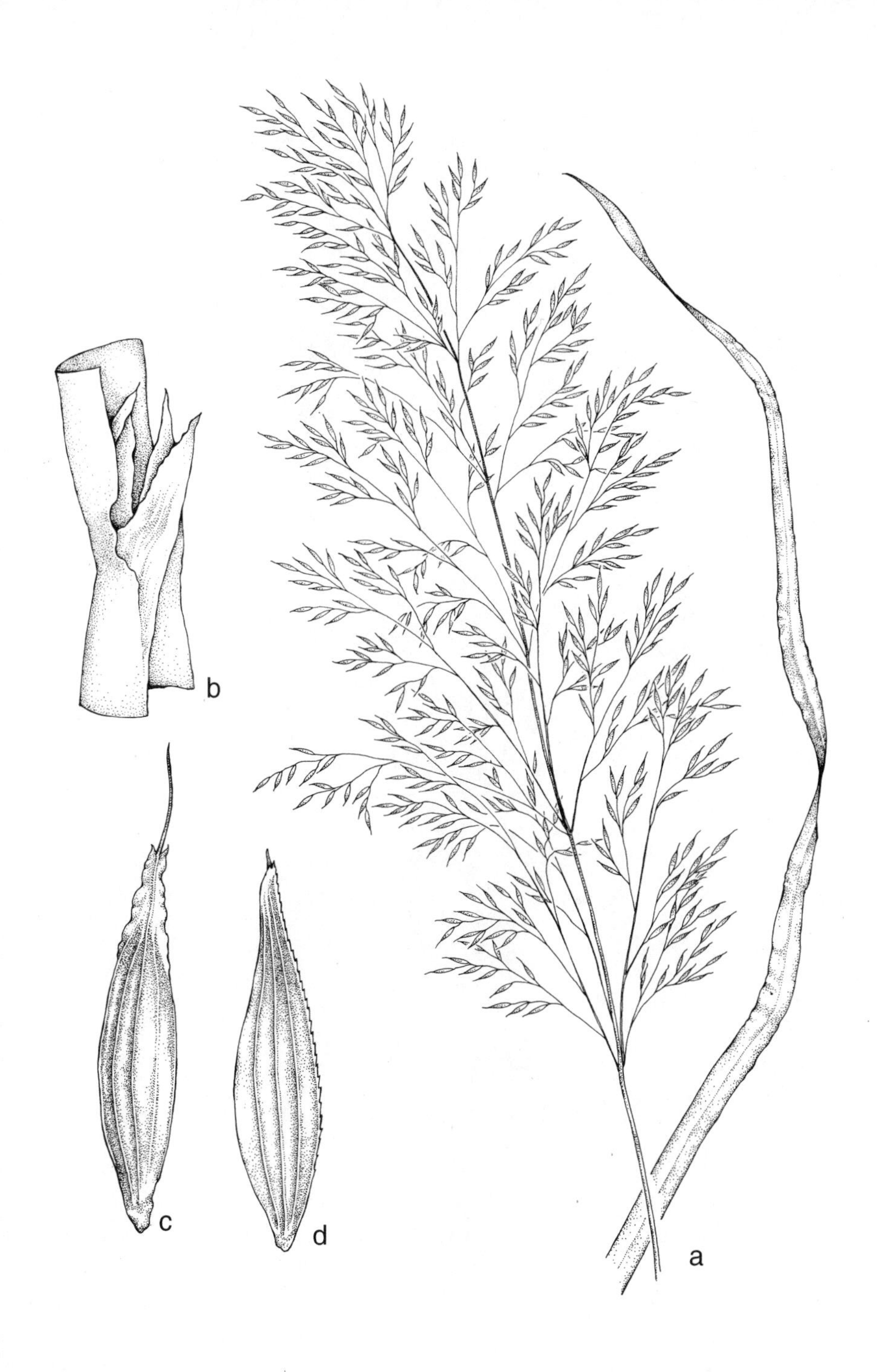

172. *Zizaniopsis miliacea* (Southern wild rice).

a. Inflorescence.
b. Sheath, with ligule.
c. Pistillate lemma.
d. Staminate lemma.

173. *Eichhornia crassipes* (Water hyacinth). Habit.

174. *Heteranthera limosa* (Mud plantain). a. Habit. b. Capsule, with spathe.

3, two of them short and with ovate anthers, one of them longer and with a sagittate anther; ovary partially 3-celled; fruit a many-seeded, dehiscent capsule.

There are about ten species in this genus, found worldwide. Four species occur in the central Midwest.

1. Stems short, erect, with the leaves clustered at the tip; leaves truncate to cuneate .... 1. *H. limosa*
1. Stems creeping, with scattered leaves; leaves cordate.
  2. Flower solitary; leaves longer than broad .... 4. *H. rotundifolia*
  2. Flowers 1–3 in a cluster; leaves as broad as long or broader.
    3. Flowering spathes more or less sessile; perianth purple .... 2. *H. multiflora*
    3. Flowering spathes on short stalks; perianth white .... 3. *H. reniformis*

1. **Heteranthera limosa** (Sw.) Willd. Ges. Naturf. Freunde Berlin Neue Schrif. 3:439. 1801. Fig. 174.
Pontederia limosa Sw. Prodr. 57. 1788.
Leptanthus ovalis Fl. Bor. Am. 1:25. 1803.

Perennial from rhizomes; leaves lanceolate to ovate, obtuse or subacute at the apex, tapering, rounded, or subcordate at the base, to 5 cm long, to 4 cm broad, usually much smaller, with a petiole to 17 cm long; flower solitary, blue marked with one or more white spots, borne from a spathe; spathe long-acuminate, 2–4 cm long; perianth tube 2–4 cm long, the segments linear-lanceolate, all the same width; filaments glabrous; stigma 3-lobed; capsule elongated, 12–15 mm long. June–July.

Muddy shores, shallow water.

IA, IL, KS, KY, MO, NE (OBL).

Mud plantain.

The solitary, axillary flower relates this species to *Zosterella dubia*, but this latter species has elongated, linear leaves, uniform stamens, and a few-seeded, indehiscent capsule.

2. **Heteranthera multiflora** (Griseb.) C. N. Horn, Phytologia 59 (4):290. 1986. Fig. 175.
Heteranthera reniformis Ruiz & Pav. var. *multiflora* Griseb. Abh. Kon. Ges. Wiss. Gott. 24:323. 1879.

Perennial from rhizomes; leaves orbicular, subacute at the apex, cordate at the base, to 5 cm long, nearly as broad, with a petiole to 15 cm long; inflorescence spicate, 2- to 8-flowered, the flowers purple; spathe sessile, short-acuminate, 1–3 cm long; perianth tube 5–10 cm long, the inner segments linear-lanceolate, the outer segments narrower; filaments pilose; stigma capitate; capsule oblongoid, 5–9 mm long. July–August.

Muddy shores; shallow water.

IL, KS, MO, NE (OBL).

Mud plantain.

This species differs from the very similar *H. reniformis* by its purple flowers and its sessile spathes.

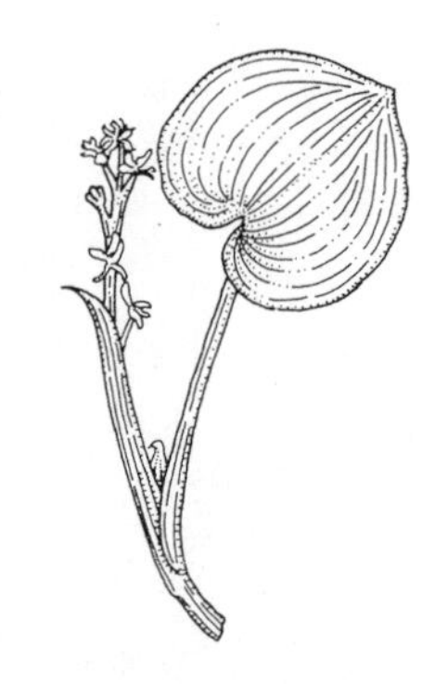

175. *Heteranthera multiflora* (Mud plantain). Habit.

3. **Heteranthera reniformis** Ruiz & Pav. Fl. Peruv. 1:43. 1798. Fig. 176.

Perennial from rhizomes; leaves orbicular, subacute at the apex, cordate at the base, to 5 cm long, nearly as broad, with a petiole to 15 cm long; inflorescence spicate, 2- to 8-flowered, the flowers white; spathe short-stalked, short-acuminate, 1–3 cm long; perianth tube 5–10 cm long, the inner segments linear-lanceolate, the outer segments narrower; filaments pilose; stigma capitate; capsule oblongoid, 5–9 cm long. July–August.

Muddy shores, shallow water.

IL, IN, KY, MO, OH (OBL).

Mud plantain.

The closely related *H. multiflora* has purple flowers and a sessile spathe.

4. **Heteranthera rotundifolia** (Kunth) Griseb. Cat. Pl. Cub. 252–253. 1866. Fig. 177.
Heteranthera limosa (Sw.) Willd. var. *rotundifolia* Kunth, Enum. Pl. Omn. Huc. Cog. 4:122. 1843.

Annual with creeping stems; leaves dimorphic: rosette leaves linear; emergent leaves to 5 cm long, lanceolate to ovate, obtuse at apex, truncate or subcordate at base, long-petiolate; flower borne singly, purple to white, 2-lipped, the lobes linear, to 18 mm long, the lateral lobes spreading, the middle lobe descending with a yellow spot at base, the tube to 30 mm long; stamens unequal, the filaments glandular-hairy. July–August.

Muddy shores, shallow water.

IL, KS, MO, NE (OBL).

Mud plantain.

This species is distinguished by its solitary flower and its leaves longer than broad.

3. **Pontederia** L.—Pickerelweed

Perennial from creeping rhizomes; leaves basal and cauline, petiolate; inflorescence a spikelike panicle with more than 8 flowers; flowers perfect; perianth funnelform, 2-lipped, with 6 segments free above, blue, with the uppermost segment marked with yellow; stamens 6, 3 of them exserted and fertile, 3 of them included and sometimes sterile; ovary 3-celled; fruit a utricle.

Five species comprise this genus, with only the following species occurring in the central Midwest.

1. **Pontederia cordata** L. Sp. Pl. 1:288. 1753. Fig. 178.
Pontederia angustifolia Pursh, Fl. Am. Sept. 1:224. 1814.
Pontederia cordata L. var. *angustifolia* (Pursh) Torr. Fl. U.S. 1:343. 1824.
Pontederia cordata L. f. *angustifolia* (Pursh) Solms-Laub. in DC. Monog. Phan. 4:532. 1883.

Rhizomes thick, widely creeping; lowest stem leaf and basal leaves similar, ovate to narrowly lanceolate, obtuse to acute at the apex, cordate or tapering to the base, to 20 cm long; petiole rather stout, to 7 cm long; inflorescence subtended by a bladeless sheath, with over 8 flowers in a spikelike panicle; flowers more or less villous; perianth purple, the tube 5–7 mm long, the lobes 6–10 mm long; fruit toothed, beaked, enclosed by the persistent perianth tube, to 1 cm long; seeds 3.5–4.5 mm long. May–September.

176. *Heteranthera reniformis* (Mud plantain). a. Habit. b. Capsule.

177. ***Heteranthera rotundifolia*** **(Mud plantain).** a. Habit. b. Flower, with upper leaves.

178. *Pontederia cordata* (Pickerelweed). a. Inflorescence and leaf. b. Flower.

Lakes, ponds, often in standing water.

IA, IL, IN, KS, KY, MO, NE, OH (OBL).

Pickerelweed.

Considerable variation exists in leaf shape, with the range extending from broadly ovate to narrowly lanceolate.

4. **Zosterella** Small—Water Star Grass

Perennial from rhizomes; leaves cauline, linear; flower solitary, borne from a spathe; perianth salverform, actinomorphic; stamens 3, uniform; ovary 1-celled; fruit a few-seeded, indehiscent capsule.

Zosterella consists of the following species:

1. **Zosterella dubia** (Jacq). Small in Small & Carter, Fl. Lanc. Co. 68. 1913. Fig. 179.
Commelina dubia Jacq. Obs. Bot. 3:9. 1768.
Schollera graminifolia Willd. Neue Schr. Ges. Naturf. Fr. Berlin 3:438. 1801.
Leptanthus gramineus Michx. Fl. Bor. Am. 1:25. 1803.
Heteranthera graminea (Michx). Vahl, Enum. 2:45. 1805.
Schollera graminea (Michx.) Raf. Am. Mo. Mag. 2:175. 1818.
Heteranthera dubia (Jacq.) MacM. Metasp. Minn. Valley 138. 1892.

Usually aquatic perennial from rather slender rhizomes; leaves linear, obtuse to subacute, to 15 cm long, to 6 mm broad, translucent; flower pale yellow, solitary, borne from a spathe 2–5 cm long; perianth tube to 6 cm long, very slender; capsule ovoid, 8–10 mm long. July–August.

Shallow water, muddy shores.

IA, IL, IN, KS, KY, MO, NE, OH (OBL).

Water star grass.

This species is often placed in the genus *Heteranthera*, but it seems best to follow Small in segregating it into the genus *Zosterella*. The major differences between the two genera are summarized below.

Zosterella	*Heteranthera*
Stamens uniform	Stamens of two kinds
Ovary 1-celled	Ovary partially 3-celled
Fruit indehiscent	Fruit dehiscent
Fruit few-seeded	Fruit many-seeded
Leaves linear	Leaves lanceolate to orbicular
Flowers yellow	Flowers blue, white, or purple

28. POTAMOGETONACEAE—PONDWEED FAMILY

Perennial aquatic herbs; stems simple or branched; leaves 2-ranked, alternate, sometimes di- or tri-morphic, entire or denticulate; stipules sheathing; spikes pedunculate, continuous to moniliform; perianth absent (some botanists consider 4 sepals to be present); stamens 4, with sepaloid connectives; carpels 4, each with a single ovule; fruit drupaceous when fresh, but appearing like a beaked achene after drying.

179. *Zosterella dubia* (Water star grass). a. Habit. b. Habit, with flower.

This family consists of three genera and ninety species, found throughout the world. Two genera occur in the central Midwest.

1. Leaves not septate, some or all over 0.5 mm broad; peduncles rigid .... 1. *Potamogeton*
1. Leaves septate, 0.3–0.5 (–2.0) mm broad; peduncles flexible .... 2. *Stuckenia*

1. **Potamogeton** L.—Pondweed

Perennial aquatic herbs; stems simple or branched, with glands sometimes present on the nodes; leaves 2-ranked, alternate, not septate, sometimes di- or tri-morphic, entire or denticulate; stipules sheathing; spikes pedunculate, continuous to moniliform; perianth absent; stamens 4, with sepaloid connectives; carpels 4, each with a single ovule; fruit drupaceous when fresh, but appearing like a beaked achene after drying; turions sometimes produced.

The species of the genus *Potamogeton* are identified more easily from fresh material than from dried. Characters that should be noted particularly when the specimens are collected are the nature of the stipules and of the submersed leaves. Characters of the fruit may be studied just as well from dried specimens. There are about one hundred species in the genus.

1. Leaves uniform, all submersed.
   2. Stipules adnate to the leaf base.
      3. Leaves auriculate at base; leaves with more than 20 veins .... 19. *P. robbinsii*
      3. Leaves rounded, tapering, or cordate at base, usually not auriculate; leaves with fewer than 20 veins.
         4. Apex of submersed leaves obtuse.
            5. Fruits 4–13 mm long, 1.3–2.4 mm broad, beakless .... 20. *P. spirillus*
            5. Fruits 1.0–2.1 mm long, 0.9–2.0 mm broad, with a minute beak.
               6. Keel of fruits with sharp wings; fruit with minute beak; floating leaves acute or obtuse .... 5. *P. diversifolius*
               6. Keel of fruits without sharp wings; fruit beakless; floating leaves acute to acuminate .... 3. *P. bicupulatus*
         4. Apex of submersed leaves acute .... 22. *P. tennesseensis*
   2. Stipules free from the leaf base (or sometimes adnate in *P. diversifolius*).
      7. Leaves sharply serrulate, crisped along the margin; fruits 5–6 mm long .... 4. *P. crispus*
      7. Leaves entire or minutely denticulate, usually flat; fruits 1.6–5.0 mm long.
         8. Submersed leaves 5–75 mm broad.
            9. Stems compressed; fruits 3-keeled .... 6. *P. epihydrus*
            9. Stems terete; fruits rounded on the sides or with only one prominent keel.
               10. Submersed leaves rounded or cordate at base, or even somewhat auriculate, some of them partly clasping the stem.
                  11. Leaves entire, boat-shaped at tip; fruits 4–5 mm long, dorsally keeled; stems zigzag .... 15. *P. praelongus*
                  11. Leaves minutely denticulate, not boat-shaped at tip; fruits 2.0–3.5 mm long, not keeled; stems not zigzag.
                     12. Tip of leaves acute; stipules fibrous-shredded .... 18. *P. richardsonii*
                     12. Tip of leaves obtuse; stipules entire .... 14. *P. perfoliatus*
               10. Submersed leaves petiolate or sessile, not clasping.
                  13. Submersed leaves more or less crispate.
                     14. Middle and upper submersed leaves broadly lanceolate to ovate, with 19–49 veins; fruiting spike 4 cm long or longer, the fruit tapering to base .... 1. *P. amplifolius*

14. Middle and upper submersed leaves linear to linear-lanceolate, with 7–19 veins; fruiting spikes 2.0–3.5 cm long, the fruit rounded at base .... 16. *P. pulcher*

13. Submersed leaves usually flat.

15. Submersed leaves petiolate.

16. Some or all the submersed leaves acuminate at apex .... 11. *P. illinoensis*

16. All the submersed leaves acute at apex.............. .... 13. *P. nodosus*

15. Submersed leaves sessile.

17. Primary stems up to 1 mm in diameter; leaves with 3–9 veins; fruits 1.9–2.3 mm long .... 9. *P. gramineus*

17. Primary stems usually more than 1 mm in diameter; leaves with 7–19 veins; fruits 2.5–3.6 mm long .... 11. *P. illinoensis*

8. Submersed leaves less than 5 mm broad.

18. Veins of submersed leaves 5–35.

19. Veins of submersed leaves 15–35; fruits quadrate, 4–5 mm long .... 24. *P. zosteriformis*

19. Veins of submersed leaves 5–9; fruits ovoid to obovoid, 1.5–2.5 mm long.

20. Stipules fibrous, whitish .... 8. *P. friesii*

20. Stipules entire, green or brown .... 9. *P. gramineus*

18. Veins of some or all of the leaves 1–3.

21. Stipules fibrous, often whitish .... 21. *P. strictifolius*

21. Stipules entire, green or brown, occasionally white.

22. Submersed leaves with a bristle tip .... 10. *P. hillii*

22. Submersed leaves obtuse.

23. Spikes subcapitate, 2–5 mm long; fruits keeled; sepaloid connectives 0.5–1.0 mm long; leaves glandless at base .... 7. *P. foliosus*

23. Spikes elongate, more or less interrupted, 6–15 mm long; fruits rounded on back; sepaloid connectives 1.2–2.5 mm long; leaves sometimes biglandular at base.

24. Stipules connate, forming cylinders with united margins at least 2/3 their length; spikes interrupted .... 17. *P. pusillus*

24. Stipules not connate, either flat or convolute; spikes continuous .... 2. *P. berchtoldii*

1. Leaves of 2 kinds, the floating usually shorter and broader than the submersed ones.

25. Leaves entire (submersed leaves sometimes crispate in *P. pulcher*).

26. Submersed leaves up to 2 mm broad, 1- to 5-nerved.

27. Submersed leaves 0.1–0.5 mm broad, 1- to 3-nerved; fruits 1.6–2.2 mm long; floating leaves elliptic to narrowly obovate .... 23. *P. vaseyi*

27. Submersed leaves 0.5–2.0 mm broad, 3- to 5-nerved; fruits 3–5 mm long; floating leaves ovate .... 12. *P. natans*

26. Submersed leaves 5–75 mm broad, 7- to 21-nerved.

28. Stems flattened; submersed leaves linear, 5–10 mm broad; fruit with 3 sharp keels .... 6. *P. epihydrus*

28. Stems terete; submersed leaves lanceolate to ovate, 10–75 mm broad; fruit not sharply 3-keeled.

29. Middle and upper submersed leaves broadly lanceolate to ovate, never crispate; fruit tapering to base .... 1. *P. amplifolius*

29. Middle and upper submersed leaves lanceolate to linear-lanceolate, sometimes crispate; fruit rounded at base .... 6. *P. pulcher*

25. Leaves minutely denticulate.

30. Submersed leaves petiolate.

31. Larger submersed leaves acute; fruits reddish .... 12. *P. nodosus*
31. Larger submersed leaves acuminate; fruits greenish .... 11. *P. illinoensis*
30. Submersed leaves sessile.
32. Submersed leaves with 3–9 veins .... 9. *P. gramineus*
32. Submersed leaves with 7–19 veins .... 11. *P. illinoensis*

1. **Potamogeton amplifolius** Tuckerm. Am. Journ. Sci. 6:225. 1848. Fig. 180.

Rootstocks thick, to 4 mm in diameter, whitish or reddish, with obtuse, black scales; stems terete, somewhat less in diameter than the rootstocks, simple or becoming somewhat branched at maturity; leaves trimorphic; floating leaves ovate to elliptic, obtuse and sometimes mucronate at the apex, cuneate or rounded at the base, coriaceous, 5–10 cm long, 2.5–5.0 cm broad, 21- to 51-nerved, with about one-fourth of the nerves more conspicuous than the others, with a petiole to 20 cm long, with the stipules persistent, fibrous, 2-keeled, to 20 cm long, 30- to 40-nerved; submersed leaves on lower part of stem lanceolate, arcuate, obtuse to subacute at the apex, to 20 cm long, 2.5–7.5 cm broad, with a petiole to 6 cm long; submersed leaves on the upper part of the stem broadly lanceolate to ovate, arcuate, obtuse to subacute at the apex, to 20 cm long, to 7.5 cm broad, 19- to 49-nerved, with the petioles to 6 cm long, with the stipules persistent, fibrous, obscurely keeled, to 11 cm long; spike continuous, composed of 9–16 whorls of flowers, in fruit 4–8 cm long, 1.0–1.5 cm thick; flowers sessile; sepaloid connectives suborbicular, 2–4 mm long, clawed; fruit obovoid, tapering to base, rounded and usually prominently keeled on the back, 3.5–5.0 mm long (excluding the beak), 2.5–4.0 mm broad, the usually prominent beak about 1 mm long.

Lakes, ponds, rivers, streams.

IA, IL (OBL).

Broad-leaved pondweed.

This species is readily distinguished by its large, arcuate submersed leaves with many veins.

2. **Potamogeton berchtoldii** Fieber, Oekon. Fl. Boch. II 1:277. 1838. Fig. 181.
Potamogeton pusillus L. var. *tenuissimus* Mert. & Koch. Roehling, Deutsch. Fl. 1:857. 1823.
Potamogeton berchtoldii Fieber var. *tenuissimus* (Mert. & Koch) Fern. Rhodora 42:246. 1940.

Rootstock slender; stems filiform, little to much branched; leaves uniform, submersed, linear, subacute to acute at the apex, tapering to the base, to 5 cm long, 0.5–1.5 mm broad, 3-nerved, with 1–5 pairs of lacunae along the midrib, with a pair of translucent glands at the base; stipules free, but the margins inrolled, obtuse, to 1.5 cm long; spikes continuous, subglobose, composed of 1–3 whorls of flowers, in fruit 2–8 mm long, nearly as thick; flowers sessile; sepaloid connectives suborbicular, about 1.5 mm long; fruits ovoid, rugulose, rounded on the sides, 2.0–2.5 mm long (excluding the beak), 1.2–2.0 mm broad, with a beak about 0.5 mm long, dark greenish brown.

Wet ditches, lakes, streams.

IA, IL, IN, KY, OH (OBL).

180. *Potamogeton amplifolius* (Broad-leaved pondweed). a. Habit. b. Fruit.

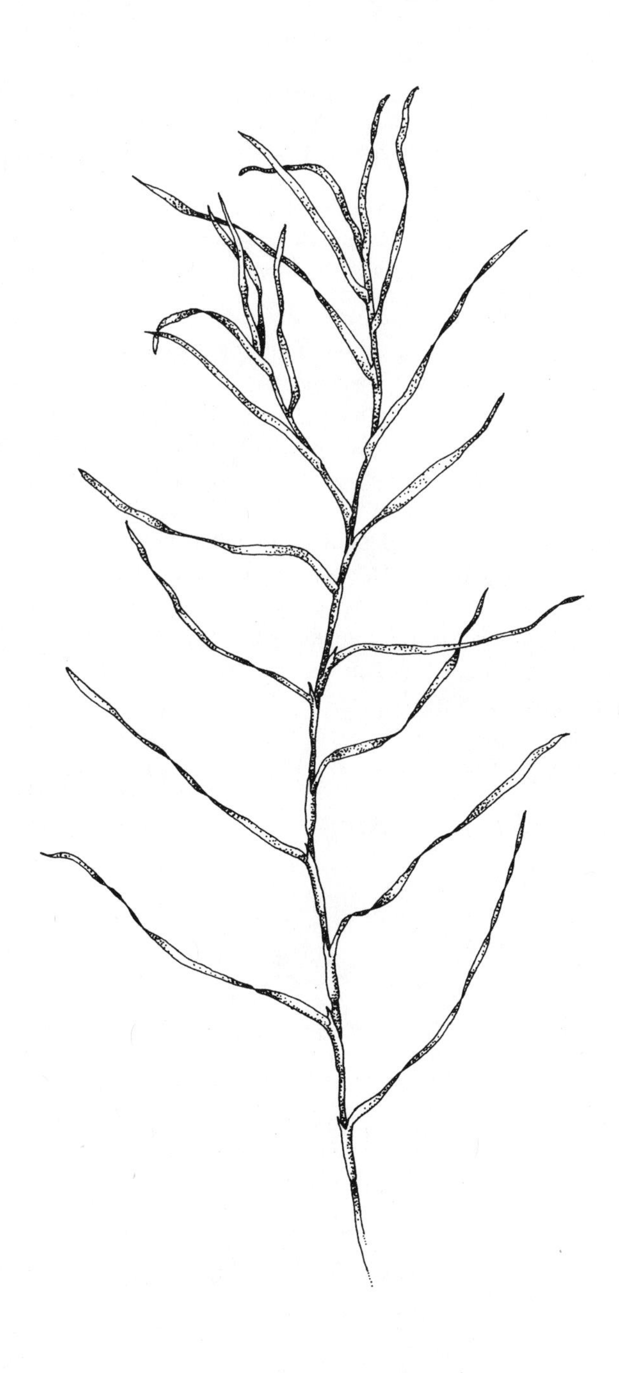

181. *Potamogeton berchtoldii* (Berchtold's pondweed). Upper part of plant.

Berchtold's pondweed.

This species is sometimes considered to be a variety of *P. pusillus*, where it is known as *P. pusillus* var. *tenuissimus*. It differs from *P. pusillus* by its ovoid fruits with rounded sides, by its continuous inflorescence, and by 1–5 rows of lacunae along the midribs of the leaves.

3. **Potamogeton bicupulatus** Fern. Mem. Amer. Acad. Arts, n. s. 17:112. 1932. Fig. 182.
Potamogeton diversifolius Raf. var. *trichophyllus* Morong, Mem. Torrey Club 3 (2):49. 1893.

Rootstocks slender, fibrous; stems profusely branched, compressed, occasionally simple, slender; leaves dimorphic, although the floating ones sometimes absent; floating leaves elliptic, oval, or narrowly obovate, acute to long-tapering at the apex, rounded at the base, 1.2–3.5 cm long, 3–11 mm broad, 3- to- 7-nerved, the compressed petiole 1.7–2.0 cm long, adnate to the stipules; stipules to 3 cm long, not becoming fibrous; submersed leaves linear, obtuse at the apex, to 45 mm long, 0.1–0.6 mm wide, sessile, lax, the stipules 2–15 mm long, sheathing for one-half of their length; spikes continuous, subglobose to elongate, 0.6–1.2 (–2.0) cm long, those in the axils of the submersed leaves subglobose, those in the axils of the floating leaves elongate; flowers sessile; sepaloid connectives suborbicular to reniform; fruits smooth and with one strong and two weak keels without sharp tips on the back, 1.1–2.1 mm long, 1.1–2.0 mm broad, beakless, green to greenish brown.

Acidic shallow water.

IN (OBL).

Slender-leaved pondweed.

This species is sometimes treated as a variety of *P. diversifolius* but differs by its acute submersed and floating leaves.

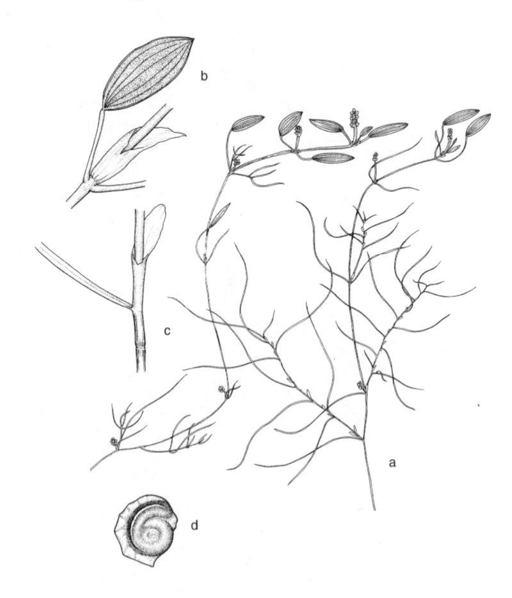

182. ***Potamogeton bicupulatus*** **(Slender-leaved pondweed).** a. Habit. b. Sheath with emersed leaf. c. Sheath with submersed leaf. d. Achene.

4. **Potamogeton crispus** L. Sp. Pl. 126. 1753. Fig. 183.

Rhizomes stout, creeping, unspotted; stems compressed, usually branched, sometimes simple, 0.5–2.5 mm in diameter; leaves uniform, submersed, broadly linear to oblong, broadly rounded to subacute at the apex, tapering to the subclasping base, 1.3–7.5 (–10.0) cm long, 0.2–1.1 cm broad, reddish green, the margins finely and irregularly dentate; stipules slightly adnate at base, papery, becoming frayed early above; spikes compact at first, becoming somewhat interrupted, cylindrical, composed of 3–5 whorls of flowers, in fruit 1–2 cm long, 1.0–1.3 cm thick; flowers sessile or short-pedicellate, the pedicels never exceeding 0.5 mm; sepaloid connectives orbicular, (0.6–) 1.2–2.0 mm long, short-clawed, anthers 0.7–1.3 mm long; fruits ovoid, strongly and obtusely keeled with a small tooth near the base, 5–6 mm long, 1.4–2.8 mm wide, with a straight or incurved beak 2–3 mm long, greenish or brownish; winter buds firm, 1–2 cm thick.

Muddy to calcareous ponds and streams.

IA, IL, IN, KS, KY, MO, NE, OH (OBL).

Curly pondweed.

This is the only species of *Potamogeton* in the central Midwest that is not native. In some bodies of water, it may become very aggressive.

The irregularly dentate, crisped margins of the leaves readily distinguish this species.

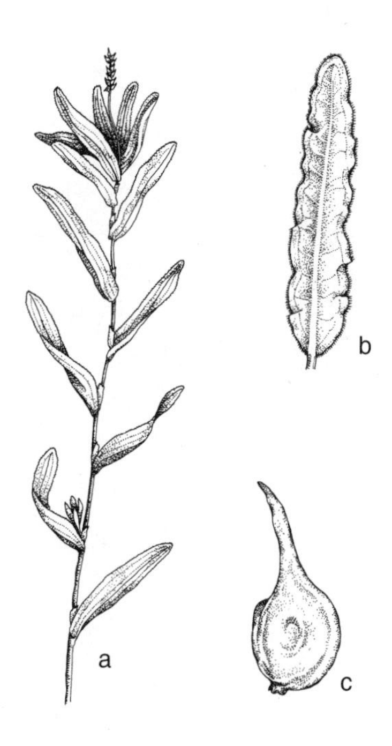

183. *Potamogeton crispus* (Curly pondweed). a. Habit. b. Leaf. c. Fruit.

5. **Potamogeton diversifolius** Raf. Med. Repos. II 5:354. 1808. Fig. 184.

Rootstocks slender, fibrous; stems profusely branched, occasionally simple, slender; leaves dimorphic; floating leaves elliptic, oval, or narrowly obovate, obtuse to acute at the apex, rounded or tapering at the base, 1.2–3.5 cm long, 0.6–1.3 cm broad, the compressed petiole 1.7–2.0 cm long, adnate to the stipules, the stipules to 3 cm long, not becoming fibrous; submersed leaves linear, obtuse at the apex, narrowed to the base, to 45 mm long, to 1.5 mm broad, sessile, the stipules 2–15 mm long, sheathing for one-half of their length; spikes continuous, subglobose to elongate, 0.6–1.2 (–2.0) cm long, those in the axils of the floating leaves elongate; flowers sessile; sepaloid connectives suborbicular, 0.7–1.0 mm long, short-clawed; fruits suborbicular to reniform, smooth and with one strong and two weak keels with sharp tips on the back, 1.0–1.5 mm long (excluding the beak), 0.9–2.0 mm broad, with a minute beak, green to greenish brown.

Quiet waters.

IA, IL, IN, KS, KY, MO, NE, OH (OBL).

Pondweed.

This is one of the most common species of *Potamogeton* in the central Midwest. It is distinguished by its adnate stipules and its acute-tipped submersed leaves.

184. ***Potamogeton diversifolius*** **(Pondweed).** a. Habit. b. Fruit.

6. **Potamogeton epihydrus** Raf. Med. Repos. II 5:354. 1808. Fig. 185.
Potamogeton claytonii Tuckerm. Am. Journ. Sci. 45:38. 1843.
Potamogeton epihydrus Raf. var. *typicus* Fern. Mem. Am. Acad. Arts & Sci. 17:114. 1932.

Rootstocks extensively creeping, slender; stems compressed, simple or little branched; leaves dimorphic; floating leaves usually opposite, elliptic to oblong-lanceolate, obtuse and sometimes cuspidate, to 8 cm long, to 3.5 cm broad, 11- to 41-nerved, coriaceous, with compressed petioles, with subcoriaceous, attenuated stipules; submersed leaves linear-elongate, to 20 cm long, 0.5–1.0 cm broad, 3- to 13-nerved, the free stipules membranous, obtuse, 2.0–4.5 cm long; transitional leaves usually present; spikes continuous, cylindrical, to 4 cm long; flowers sessile; sepaloid connectives flabellate, 1.5–3.0 mm long, short-clawed; fruits sessile, greenish brown, rhombic-obovoid, laterally compressed, 3-keeled, pitted, 3.0–4.5 mm long (excluding the beak), 3.0–3.5 mm broad, with a merely toothlike beak.

Quiet ponds and lakes, streams, rivers.

IA, IL, IN, KY, OH (OBL).

Pondweed.

This species has both floating and submersed leaves, and the submersed leaves are linear-elongate.

7. **Potamogeton foliosus** Raf. Med. Rep. II 5:354. 1808. Fig. 186.
Potamogeton niagarensis Tuckerm. Am. Journ. Sci. 7:354. 1849.
Potamogeton pauciflorus Pursh var. *niagarensis* (Tuckerm.) Gray, Man. Bot. 435. 1856.
Potamogeton foliosus Raf. var. *niagarensis* (Tuckerm.) Morong, Mem. Torrey Club 3:39. 1893.
Potamogeton foliosus Raf. var. *genuinus* Fern. Mem. Acad. Arts & Sci. 17:43. 1932.
Potamogeton foliosus Raf. var. *macellus* Fern. Mem. Acad. Arts. & Sci. 17:46. 1932.

Rhizomes filiform, fibrous, rooting at the nodes; stems filiform, flattened, simple to much branched; leaves uniform, submersed, elongate, linear, subacute at apex, subcuneate at base, 2–14 cm long, up to 2.5 mm broad, green or bronze, 1- to 5-nerved; stipules connate at first to form tubular sheaths up to 17 mm long, at length splitting and falling away; spikes continuous, subcapitate, cylindric, composed of 2–3 whorls of flowers, in fruit up to 5 mm long, 2–5 mm thick; flowers sessile; sepaloid connectives suborbicular to rhombic, 0.6–1.0 mm long, short-clawed, green or brownish; fruits obovoid, compressed, dentate and strongly one-keeled on the back, 1.8–2.5 mm long (excluding the beak), nearly as broad, with a short beak less than 0.5 mm long, greenish brown.

Ponds, lakes, rivers, streams.

IA, IL, IN, KS, KY, MO, NE, OH (OBL).

Leafy pondweed.

This is the only *Potamogeton* in the central Midwest with linear submersed leaves and sharply keeled fruits.

8. **Potamogeton friesii** Rupr. Beitr. Pfl. Russ. Reich. 4:43. 1845. Fig. 187.
Potamogeton pusillus L. var. *major* Fries, Nov. Fl. Suec. 48. 1828.
Potamogeton mucronatus Schrad. ex Reich. Ic. Fl. Germ. Helv. 7:15. 1845, in synon.

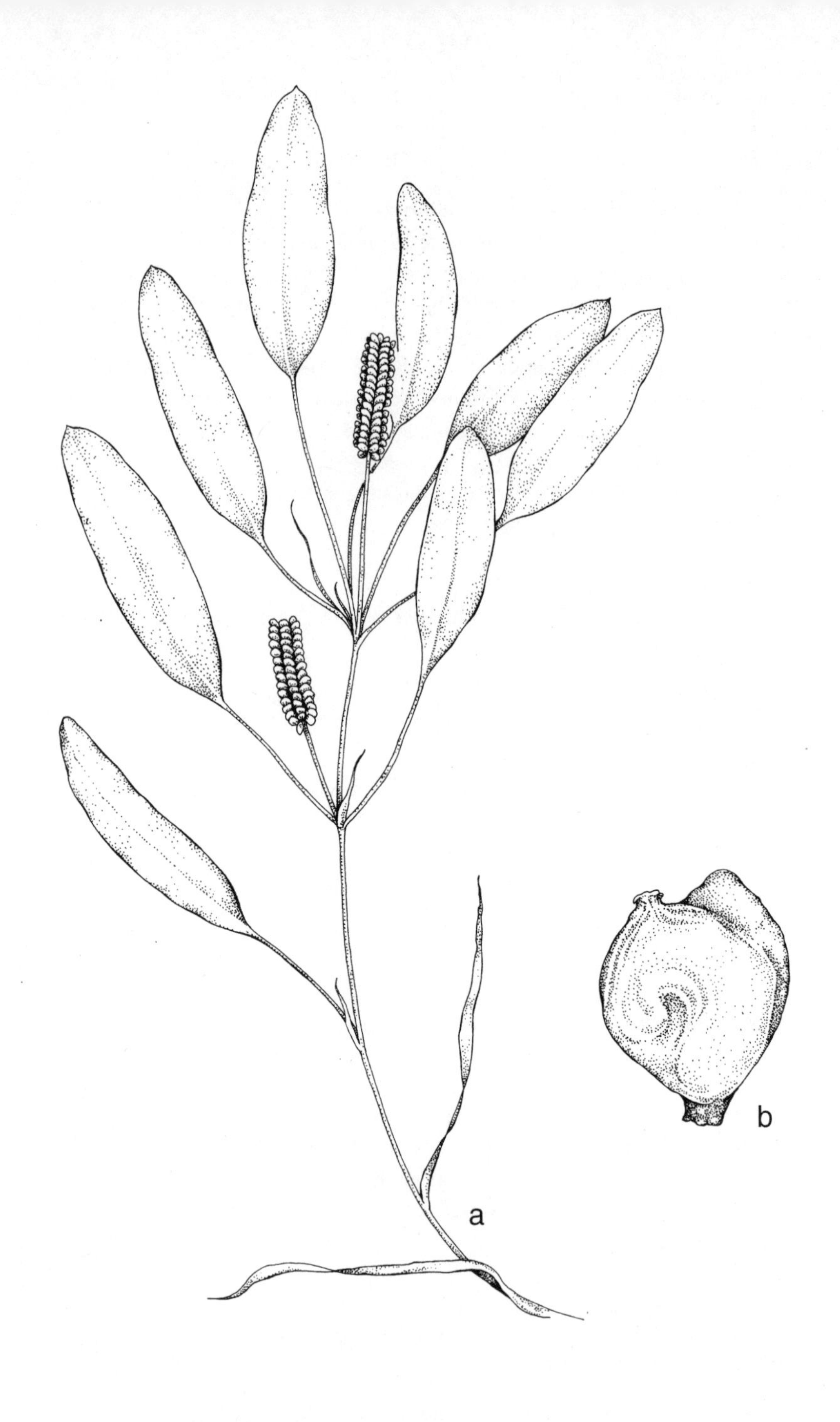

185. *Potamogeton epihydrus* (Pondweed). a. Habit. b. Fruit.

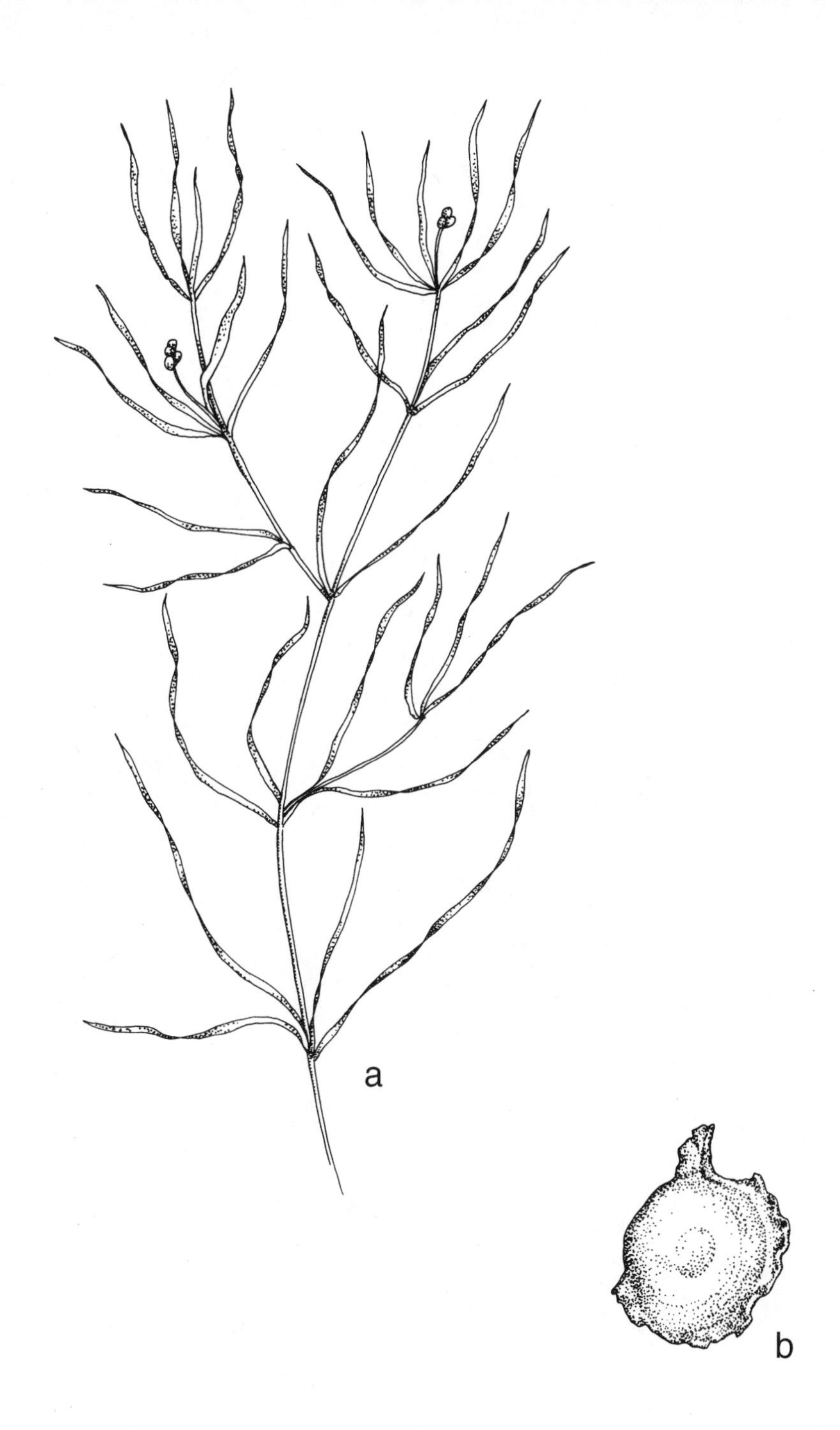

186. *Potamogeton foliosus* (Leafy pondweed). a. Habit. b. Fruit.

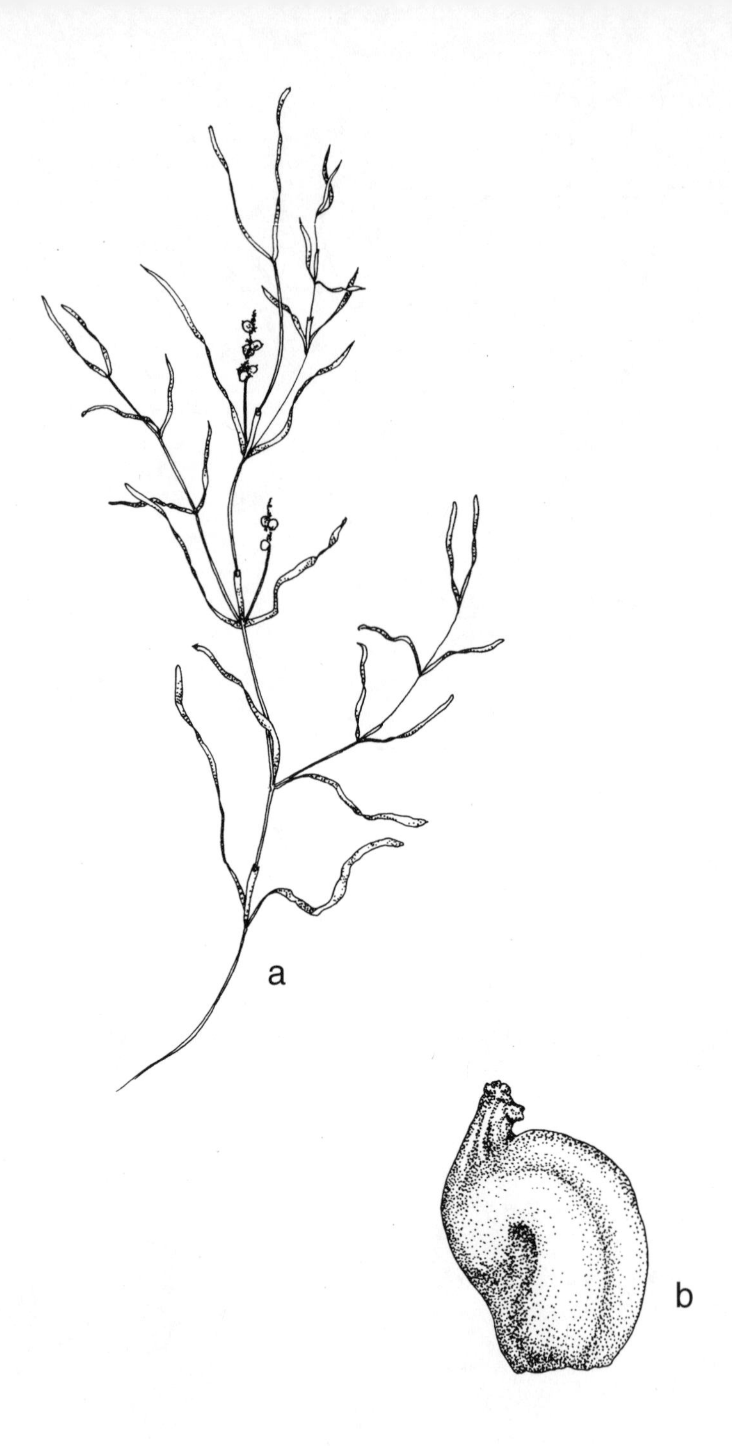

187. *Potamogeton friesii* (Fries' pondweed).   a. Habit.   b. Fruit.

Rhizomes absent; stems simple to sparsely branched, flattened; leaves uniform, submersed, translucent, filiform to linear, obtuse to subacute to acute at apex, cuneate at base, 2.0–10.4 cm long, 0.7–3.5 mm broad, 5- to 9-nerved, with two glands at base; stipules at first connate forming a tubular sheath 7–16 mm long, whitish, fibrous; spikes rarely interrupted, cylindric, composed of 3–4 whorls of flowers, in fruit up to 2.7 cm long; flowers sessile; sepaloid connectives suborbicular to rhombic, 1.5–2.5 mm long, short-clawed; fruits obovoid, rounded and 3-keeled on the back, 2–3 mm long (excluding the beak), with a beak less than 1 mm long, brownish.

Calcareous waters of lakes and streams.

IA, IL, IN, KY, NE, OH (OBL).

Fries' pondweed.

The white, fibrous stipules are distinctive for this species.

9. **Potamogeton gramineus** L. Sp. Pl. 1:127. 1753. Fig. 188.
Potamogeton gramineus L. var. *graminifolius* Fries, Nov. Fl. Suec. 36. 1828.
Potamogeton heterophyllus L. f. *graminifolius* (Fries) Morong, Mem. Torrey Club 3 (2):24. 1893.
Potamogeton gramineus L. var. *typicus* Ogden, Rhodora 45:143. 1943.

Rootstocks thin or thickish, frequently red-spotted; stems much branched, to 1 mm in diameter; leaves dimorphic; floating leaves elliptic to ovate, obtuse and often mucronate at the apex, tapering or rounded at the base, coriaceous, to 7 cm long, 1–3 cm broad, 13- to 23-nerved, with a petiole to 15 cm long; submersed leaves sessile, narrowly elliptic to oblanceolate, acute at the apex, tapering to the base, 3- to 9-nerved, denticulate, to 7 cm long, 2–8 mm broad; stipules persistent, obtuse at the apex, to 3 cm long, 1–5 mm broad, obscurely 2-keeled; spike generally compact, comprised to 5–10 whorls of flowers, in fruit 1–3 cm long, 5–10 mm broad; flowers sessile or on pedicels up to 0.5 mm long; sepaloid connectives suborbicular, 0.7–2.5 mm long and broad, short-clawed; fruit obovoid, prominently keeled, 1.5–3.0 mm long (excluding the beak), 1.5–2.5 mm broad, with a short curved beak, greenish.

Lakes, ponds, streams.

IA, IL, IN, KS, KY, NE, OH (OBL).

Grass-leaved pondweed.

The species varies with respect to the shape and size of the submersed leaves. It also hybridizes with *P. richardsonii* to form *P. X hagstroemii* A. Bennett (Fig. 189) and with *P. illinoensis* to form *P. X spathulaeformis* (J. W. Robbins) Morong (Fig. 190).

Potamogeton gramineus differs from the similar *P. illinoensis* by its small fruits and fewer nerves of the submersed leaves.

10. **Potamogeton hillii** Morong, Bot. Gaz. 6:290. 1881. Fig. 191.
Potamogeton porteri Fern. Mem. Gray Herb. 3:73. 1932.

Rhizomes usually absent; stems filiform, branched, to 1 m long; leaves uniform, submersed, elongated, linear, acute to cuspidate at the apex with a bristle tip, to 8 cm long, 0.5–2.0 mm broad, 3- to 5-nerved; stipules free from the blades, white or pale

188. *Potamogeton gramineus* (Grass-leaved pondweed). a. Habit. b. Fruit.

189. *Potamogeton* X *hagstroemii* (Hagstroem's pondweed). Habit.

190. *Potamogeton* **X** *spathulaeformis* **(Pondweed).** Upper part of plant.

brown, to 15 mm long, at length splitting and falling away; peduncles axillary, to 15 mm long, usually recurved; spikes continuous, capitate, 4–7 mm thick; flowers sessile; fruits brown to greenish brown, ovoid to orbicular, sessile, 2.3–4.0 mm long, 2.0–3.2 mm broad, with 1 (–2) keels on the back and a pair of lateral keels, with a beak about 0.5 mm long.

In clear, cold, alkaline waters.

OH (OBL).

Hill's pondweed.

This northern pondweed is readily recognized by its bristle-tipped leaves with 3 or 5 veins.

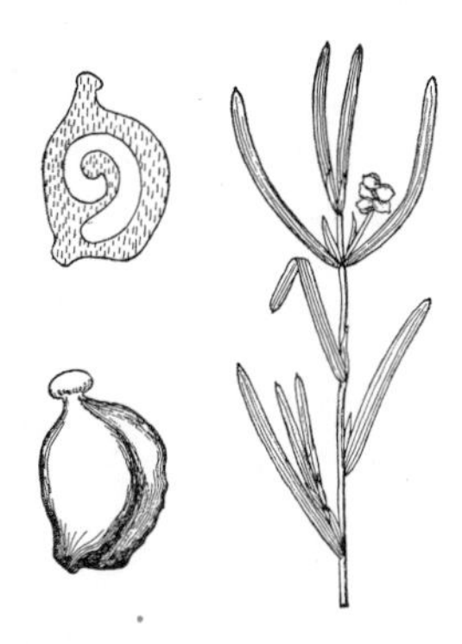

191. *Potamogeton hillii* **(Hill's pondweed).** Habit (right). Achenes (left).

11. **Potamogeton illinoensis** Morong, Bot. Gaz. 5:50. 1880. Fig. 192.

Rootstocks stout; stems slender to rather stout, up to 5 mm in diameter, simple to little branched; leaves dimorphic; floating leaves (often absent) elliptic to ovate, obtuse and mucronate, rounded or tapering to the base, denticulate, to 15 (–20) cm long, 2–5 (–6) cm broad, 13- to 29-nerved, with a petiole shorter than the blade; submersed leaves petiolate or sessile, linear to lanceolate to narrowly ovate, acute to acuminate and usually mucronate at the apex, rounded or tapering to the base, denticulate, 7- to 19-nerved, to 20 cm long, to 4.5 cm broad, with a petiole somewhat compressed, up to 3.5 cm long; stipules obtuse, to 5 (–8) cm long, two-keeled; spike compact, continuous, composed of 8–15 whorls of flowers, in fruit to 7 cm long, about 1 cm thick; flowers sessile; sepaloid connectives suborbicular, 1.5–3.0 mm long, short-clawed; fruits ovoid to suborbicular, with one prominent and two usually obscure keels, 2.5–3.5 mm long, 2–3 mm broad, with the beak about 0.5 mm long, brown.

Ponds, lakes, and streams, usually in alkaline water.

IA, IL, IN, KS, KY, MO, NE, OH (OBL).

Illinois pondweed.

Potamogeton illinoensis is similar to *P. gramineus*, but differs by its little-branched stems, its prominently keeled stipules, and its generally larger spikes, leaves, and sepaloid connectives. It hybridizes with *P. gramineus* to form *P. X spathulaeformis* (J. W. Robbins) Morong (Fig. 190).

12. **Potamogeton natans** Sp. Pl. 1:126. 1753. Fig. 193.

Rootstocks thin, fibrous, white with reddish spots; stems slender, simple to rarely branched, with ridges; leaves dimorphic, coriaceous; floating leaves lanceolate to elliptic to ovate, obtuse at the apex, rounded to cordate at the base, 4–15 cm long, 2–6 cm broad, 13- to 37-nerved, with about one-third of the nerves more prominent, with petioles 3–15 cm long; submersed leaves long-linear, obtuse at the apex, tapering to the sessile base, 10–20 cm long, 0.5–2.0 mm broad, obscurely 3- to 5-nerved; stipules clasping, persistent, fibrous, whitish, linear to lanceolate, 3–11 cm long, 2-keeled; spikes continuous, cylindrical, composed of 8–14 whorls of flowers,

192. *Potamogeton illinoensis* (Illinois pondweed). a. Habit. b. Fruit.

193. *Potamogeton natans* (Pondweed). a. Habit. b. Fruit.

in fruit 2.5–6.0 cm long, 0.9–1.2 cm thick; flowers sessile or nearly so; sepaloid connectives suborbicular, 1.6–2.8 mm broad, short-clawed; fruits obovoid, smooth and without a keel on the back, 3–5 mm long (excluding the beak), 2.0–3.5 mm broad, with a broad beak less than 1 mm long.

Lakes and streams.

IA, IL, IN, KS, KY, MO, NE, OH (OBL).

Pondweed.

This species is distinguished by its dimorphic leaves, the submersed ones 0.5–2.0 mm broad. It differs from the similar *P. vaseyi* by its larger fruits and its 3- to 5-nerved submersed leaves.

13. **Potamogeton nodosus** Poir. Lam. Encycl. Meth. Bot. Suppl. 4:535. 1816. Fig. 194.
Potamogeton occidentalis Sieb. ex. Cham. & Schlecht. Linnaea 2:224. 1827.
Potamogeton americanus Cham. & Schlecht. Linnaea 2:226. 1827.

Rootstocks thick, reddish spotted; stems terete, simple, to 2 mm in diameter; leaves dimorphic; floating leaves elliptic, obtuse to subacute at the apex, cuneate or rounded at the base, 9- to 21-nerved, coriaceous, 3–15 cm long, 1.5–5.2 cm broad; submersed leaves petiolate, linear-lanceolate to elliptic, acute at the apex, tapering to the base, denticulate (at least when young), thin, to 20 cm long, to 3.5 cm broad, 7- to 15-nerved, with the compressed petiole 2–13 cm long; stipules linear, acute or obtuse, delicate, brownish, 2–9 cm long, usually two-keeled; spike more or less continuous, cylindrical, composed to 10–17 whorls of flowers, in fruit 3–10 cm long, about 1 cm thick; flowers sessile; sepaloid connectives suborbicular, 1.4–2.6 mm long, short-clawed; fruits obovoid, smooth and prominently one-keeled on the back, 2.0–2.4 (–4.3) mm long (excluding the beak), 2.5–3.0 mm broad, with a short beak less than 1 mm long, brownish red.

Ponds and streams.

IA, IL, IN, KS, KY, MO, NE, OH (OBL).

Pondweed.

Potamogeton nodosus is distinguished by its dimorphic leaves that are denticulate, at least when young, and by its large fruits more then 3.5 mm long, including the beak. This species hybridizes with *P. richardsonii* to form *P. X rectifolius* A. Bennett (Fig. 195).

14. **Potamogeton perfoliatus** L. Sp. Pl. 1:126. 1753. Fig. 196.

Rhizomes rather slender; stems terete, not punctate, lacking nodal glands; leaves all alike, submersed, sessile, lax; stipules inconspicuous or absent, soon deteriorating, free from the blade, 3.5–6.5 cm long, not shredding at the apex, the blade olive-green, broadly lanceolate, orbiculate to ovate, not curved, 1–10 cm long, 7–40 mm broad, rounded at the base, clasping, the margins entire, sometimes crispate, the apex usually more or less obtuse; spikes unbranched, terminal or axillary, cylindric, to 5 cm long; fruits sessile, greenish, obovoid, not keeled, 1.5–3.0 mm long, 1.5–2.2 mm wide, the beak erect, to 0.6 mm long.

Lakes, rivers, streams.

OH (OBL).

194. *Potamogeton nodosus* (Pondweed). a. Habit. b. Fruit.

195. ***Potamogeton X rectifolius*** **(Pondweed).** Upper part of plant.

Clasping pondweed.

This species is readily distinguished by the large leaves that are clasping at the base, differing from *P. richardsonii* by its obtuse leaf tips.

196. ***Potamogeton perfoliatus* (Clasping pondweed).** Habit. Achene (lower left).

15. **Potamogeton praelongus** Wulfen. Roem. Arch. Bot. 3:331. 1805. Fig. 197.

Rhizomes rather stout, whitish, suffused with red; stems slender, to 4 mm in diameter, simple or sparingly branched, flexuous, whitish or olive-green; leaves uniform, all submersed, broadly lanceolate, obtuse and cucullate at the apex, rounded or cordate at the partly clasping base, 2–18 (–36) cm long, 0.8–3.0 cm broad, 13- to 25-nerved, with 3–7 of the nerves more prominent, sessile, the margins entire; stipules mostly persistent, more or less oblong, 2.5–6.0 (–10.0) cm long, white, strongly nerved, without keels; spikes interrupted, sometimes moniliform, cylindrical, with 5–10 (–12) whorls of flowers, in fruit 3–5 cm long, 1.0–1.5 cm broad; flowers sessile; sepaloid connectives suborbicular to rhombic, 1.7–3.0 mm broad, short-clawed; fruits obovoid, rounded and prominently 1-keeled on the back, 4–5 mm long (excluding the beak), 2.5–4.0 mm broad, with a thick beak up to 1 mm long.

Lakes, rivers.

IA, IL, IN, NE (OBL).

Pondweed.

This species is characterized by its whitish, flexuous stems, its large fruits, its large, cucullate, ovate leaves, and its conspicuous whitish stipules.

16. **Potamogeton pulcher** Tuckerm. Am. Journ. Sci. 45:38. 1843. Fig. 198.

Rootstocks slender, up to 1 mm in diameter, usually red-spotted; stems terete, to 2.5 mm in diameter, simple, black-punctate; leaves trimorphic; floating leaves ovate to orbicular, obtuse and sometimes bluntly mucronate at the apex, cordate or rounded at the base, to 7 (–11) cm long, 1.5–5.5 (–8.5) cm broad, 19- to 35-nerved, coriaceous, with the petiole up to 1.5 cm long, with the stipules early decaying; lower leaves oblong, obtuse, subcoriaceous; upper leaves lanceolate to lance-linear, acute or subacute, translucent; transitional leaves present; spikes continuous, cylindrical, composed of about 10 whorls of flowers, in fruit 2.0–3.5 cm long, about 1 cm thick; flowers sessile or nearly so; sepaloid connectives suborbicular, 1.5–4.0 mm long, clawed; fruits obliquely ovoid, rounded at the base, rounded on the back with one prominent and two rather obscure keels, 2–4 mm long (excluding the beak), 2.0–3.5 broad, with a prominent beak up to 0.8 mm long, light brown to olive-green.

Shallow water.

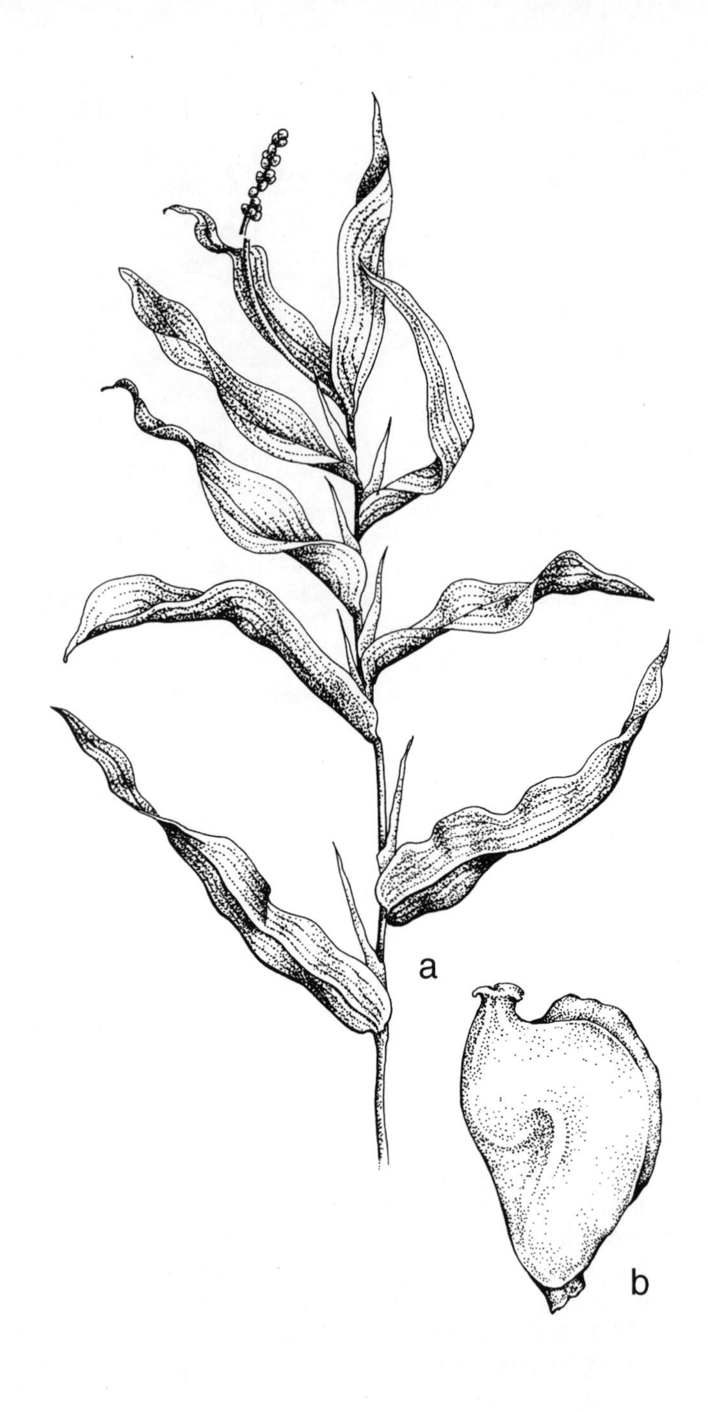

197. *Potamogeton praelongus* (Pondweed). a. Habit. b. Fruit.

198. ***Potamogeton pulcher*** **(Pondweed).** a. Habit. b. Fruit.

IL, IN, KY, MO, OH (OBL).

Pondweed.

The conspicuously punctate stems, the large cordate floating leaves, and the lanceolate submersed leaves abruptly tapering at the base serve to distinguish this pondweed.

17. **Potamogeton pusillus** L. Sp. Pl. 1:127. 1753. Fig. 199.
Potamogeton panormitanus Biv. var. *minor* Biv. Figlio Andrea 6. 1838.
Potamogeton panormitanus Biv. var. *major* Fischer, Ber. Bayer Bot. Gesells. 11:109. 1907.
Potamogeton pusillus L. var. *minor* (Biv.) Fern. & Schub. Rhodora 50:154. 1948.

Rootstocks very slender; stems filiform, compressed, much branched, with green nodal glands; leaves uniform, submersed, linear, obtuse to acute at the apex, tapering to the base, 3–21 mm long, 0.5–2.0 mm broad, 1- to 3-nerved, usually bearing two translucent glands near the base; stipules connate to about the middle, membranous, splitting at maturity but not becoming fibrous, to 1.7 cm long; spikes interrupted, subcylindric, composed to 2–5 whorls of flowers, in fruit 4–15 mm long; flowers sessile; sepaloid connectives suborbicular, 1.3–1.7 mm long, short-clawed; fruits obovoid to reniform, smooth and very slightly one-keeled on the back, 1.0–1.8 mm long (excluding the beak), 1.0–1.8 mm broad, the beak about 0.5 mm long, greenish brown.

Ponds, lakes, streams.

IA, IL, IN, KS, KY, MO, NE, OH (OBL).

Small pondweed.

Potamogeton pusillus is distinguished from *Stuckenia pectinata* by its nonseptate, shorter leaves, its much interrupted spikes, and its smaller fruits. From *P. friesii* and *P. strictifolius* it differs by its nonfibrous stipules. Size of individual plants and their parts is variable. Those specimens with more robust features have been segregated as var. *major*, while those with leaves less than 1 mm broad have been called var. *minor*.

18. **Potamogeton richardsonii** (Benn.) Rydberg, Bull. Torrey Club 32:599. 1905. Fig. 200.
Potamogeton perfoliatus L. var. *lanceolatus* Robbins. Gray, Man. Bot. 5:488. 1867, non Blytt (1861).
Potamogeton perfoliatus L. var. *richardsonii* Benn. Journ. Bot. 27:25. 1889.

Rhizomes thick, whitish or yellowish, unspotted; stems slender, to 2.5 mm in diameter, simple to branched; leaves all submersed, the lower ovate to ovate-lanceolate, the upper becoming lanceolate, acute at the apex, cordate at the partly clasping base, 1.5–14.0 cm long, 0.5–3.5 cm broad, 7- to 33-nerved, with 3–7 nerves more prominent, with the margins denticulate; stipules ovate to lanceolate, obtuse, 1.0–2.7 cm long, whitish, strongly nerved, without keels, fibrous-shredded; spikes mostly interrupted and moniliform, sometimes continuous, cylindrical, composed of 6–12 whorls of flowers, in fruit 1.5–5.0 cm long, to 1 cm broad; flowers sessile; sepaloid connectives suborbicular to rhombic, 1.5–2.5 mm broad, short-clawed; fruits obovoid, rounded and usually keel-less on the back, 2.5–3.5 mm long (excluding the beak), 2.0–3.3 mm broad, with a beak up to 1 mm long, brownish.

Lakes and rivers.

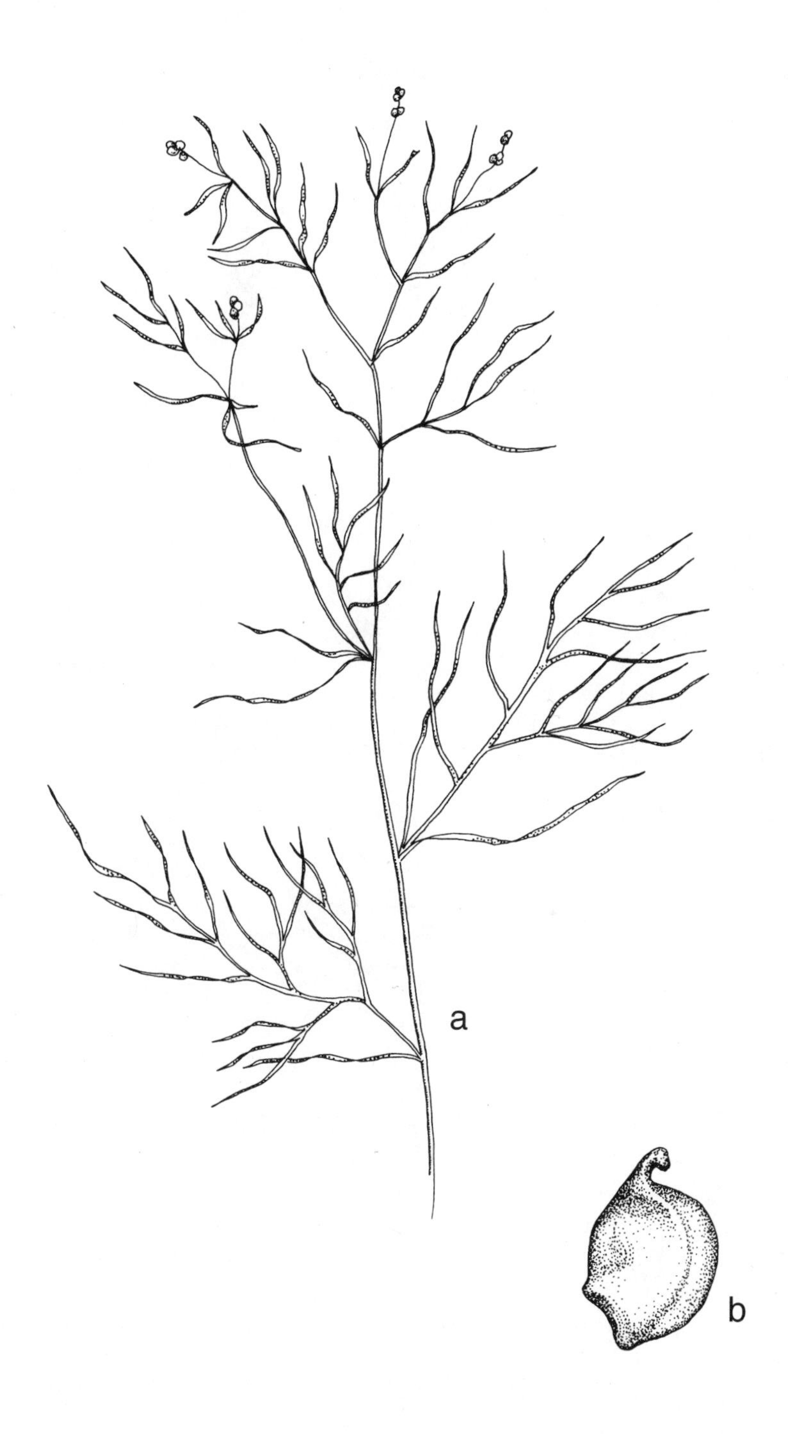

199. *Potamogeton pusillus* (Small pondweed). a. Habit. b. Fruit.

200. *Potamogeton richardsonii* (Richardson's pondweed). a. Habit. b. Fruit.

201. *Potamogeton robbinsii* (Robbins' pondweed). a. Habit. b. Fruit.

IA, IL, IN, NE, OH (OBL).

Richardson's pondweed.

This species differs from *P. perfoliatus,* one of the other species with clasping submersed leaves, by its acute leaf tips, and from *P. robbinsii,* another species with clasping submersed leaves, by its stipules free from the leaf base and its unbranched spikes. *Potamogeton richardsonii* hybridizes with *P. nodosus* to form *P. X rectifolius* A. Bennett (Fig. 195).

19. **Potamogeton robbinsii** Oakes, Hovey's Mag. 7:180. 1841. Fig. 201.

Rhizomes very slender and long-fibrous, not tuber-producing; stems simple or profusely branched, sometimes attaining a length of over one meter, rooting at the nodes; leaves submersed, adnate to the lower half of the stipules, those of the sterile stems crowded, linear to lanceolate, stiff, acute at the apex, auriculate at the base, 2–80 cm long, 1–8 mm broad, 20- to 60-nerved, with a serrulate margin, those of the flowering stems remote and smaller; stipules 1.0–2.5 mm long, sheathing for one-half of their length, whitish, becoming fibrous; spikes interrupted, cylindrical, branching, to 2 cm long; flowers sessile; sepaloid connectives suborbicular to rhombic, short-clawed; fruits flattened, obovoid, smooth, with a prominent keel on the back, 4–5 mm long (excluding the beak), 2.7–3.3 mm broad, with a central beak up to 1 mm long.

Ponds, lakes, rivers.

IL, IN, OH (OBL).

Robbins' pondweed.

This pondweed is the only one with branching spikes, but the plants very seldom flower. The leaves have definite auricles at the base that protrude beyond the stem.

20. **Potamogeton spirillus** Tuckerm. Am. Journ. Sci. & Arts, ser. 2, 6:228. 1848. Fig. 202.

Rhizomes rather slender; stems flattened, not punctate, to 40 cm long; leaves submersed and floating, or floating leaves sometimes absent, spirally arranged; submersed leaves sessile, linear, not curved, to 8 cm long, 0.5–2.0 mm wide, tapering to the base, entire, obtuse at the apex, with broad lacunae extending from the midvein to the margin, with 1–3 veins; stipules more or less persistent, adnate to the blade for one-half its length, not fibrous; floating leaves petiolate, light green, oblong to obovate, to 3.5 cm long, 2–13 mm wide, obtuse at the apex, tapering or rounded at the base, with 5–15 veins; spikes unbranched, with dimorphic peduncles, some of them submersed, axillary or terminal, to 2.5 cm long, some submersed and capitate or emersed and ellipsoid, to 1.2 cm long; fruits sessile, greenish brown, more or less orbicular,

202. *Potamogeton spirillus* **(Snailseed pondweed).** Habit. Achene (above and lower right).

compressed, winged with blunt-tipped wings, to 4–13 mm long, to 1.3–2.4 mm broad, without a beak.

Ponds, lakes, streams.

IA, NE, OH (OBL).

Snailseed pondweed.

The common name is derived from the embryo within the seed, which has a full spiral resembling a snail. The broad lacunae between the midvein and the margins of the submersed leaves are diagnostic.

21. **Potamogeton strictifolius** Benn. Journ. Bot. 40:148. 1902. Fig. 203.

Rootstocks slender; stems filiform, flattened, simple to sparsely branched, the branchlets rigid; leaves firm, uniform, submersed, linear, obtuse and sometimes mucronate, to 2.5 mm broad, 3-nerved, the margins more or less revolute; stipules connate at first, whitish, chartaceous, splitting and becoming fibrous at maturity, to 2 cm long; spikes interrupted, cylindric, composed of 3–4 whorls of flowers, in fruit 0.6–1.5 cm long; flowers sessile; sepaloid connectives suborbicular, (1.3–) 1.5–1.8 mm long, short-clawed; fruits obovoid to ovoid, rounded on the back, 2–3 mm long (excluding the beak), 1.5–2.0 mm broad, the marginal beak less than 1 mm long.

Lakes, slow-moving streams.

IL, IN, NE, OH (OBL).

Stiff pondweed.

The leaves are often revolute in this species. It is somewhat similar to *P. pusillus* but has fibrous stipules and slightly larger fruits. The submersed leaves are firm and stiff.

22. **Potamogeton tennesseensis** Fern. Rhodora 38:167, Plate 412. 1936. Fig. 204.

Rhizomes slender; stems terete, not punctate, to 35 cm long; leaves dimorphic, the floating ones sometimes absent, spirally arranged, acute at the apex; submersed leaves sessile, linear-filiform, not curved, to 10 cm long, 0.2–1.0 (–2.0) mm broad, acute at the apex, tapering to the base, the margins entire, long-tapering at the tip, with numerous broad lacunae, 1- to 3-veined, the stipules persistent, adnate to the blade for less than one-fourth its length, not fibrous; floating leaves petiolate, lance-oblong, to 5 cm long, 5–13 mm broad, tapering to the base, acute at the apex, 9- to 23-nerved; spikes cylindric, continuous, to 2 cm long; fruits sessile, greenish brown, more or less orbicular, keeled, 2.5–3.0 mm long, 2.0–2.5 mm broad, with a beak 0.5 mm long.

Rivers, streams.

KY, OH (OBL).

Tennessee pondweed.

This rare species differs from *P. bicupulatus* by its narrower submersed leaves with many broad lacunae and by its floating leaves with 9–23 nerves. It differs from *P. epihydrus* by its fewer veins on its submersed leaves, by its stipules at least partly adnate to the leaves, and by its acute floating leaves.

23. **Potamogeton vaseyi** Robbins. Gray, Man. Bot. 485. 1867. Fig. 205.

Rootstocks slender, fibrous, much branched, those with flowers and fruits bearing expanded leaves, those without flowers and fruits bearing unexpanded leaves; leaves

203. *Potamogeton strictifolius* (Stiff pondweed). a. Habit. b. Fruit.

204. *Potamogeton tennesseensis* (Tennessee pondweed). Habit.

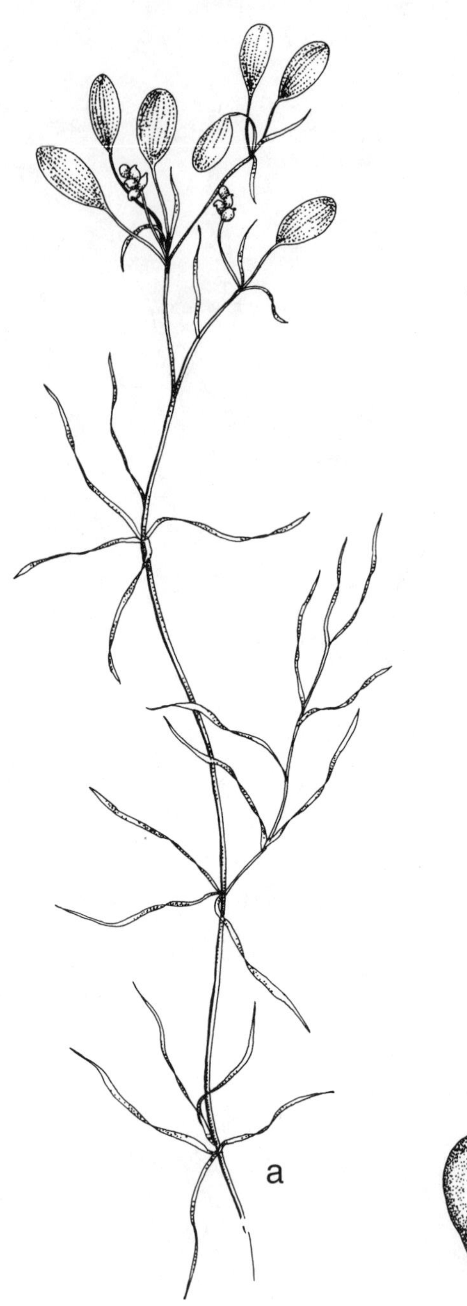

205. *Potamogeton vaseyi* (Vasey's pondweed). a. Habit. b. Fruit.

dimorphic; floating leaves elliptic to oval, obtuse at the apex, tapering to the base, 0.6–1.5 cm long, 3–7 mm broad, 5- to 9-nerved, the margins entire, the petiole up to twice as long as the blade; submersed leaves linear-filiform, delicate, mucronulate at the apex, tapering to the sessile base, 3.0–9.5 mm long, up to 0.5 mm broad; stipules slender, delicate, to 1.2 cm long; spikes interrupted, cylindric, composed of 3–8 whorls of flowers, in fruit 6–9 mm long; flowers sessile; sepaloid connectives suborbicular, about 1 mm long, short-clawed; fruits suborbicular, flattened, strongly keeled on the back, 0.9–2.0 mm in diameter, with a marginal recurved beak about 0.5 mm long, greenish.

Lakes, ponds.

IA, IL, IN, OH (OBL).

Vasey's pondweed.

Floating leaves are present only on fertile stems. The flattened fruits are strongly keeled on the back.

24. **Potamogeton zosteriformis** Fern. Mem. Amer. Acad. 17:36. 1932. Fig. 206.

Rootstocks slender, fibrous; stems usually branched, flattened, to 3 mm broad, constricted at the nodes; leaves uniform, submersed, linear, elongate, obtuse to acuminate at the apex, clasping at the base, 6–12 (–20) cm long, 2–5 mm broad, 15- to 35-nerved, with three nerves more prominent, sessile; stipules firm, 1.5–4.0 cm long; flowers sessile; sepaloid connectives suborbicular to rhombic, 2.0–2.6 mm long, short-clawed; fruits suborbicular to quadrate, with a winged, dentate keel on the back, usually umbonate at the base, 1.5–3.0 mm long (excluding the beak), 1.5–3.0 mm broad, with a marginal beak to 1 mm long, brown.

Ponds, lakes, streams.

IA, IL, IN, KS, NE, OH (OBL).

Flat-stemmed pondweed.

This species is readily recognized by its broad, flattened stems.

2. **Stuckenia** Borner—Linear Pondweeds

Rhizomes present; stems terete; leaves submersed, sessile, linear, tapering to the base, obtuse to acute at the apex, the margins entire, 1- to 5-nerved; stipules adnate to the base of the leaves; spikes capitate or cylindric, submersed, on flexible peduncles; fruits rounded, with or without a beak, turgid. The species is this genus have usually been included in the genus *Potamogeton*. *Stuckenia* species differ by their submersed fruits on flexible peduncles.

Six species comprise the genus, found in most parts of the world. Two species occur in the central Midwest.

1. Apex of leaves obtuse, notched; stems sparsely branched; fruit beakless .... 1. *S. filiformis*
1. Apex of leaves acute, apiculate; stems much branched; fruit beaked .... 2. *S. pectinata*

1. **Stuckenia filiformis** (Persoon) Borner, Fl. Deut. Volk 713. 1912. Fig. 207.
Potgamogeton filiformis Persoon, Syn. Pl. 1:152. 1805.

Rhizomes very slender; stems sparsely branched, more or less terete, to 30 cm long; leaves uniform, submersed, filiform or a little broader, to 15 cm long, 0.2–

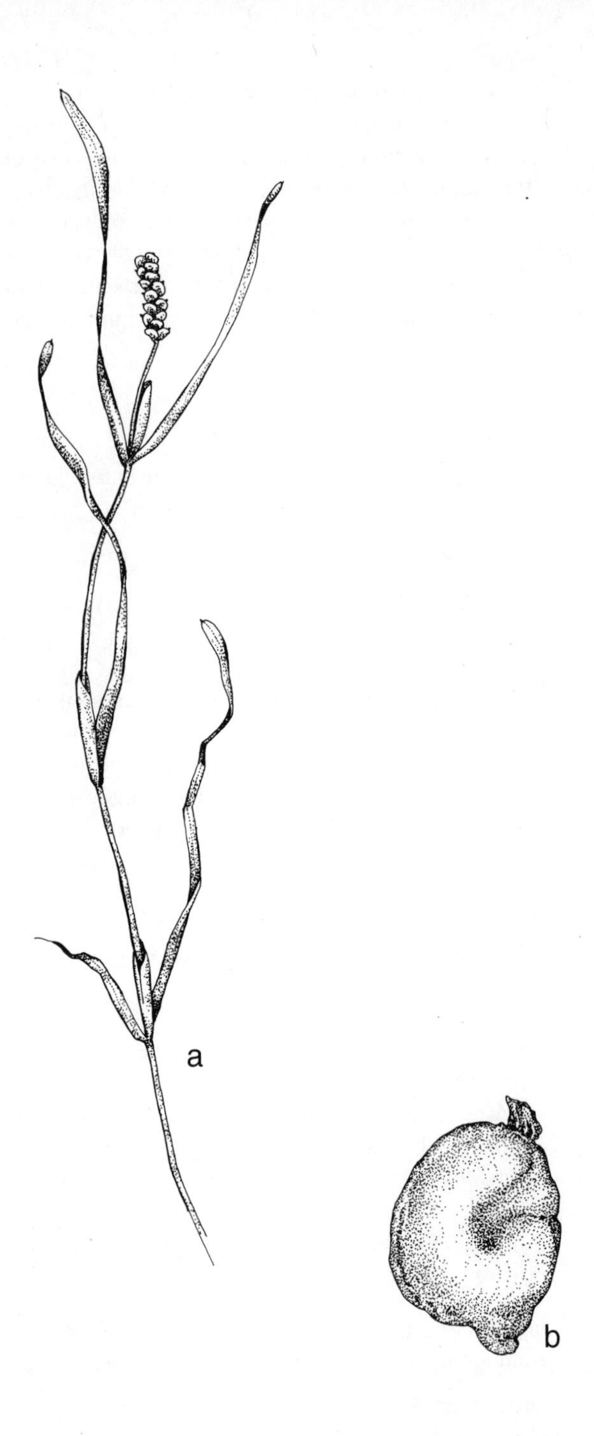

206. *Potamogeton zosteriformis* (Flat-stemmed pondweed). a. Habit. b. Fruit.

0.5 mm broad, obtuse and notched at the apex, 1- to 3-nerved; stipules to 4 (–9) cm long, about the same width as the stem; spikes cylindric to moniliform, submersed, to 5.5 cm long, with 2–9 whorls of flowers; fruits dark brown, obovoid, 2–3 mm long, 1.5–2.5 mm broad, without a beak.

Ponds, lakes, streams.

OH (OBL).

Thread-leaved pondweed.

This species is readily recognized by its filiform leaves that are notched at the tip and its beakless fruits.

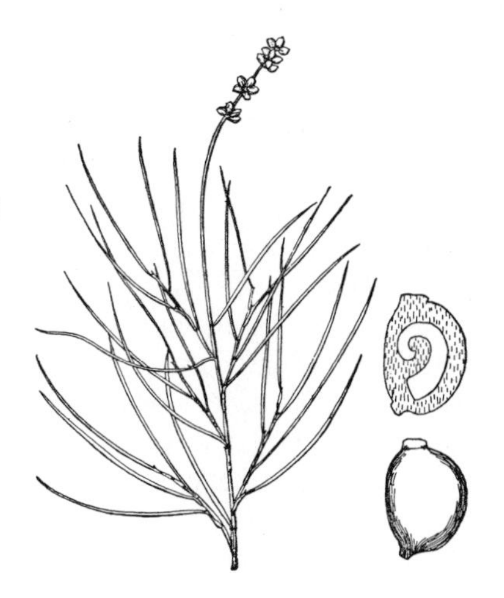

207. *Stuckenia filiformis* **(Thread-leaved pondweed).** Upper part of plant in flower (left). Achenes (far right).

2. **Stuckenia pectinata** (L.) Borner, Fl. Deut. Volk 713. 1912. Fig. 208.
Potamogeton pectinatus L. Sp. Pl. 1:127. 1753.

Rhizomes very slender, fibrous, bearing small tubers; stems filiform, much branched, up to 1 m long; leaves uniform, submersed, filiform, acute and mucronate at the apex, tapering to the base, to 7.5 cm long, 0.3–1.0 mm broad, the margins entire; stipules connate more than half their length, 2–5 cm long; spikes interrupted, moniliform, composed of 2–6 whorls of flowers, in fruit 1.5–5.0 cm long; flowers sessile; sepaloid connectives suborbicular to broadly rhombic, 0.7–1.0 long, short-clawed; fruits reniform, smooth and weakly one-keeled on the back, 2.5–4.2 mm long (excluding the beak), 1.0–2.5 mm broad, with a short, curved beak near the ventral margin, brown.

Lakes, rivers, streams.

IA, IL, IN, KS, KY, MO, NE, OH (OBL).

Sago pondweed.

This common species is recognized by its filiform leaves that are acute at the apex and by its curved beak on the fruits. The tubers are a major source of food for waterfowl.

29. RUPPIACEAE—DITCH GRASS FAMILY

Only the following genus is in the family.

1. **Ruppia** L.—Ditch Grass

Aquatic herbs; leaves capillary, sheathing at the base; flowers bisexual, borne two on a spadix; perianth absent; stamens 2, sessile; carpels 4, free, 1-celled, each with a single ovule; fruit drupaceous, borne on a stipe (podogyne).

This genus, found nearly throughout the world, consists of ten species. Only the following occurs in the central Midwest.

1. **Ruppia cirrhosa** (Petagna) Grande, Bull. Orto Bot. Regia Univ. Napoli 5:58. 1918. Fig. 209.
Buccaferrea cirrhosa Petagna, Inst. Bot. 5:1826. 1787.

Slender, submersed herbs; stems filiform, branched; leaves alternate, capillary, up to 10 cm long, less than 0.5 mm broad, with membranous basal sheaths up to 1 cm long; bisexual flowers two per spadix, enclosed at anthesis by the sheathing leaf-base; perianth absent; stamens 2, each with 2 large locules; peduncles 1.5–3.0 mm long; carpels 4, free, each with a single suspended ovule; drupes slenderly and excentrically beaked, 1.5–3.0 mm long, the podogyne 10–35 mm long. June–September.

Saline waters.

IL, KS, MO, NE, OH (OBL).

Ditch grass; widgeon grass.

This species resembles a very filiform pondweed, but the fruits are borne on a long stalk, or podogyne. In the past, our plant has been called *R. maritima,* but that is a different species that does not occur in the central Midwest.

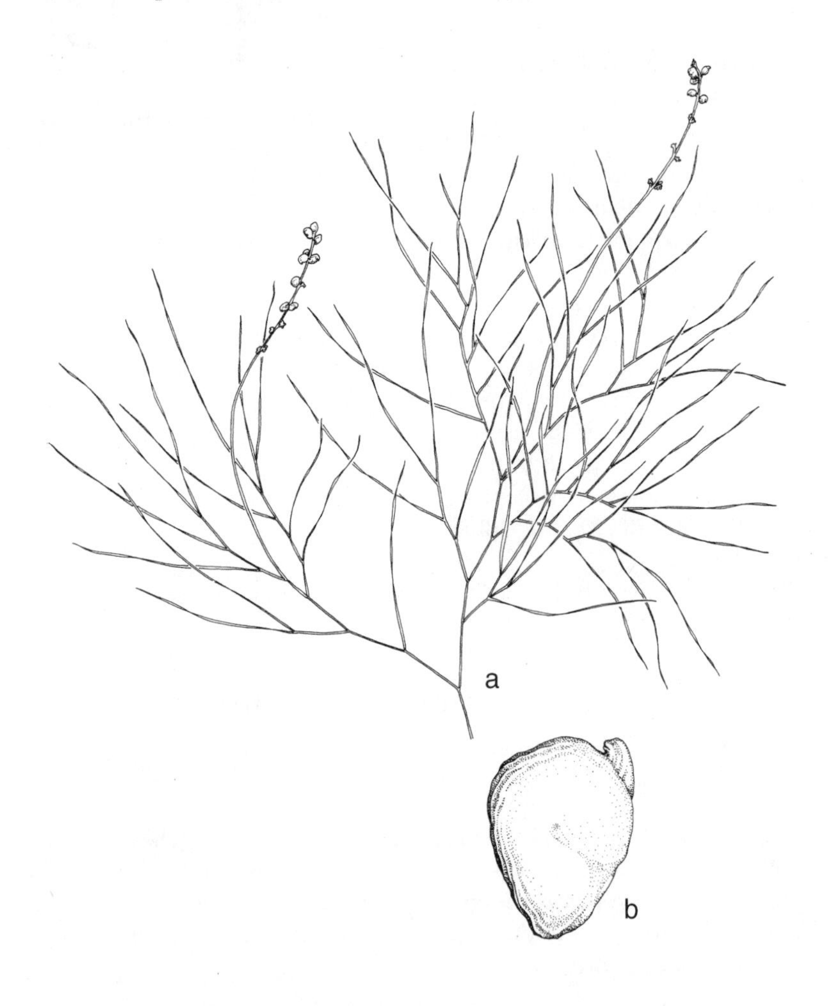

208. ***Stuckenia pectinata* (Sago pondweed).** a. Habit. b. Fruit.

209. *Ruppia cirrhosa* (Ditch grass). a. Habit. b. Fruit.

30. SCHEUCHZERIACEAE—ARROW-GRASS FAMILY

Only the following genus is in this family.

1. **Scheuchzeria** L.—Arrow-grass

Plants with creeping rhizomes; leaves terete, linear, most of them basal; inflorescence loosely racemose, bracteate; flowers perfect; sepals 3, greenish, deciduous; petals 3, greenish, persistent; stamens 6; carpels free except at base, developing into 3 follicles; ovules 2.

This genus is sometimes placed in the Juncaginaceae, but it differs from that family by its loosely racemose inflorescence that is subtended by bracts and by its basal and cauline leaves.

The following species, found in North America, Europe, and Asia, is the only one in the genus.

1. **Scheuchzeria palustris L. var. americana** Fern. Rhodora 25:178. 1923. Fig. 210.
Scheuchzeria americana (Fern.) G. N. Jones, Fl. Ill. 44. 1945.

Perennial from a creeping, jointed rootstock; leaves terete, linear, to 30 cm long, 1–3 mm broad, open at the tip, glabrous, the basal ones clustered, the cauline ones alternate, 1–3 in number, scattered; sheaths at base of leaf enlarged, partly sheathing the stem; raceme bracteate, the lowest bract leaflike, the upper sheathlike; sepals ovate-lanceolate, acute, 2–3 mm long; petals ovate-lanceolate, acute, 2–3 mm long; fruits 6–10 mm long, with a curved beak; seeds narrowly ellipsoid, 4–5 mm long, black. June–September.

Sphagnum bogs, marshes.

IA, IL, IN, OH (OBL).

Arrow-grass.

Our plants belong to var. *americana.* Var. *palustris,* which is Eurasian, has slightly smaller, nearly beakless follicles and slightly smaller seeds.

31. SPARGANIACEAE—BUR-REED FAMILY

Only the following genus comprises the family.

1. **Sparganium** L.—Bur-reed

Perennial aquatics from stout rhizomes; leaves alternate, elongate, sheathing at the base, often keeled on the back; inflorescence globose, axillary; flowers unisexual; perianth parts 3 or 6; stamens 5, with slender, free filaments; ovary 1, superior, 1- to 2-celled, each cell with 1 ovule; fruit a beaked achene subtended by the persistent perianth.

Fourteen species comprise the genus, five of them in the central Midwest.

1. Plants floating or suberect, to 30 cm long; staminate head 1; pistillate heads about 1 cm in diameter, the lowest short-pedunculate; beak of achene 0.5–1.5 mm long .... 5. *S. natans*
1. Plants erect, (5–) 30–120 cm tall; staminate heads 3–25; pistillate heads 1.5–3.5 cm in diameter, all sessile; beak of achene 2–6 mm long.
    2. Central axis bearing 1–4 pistillate heads; achene stipitate, fusiform, tapering to the summit; stigma 1.

3. At least one of the pistillate heads borne above the subtending bract (supra-axillary); inflorescence simple; achene usually greenish brown, even at maturity .... 3. *S. emersum*
3. All heads axillary in the subtending bract; inflorescence usually with 1 or more branches; achene pale or dark brown.
   4. Leaves stiff, strongly keeled; branches of inflorescence without any pistillate heads; pistillate heads 2.5–3.5 cm in diameter; body of achene 5–7 mm long, shiny, pale brown .... 2. *S. androcladum*
   4. Leaves soft, usually not strongly keeled; branches of inflorescence with 1–3 pistillate heads; pistillate heads 1.5–2.5 cm in diameter; body of achene 3–5 mm long, dull, dark brown .... 1. *S. americanum*

2. Central axis bearing only staminate heads; achene not stipitate, obpyramidal, truncate at the summit; stigmas 2 .... 4. *S. eurycarpum*

1. **Sparganium americanum** Nutt. Gen. Pl. 2:203. 1818. Fig. 211.

Stems erect, to nearly 1 m tall; leaves to 80 cm long, 4–15 mm broad, thin, usually scarcely keeled; bracts similar to the leaves, smaller; inflorescence usually branched, occasionally simple, the central axis bearing 1–4 pistillate heads and 3–10 staminate heads, the branches bearing 1–3 pistillate heads and 1–6 staminate heads; pistillate heads 1.5–2.5 cm in diameter, sessile; sepals spatulate, about two-thirds as long as the achene; stigma 1; achene fusiform, tapering to the apex, dull, dark brown, the body 3–5 mm long, the beak 2–5 mm long, the stipe 2–3 mm long. June–July.

Shallow water.

IA, IL, IN, KS, KY, MO, OH (OBL).

American bur-reed.

This species has very obviously thin leaves when compared with the leaves of the similar *S. androcladum* and *S. eurycarpum*. It differs further from *S. eurycarpum* by its single stigma and by the achene that tapers to the summit.

2. **Sparganium androcladum** (Engelm.) Morong, Bull. Torrey Club 15:78. 1888. Fig. 212.
Sparganium simplex Huds. var. *androcladum* Engelm. Gray. Man. Bot 481. 1867.
Sparganium americanum Nutt. var. *androcladum* (Engelm.) Fern. & Eames, Rhodora 9:87. 1907.
Sparganium lucidum Fern. & Eames, Rhodora 9:87. 1907.

Stems erect, to 1 m tall; leaves to 75 cm long, 5–15 mm broad, stiff, keeled; bracts similar to the leaves, smaller; inflorescence usually branched, occasionally simple, the central axis bearing 1–4 pistillate heads and 5–10 staminate heads, the branches without pistillate heads and with 3–6 staminate heads; pistillate heads 2.5–3.5 cm in diameter, sessile; sepals spatulate, about two-thirds as long as the achene; stigma 1; achene ellipsoid-fusiform, tapering to the apex, shiny, pale brown, the body 5–7 mm long, the beak 4–6 mm long, the stipe 2–4 mm long. June–July.

Shallow water

IA, IL, IN, KY, MO, OH (OBL).

Bur-reed.

The achene that tapers to the summit and the presence of but a single stigma separate this species from the similar appearing *S. eurycarpum*. From *S. ameri-*

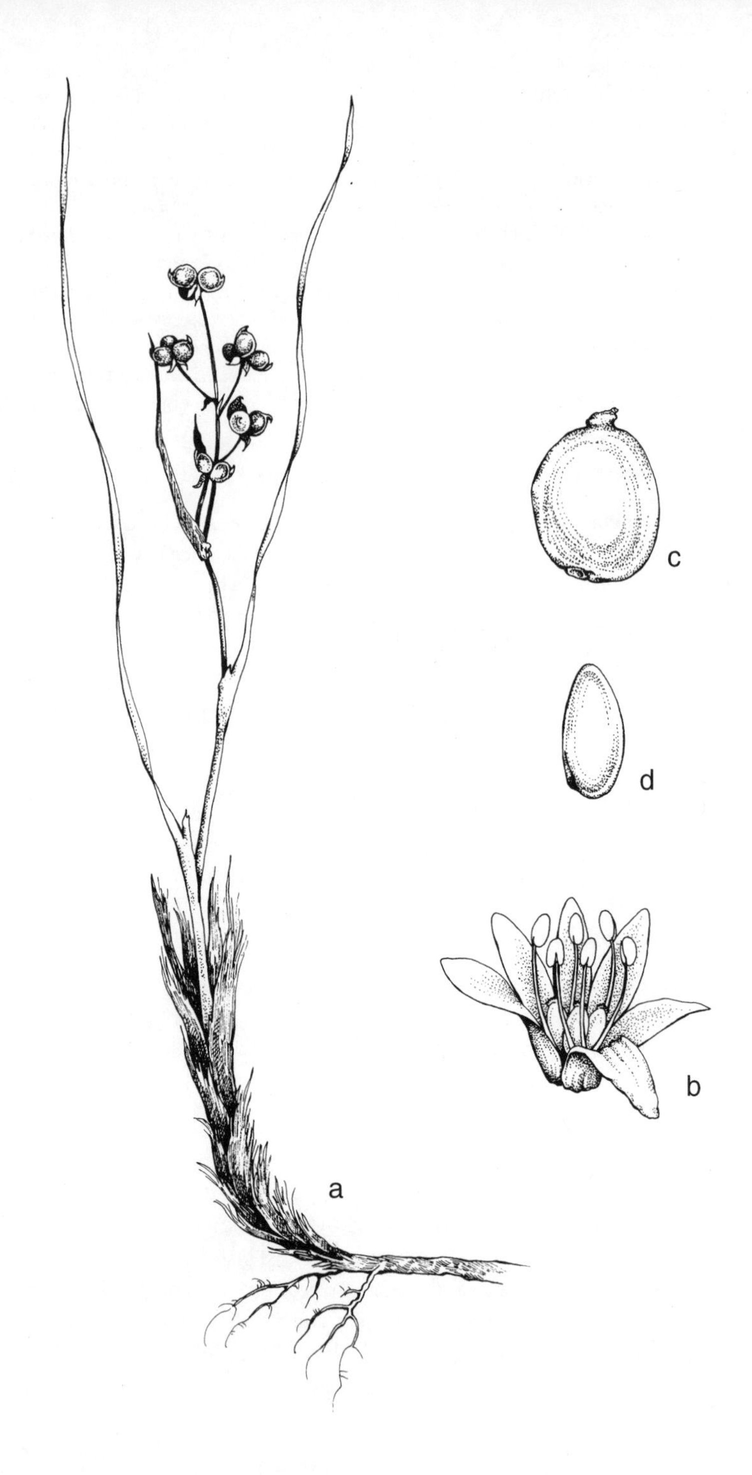

210. *Scheuchzeria palustris* (Arrow-grass).

a. Habit.
b. Flower.
c. Follicle.
d. Seed.

211. *Sparganium americanum* (American bur-reed). a. Habit. b. Achene.

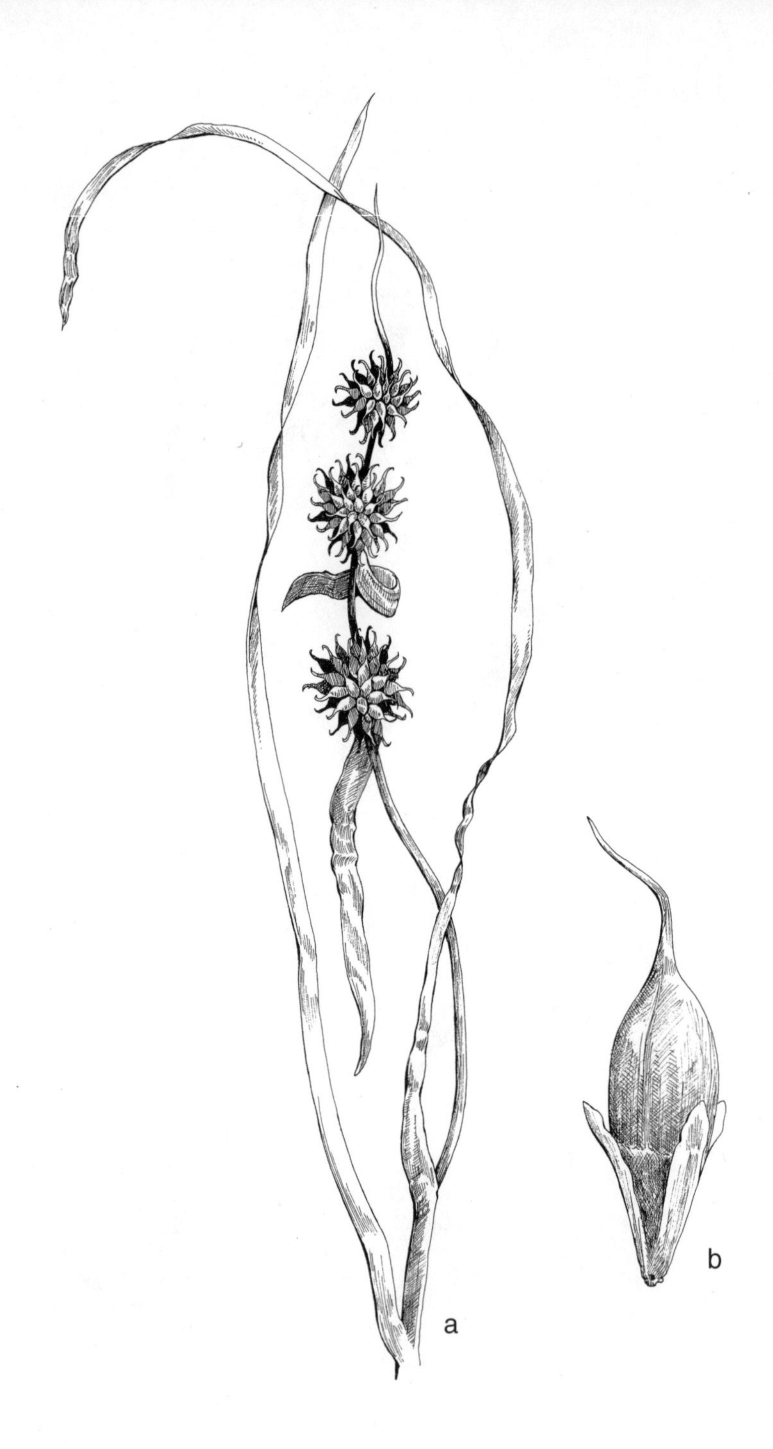

212. *Sparganium androcladum* (Bur-reed). a. Habit. b. Achene.

canum, this species differs by its larger, shiny, pale brown achenes and its stiffer, keeled leaves.

3. **Sparganium emersum** Rehm. Verh. Naturf. Vereins Brunn 10:80. 1872. Fig. 213.
Sparganium simplex Huds. var. *acaule* Beeby ex Macoun, Cat. Canad. Pl. 5:367. 1909.
Sparganium acaule (Beeby) Rydb. N. Am. Fl. 17:8. 1909.
Sparganium chlorocarpum Rydb. N. Am. Fl. 17:8. 1909.
Sparganium chlorocarpum Rydb. var. *acaule* (Beeby) Fern. Rhodora 24:29. 1922.

Stems erect, to 60 cm tall; leaves to 80 cm long, 2–10 mm wide, thin, slightly keeled; bracts similar to the leaves, smaller, ascending to erect; inflorescence simple, bearing 1–4 pistillate heads and 2–9 staminate heads; pistillate heads sessile, 1.5–2.5 cm in diameter; sepals spatulate, about two-thirds as long as the achene; stigma 1; achene fusiform, tapering to the apex, shiny, greenish to brownish, the body 4–6 mm long, the beak 2.0–4.5 mm long, the stipe 2.5–3.5 mm long. June–August.

Shallow water.

IA, IL, IN, OH (OBL).

Green-fruited bur-reed.

For many years, this species was known as *S. chlorocarpum.*

Achenes in *S. emersum* generally remain greenish, even at maturity. In addition, one or more of the pistillate heads is supra-axillary, that is, borne above the subtending bracts.

4. **Sparganium eurycarpum** Engelm. Gray, Man. Bot. 430. 1856. Fig. 214.

Stems erect, usually at least 1 m tall; leaves to 75 cm long, 6–12 mm broad, stiff, keeled; bracts similar to the leaves, shorter; inflorescence branched, the central axis bearing only staminate heads, the branches bearing 1–3 pistillate heads and 0–20 staminate heads; pistillate heads 2.0–3.5 cm in diameter, sessile; sepals spatulate, frequently falling early; stigmas 2; achene obpyramidal, truncate at the summit, rather dull, brown, the body 6–9 mm long, with a cleft beak 2.5–3.5 mm long, without a stipe. June–August.

Shallow water.

IA, IL, IN, KS, KY, MO, NE, OH (OBL).

Bur-reed.

The two styles and the cleft beaks of the achene readily distinguish this species.

5. **Sparganium natans** L. Sp. Pl. 1:971. 1753. Fig. 215.
Sparganium minimum Fries, Summa Veg. Scand. 2:560. 1849.

Stems floating or suberect, to 30 cm long; leaves to 50 cm long, 2–7 mm wide, thin, without a keel; bracts similar to the leaves, smaller; inflorescence simple, bearing 1–3 pistillate heads and 1 staminate head, the lowest pistillate head short-pedunculate; pistillate heads about 1 cm in diameter; sepals spatulate, a little more than half as long as the achene; stigma 1; achene elliptic-ovoid or fusiform, abruptly narrowed at the apex, dull, greenish or brownish, the body 3–4 mm long, the beak 0.5–1.5 mm long, the stipe 1–2 mm long. June–August.

Shallow water.

IL, IN (OBL).

213. *Sparganium emersum* (Green-fruited bur-reed). a. Habit. b. Achene.

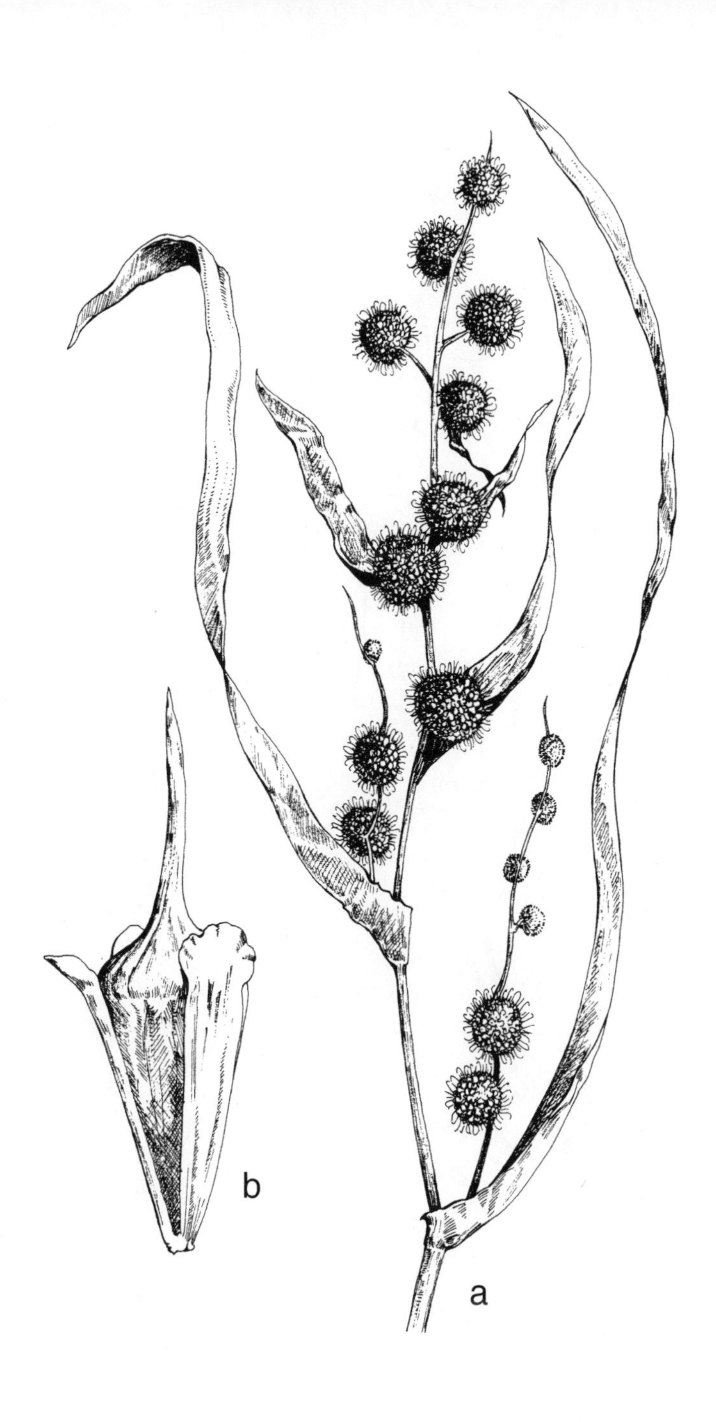

214. *Sparganium eurycarpum* (Bur-reed). a. Habit. b. Achene.

215. *Sparganium natans* **(Least bur-reed).** a. Habit. b. Achene.

Least bur-reed.

This is the only species of *Sparganium* with a floating habit and with a single staminate head. The pistillate heads are about 1 cm in diameter, smaller than in the other species.

32. TYPHACEAE—CAT-TAIL FAMILY

Only the following genus comprises this family.

1. **Typha** L.—Cat-tail

Perennials from stout rhizomes; leaves cauline, elongated, sheathing at the base; inflorescence terminal, elongated; flowers unisexual; perianth none; stamens 1–7, usually 3, the filaments connate; ovary 1, superior, 1-celled, stipitate, with 1 ovule; achene stipitate, the style persistent.

There are about ten species in the genus. Three species occur in the central Midwest, and each of them apparently hybridizes with each other. The key below, which includes the hybrids, is adapted from the key written by Galen Smith for Flora North America.

1. Pistillate bracteoles either absent or narrower than the stigmas, not evident at the spike surface; pistillate spikes green in flower when fresh, 19–36 mm thick in fruit.
    2. Pistillate bracteoles absent; pistillate spikes contiguous, with staminate spikes, in fruit 24–36 mm thick; seeds numerous; pollen in tetrads .... 3. *T. latifolia*
    2. Pistillate bracteoles present; pistillate spikes usually separated from staminate spikes by a gap, in fruit 19–25 mm thick; seeds few or none; pollen a mixture of 1–4 grains.
        3. Mucilage glands absent from blade; pistillate spikes after flowering medium to dark brown .... *T. angustifolia* X *T. latifolia*
        3. Mucilage glands usually present on adaxial surface of blade near the sheath; pistillate spikes after flowering bright orange .... *T. domingensis* X *T. latifolia*
1. Pistillate bracteoles as wide as or wider than the stigmas, evident at the spike surface; pistillate spikes brown at all stages, or white when fresh, 13–25 mm thick in fruit.
    4. Mucilage glands absent from surface of blade; pistillate bracteoles darker than stigmas; pistillate spikes medium to dark brown .... 1. *T. angustifolia*
    4. Mucilage glands present on adaxial surface of all of sheath and usually part of adjacent blade; pistillate bracteoles paler or same color as stigmas; pistillate spikes bright cinnamon- to orange-brown to medium brown.
        5. Pistillate bracteoles paler than the stigmas; pistillate spikes usually bright cinnamon- to orange-brown; mucilage glands numerous on leaf blade .... 2. *T. domingensis*
        5. Pistillate bracteoles the same color as the stigmas; pistillate spikes medium brown; mucilage glands few or absent from leaf blade .... *T. angustifolia* X *T. domingensis*

1. **Typha angustifolia** L. Sp. Pl. 1:971. 1753. Fig. 216.

Coarse perennial from stout, creeping rhizomes; stems to 1.5 m tall, with several cauline leaves; leaves elongated, somewhat convex, 4–12 mm wide, with mucilage glands at the transition between blade and sheath but absent from the blade; staminate spikes separated from pistillate spikes by 1–12 cm, to 17 cm long, to 1 cm thick at anthesis, the scales stramineous to medium brown, filiform, with pollen in irregular clusters or borne singly; pistillate spikes medium to dark brown when fresh, to 20 cm long, to 6 mm wide at anthesis, and to 2.2 cm wide in fruit, with

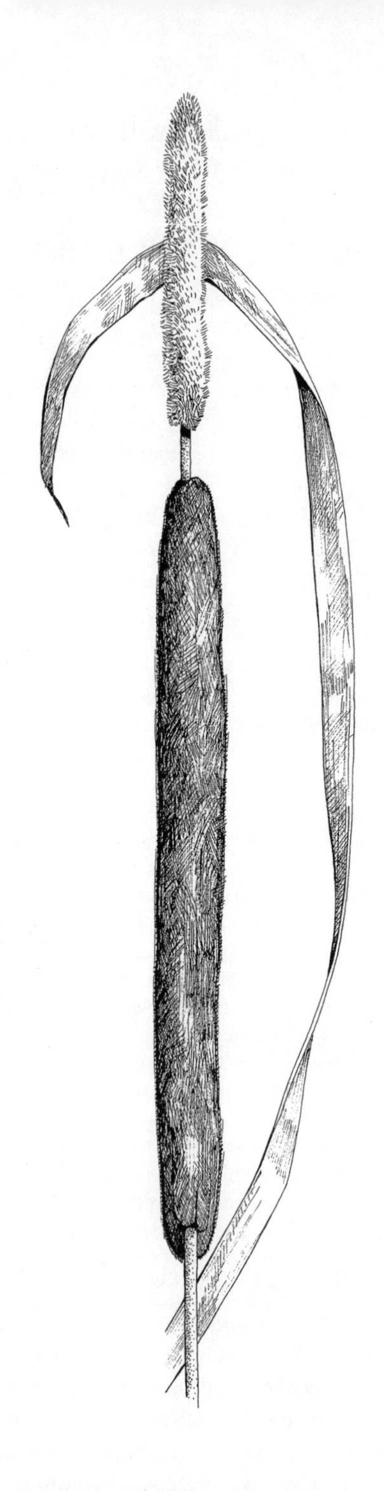

216. ***Typha angustifolia*** **(Narrow-leaved cat-tail).** Inflorescence.

bracteoles darker than the stigmas; achenes about 1 mm long, surrounded by numerous white hairs to 8 mm long. June–September.

Marshes, wet ditches.

IA, IL, IN, KS, KY, MO, NE, OH (OBL).

Narrow-leaved cat-tail.

True specimens of *T. angustifolia* will have leaves not more than 8 mm wide and pistillate spikes less than 2.2 cm broad. This species hybridizes with *T. latifolia* throughout the ranges of these two species, forming the hybrid known as *T.* X *glauca.* The hybrid usually can be identified by the characters given in the key above. *Typha angustifolia* also hybridizes with *T. domingensis.* These latter hybrids are known from Kansas, Missouri, and Nebraska in the central Midwest.

Some botanists believe that *T. angustifolia* is not native to the United States.

2. **Typha domingensis** Persoon, Syn. Pl. 2:532. 1807. Fig. 217.

Coarse perennial from stout, creeping rhizomes; stems to 4 m tall, with numerous cauline leaves; leaves elongated, more or less convex, to 18 mm wide, with mucilage glands at the sheath-blade transition and along the entire sheath and part of the blade; staminate spikes separated from the pistillate spikes by up to 8 cm, to 35 cm long, to 1 cm thick at anthesis, the scales stramineous to bright orange-brown, linear, with pollen grains borne singly; pistillate spikes bright cinnamon-brown when fresh, becoming bright orange-brown, to 35 cm long, to 6 mm wide at anthesis, to 25 mm wide in fruit, the bracteoles paler than or about the same color as the stigmas; achenes about 1 mm long, surrounded by numerous whitish hairs. June–September.

Marshes, wet ditches.

IL, KS, KY, MO, NE (OBL).

Southern cat-tail.

This species seems to have the broad leaves of *T. latifolia* but the narrower pistillate spikes of *T. angustifolia.* Its height is usually considerably greater than in the other two species of *Typha.*

Typha domingensis hybridizes with *T. latifolia* to form *T.* X *provincialis* A. Camus, and with *T. angustifolia.* The distinguishing characteristics of the hybrids may be found in the key above. The hybrid with *T. latifolia* is known from Missouri and Nebraska.

3. **Typha latifolia** L. Sp. Pl. 1:971. 1753. Fig. 218.

Coarse perennial from stout, creeping rhizomes; stems to 3 m tall, with numerous cauline leaves; leaves elongated, flat, to nearly 3 cm broad, often glaucous, with obscure mucilage glands, or the glands absent; staminate spikes usually contiguous with the pistillate spikes, to 12 cm long, to 2 cm thick at anthesis, the scales colorless or stramineous, filiform; pistillate spikes pale green at anthesis, becoming brown, to 25 cm long, to 8 mm wide at anthesis, 24–36 mm wide in fruit, the bracteoles absent; achenes about 1 mm long, surrounded by numerous white hairs up to 1 cm long. June–September.

Marshes, wet ditches.

217. *Typha domingensis* (Southern cat-tail). a. Habit. b. Inflorescence.

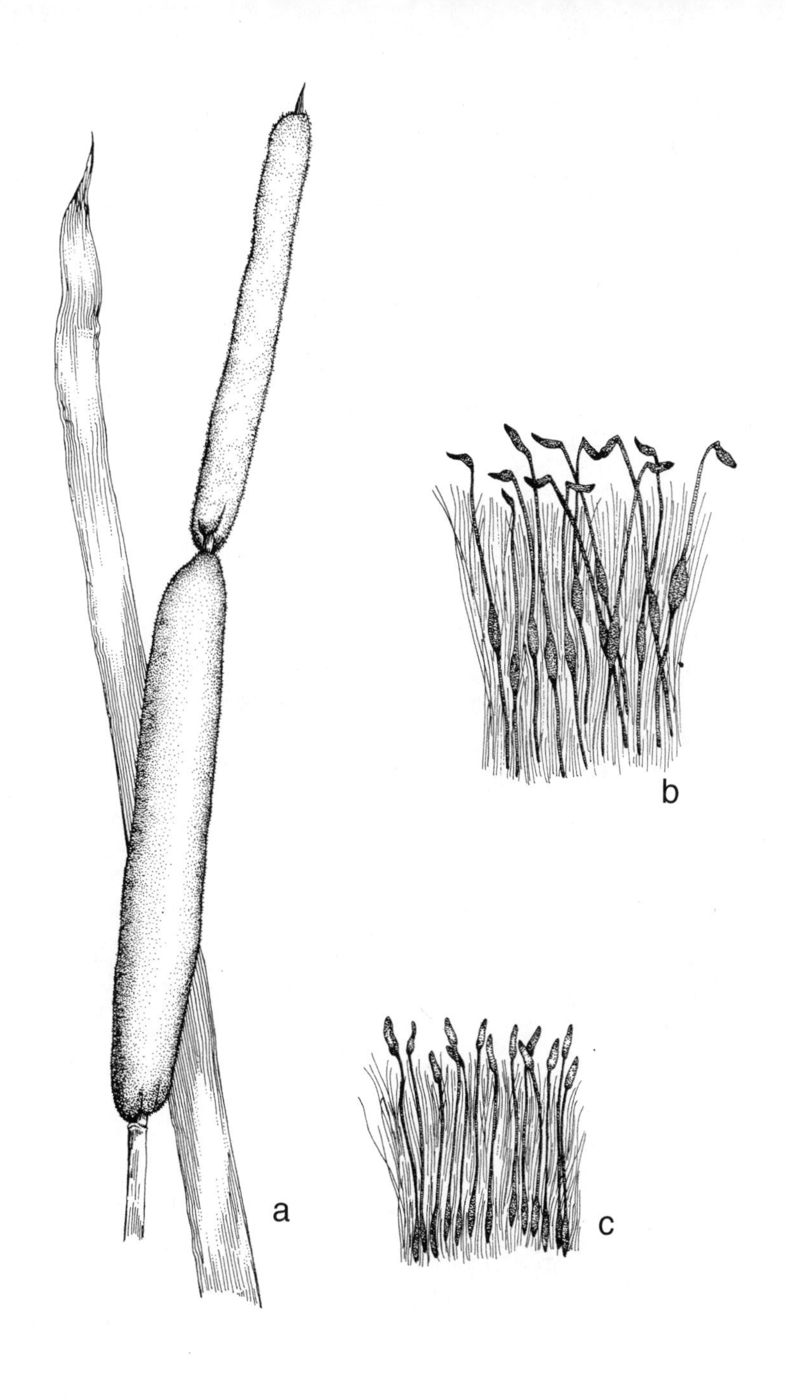

218. *Typha latifolia* (Common cat-tail).

a. Leaf and inflorescence.
b. Pistillate flowers.
c. Staminate flowers.

IA, IL, IN, KS, KY, MO, NE, OH (OBL).

Common cat-tail; broad-leaved cat-tail.

This common species usually has leaves more than 15 mm broad and fruiting spikes at least 24 mm broad. It hybridizes with *T. angustifolia* to form the rather common *T.* X *glauca*, and with *T. domingensis* to form the rarer *T.* X *provincialis*.

33. XYRIDACEAE—YELLOW-EYED GRASS FAMILY

Perennial herbs with narrow, erect, basal leaves; flowers perfect, crowded in a dense head upon an elongated scape; calyx bilaterally symmetrical, distinguishable in color from the radially symmetrical corolla; fertile stamens 3, attached to the base of the corolla; staminodia present; ovary 1-celled, superior.

Five genera comprise this family, with about three hundred species, most of them tropical or subtropical.

Only the following genus occurs in the central Midwest.

1. **Xyris** L.—Yellow-eyed Grass

Leaves basal, elongated, narrow and grasslike; flowering head one on an unbranched scape; lateral sepals keeled, firm, the anterior sepal not keeled, membranous; corolla yellow; staminodia cleft at apex; capsule ellipsoid, many-seeded.

Several species of *Xyris* occur in the southeastern United States, but only two are known from the central Midwest, both of them occasionally occurring in standing water.

1. Leaf blades twisted, to 5 mm wide; plants bulbous at base; lateral sepals strongly curved, 4.5–5.5 mm long .... 2. *X. torta*
1. Leaf blades not twisted, to 15 mm wide; plants not bulbous at base; lateral sepals only slightly curved, 5–7 mm long .... 1. *X. difforme*

1. **Xyris difforme** Chapman, Fl. S. U.S. 500. 1860. Fig. 219.

Plants not bulbous at base; leaves to 50 cm long, to 7 mm broad, deep green, rather stiff, not twisted; scape not twisted, usually bicostate, to 70 cm tall; head ovoid, 5–20 mm long, acute at the apex; bracts brown, obovate to nearly orbicular; lateral sepals 5–7 mm long, the keel scarious; seeds 0.5 mm long. June–July.

Swamps, bogs, seeps.

IN, KY, OH (OBL).

Yellow-eyed grass.

This species is readily distinguished from the other *Xyris* in the central Midwest by its nontwisted leaves and nontwisted scapes.

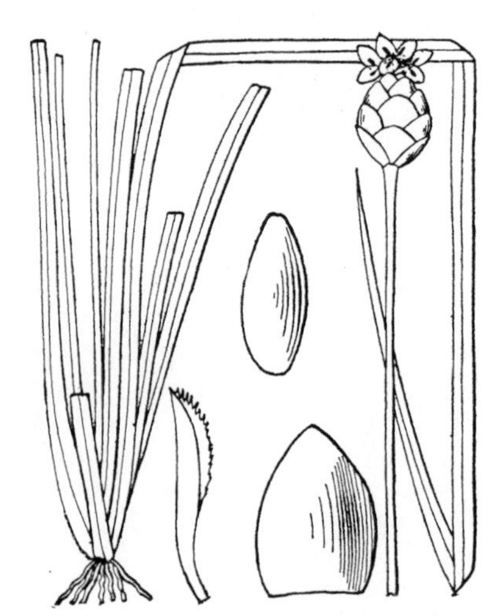

219. *Xyris difforme* (Yellow-eyed grass). Habit. Scales (center).

2. **Xyris torta** Sm. Rees, Cyclop. 39. 1818. Fig. 220.

Plants bulbous, the bulb up to 1 cm long and thick; leaves 10–30 cm long, to 5 mm broad, gray or blue-green, stiff, twisted; scape twisted, bicostate, to 80 cm tall; head globose to ovoid to ellipsoid, to 25 mm long, acute or obtuse at the apex; bracts pale brown, obovate to orbicular, with a gray-green central area; lateral sepals 4.5–5.5 mm long, the keel ciliate; seeds 0.5 mm long. June–August.

Bogs, wet sandy areas.

IA, IL, IN, KY, MO, OH (OBL).

Twisted yellow-eyed grass.

The twisted leaves and twisted scapes are the distinguishing characteristics for this species.

34. ZANNICHELLIACEAE—HORNED PONDWEED FAMILY

Aquatic herbs; leaves alternate or opposite, submersed, sessile, with linear blades; inflorescence axillary, cymose, bearing unisexual flowers, without bracts; perianth absent; stamen 1; pistils (1–) 2–8, free from each other; fruit drupaceous.

Four genera and about twelve species comprise this family that is known from most parts of the world.

1. **Zannichellia** L.—Horned Pondweed

Monoecious submersed perennial with 1 or 2 roots; leaves linear, not dilated at base, entire; flowers unisexual, reduced, axillary; perianth absent; stamen 1; pistils 2–4, free from each other, 1-celled, each cell with a single ovule; fruits in fascicles of 4, axillary, with a persistent beaklike style.

Five species make up this genus, with only one known from our area.

1. **Zannichellia palustris** L. Sp. Pl. 1:969. 1753. Fig. 221.
Zannichellia intermedia Torr. Beck, Bot. N. & M. St. 385. 1833.

Slender monoecious perennials, submersed, with 1–2 roots; stems threadlike, sparsely branched, light green; leaves opposite, linear, entire, acute, dark green, 3–7 cm long, 0.4–0.7 mm wide, the base with clasping stipules not dilated; unisexual flowers adjacent in same axil; perianth none; staminate flowers with a single stamen, the anther borne on a slender filament; pistillate flowers with 2–4 (usually 4) separate carpels, each with a single ovule, slender style, peltate stigma; fruits usually in fascicles of 4, axillary, short-pedicellate, flattish, falcate, obliquely oblongoid, dentate on convex surface, brownish, the body 2.0–2.5 mm long, with a persistent, beaklike style 0.8–1.2 mm long. June–September.

Ponds, streams, wet ditches.

IA, IL, IN, KS, KY, MO, NE, OH (OBL).

Horned pondweed.

Although *Zannichellia palustris* resembles some other submersed aquatics, the beaked fruits borne in fascicles of four are distinctive. Vegetatively, *Zannichellia* may be distinguished from *Najas* by its clasping stipules.

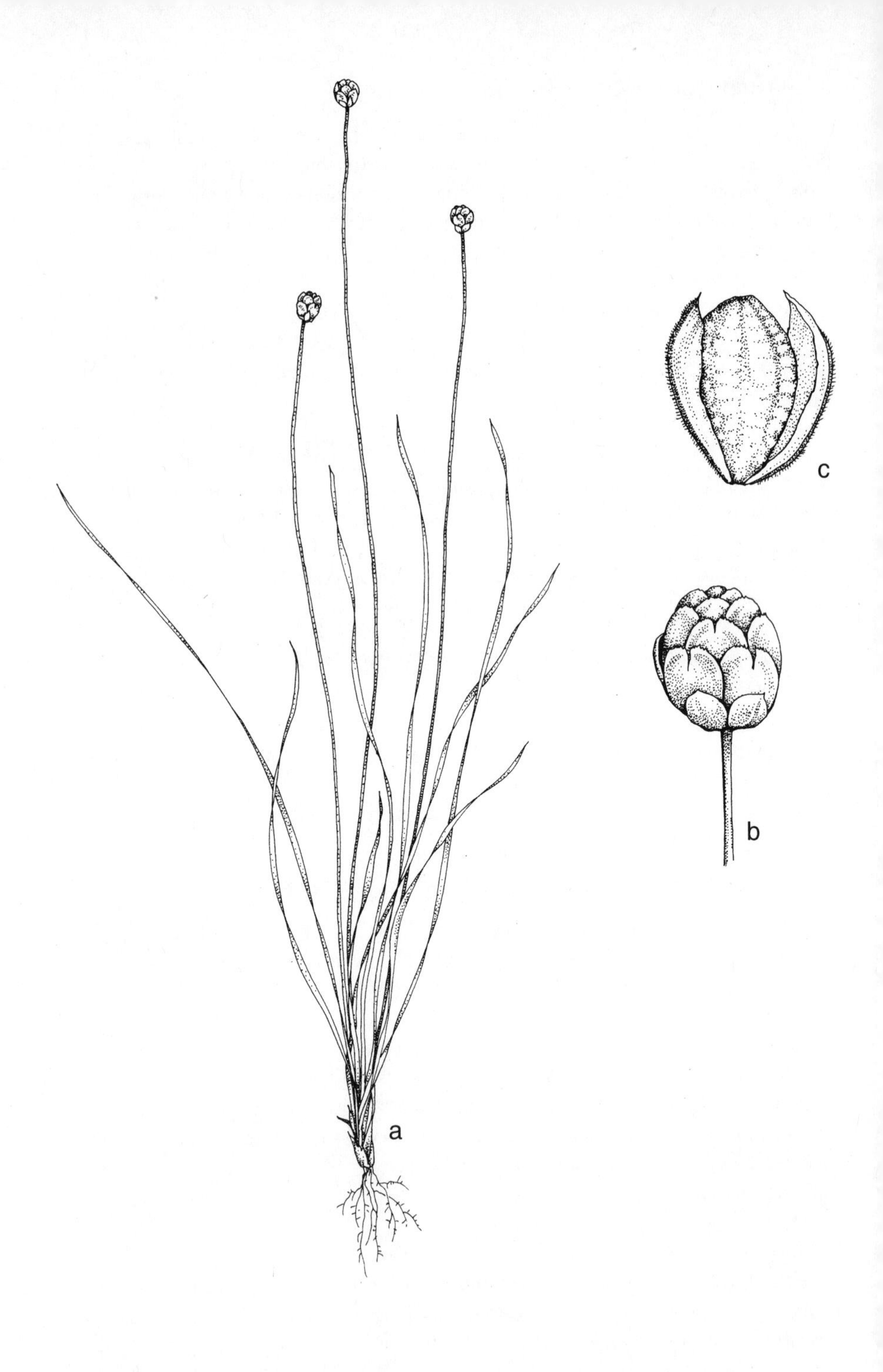

220. *Xyris torta*
(Twisted yellow-eyed grass).

a. Habit.
b. Flowering head.
c. Fruit with bracts.

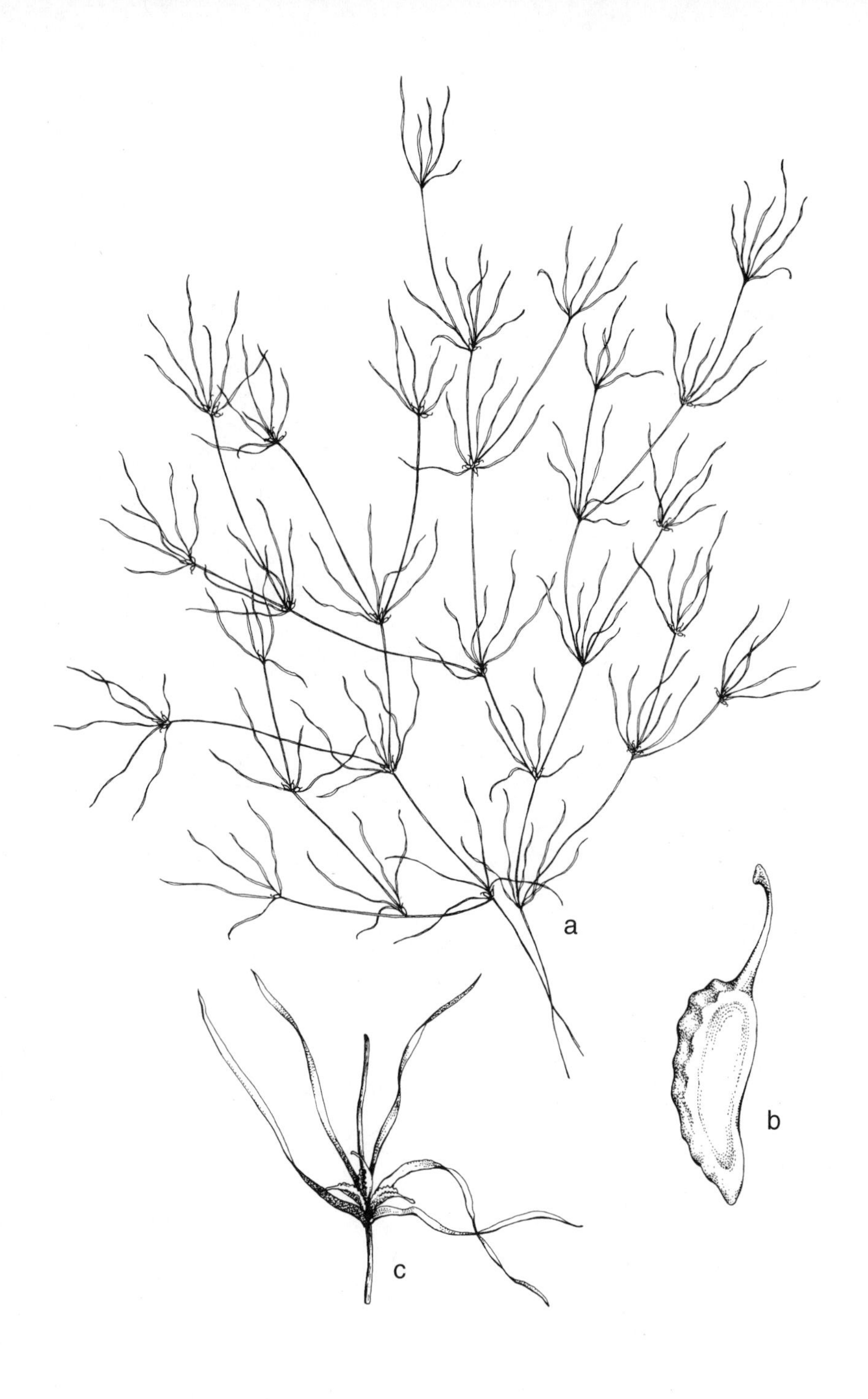

221. *Zannichellia palustris* (Horned pondweed).
a. Habit.
b. Nutlet.
c. Cluster of fruits.

Glossary
Illustration Credits
Index to Genera and Species
Index to Common Names

Glossary

acaulescent. Seemingly without aerial stems.
achene. A type of one-seeded, dry, indehiscent fruit with the seed coat not attached to the mature ovary wall.
acicular. Needlelike.
actinomorphic. Having radial symmetry; regular, in reference to in a flower.
acuminate. Gradually tapering to a point.
acute. Sharp, ending in a point.
adnate. Fusion of dissimilar parts.
alternate. Referring to the condition of structures arising singly along an axis; opposed to opposite.
anastomosing. Forming a network of cross-veins.
androphore. A stalk beneath a staminate flower.
annual. Living for a single year.
anther. The terminal part of a stamen which bears pollen.
anthesis. Flowering time.
antrorse. Projecting forward.
apical. Relating to the apex or tip.
apically. At the apex.
apiculate. Abruptly short-pointed at the tip.
apiculus. A short-pointed tip.
appressed. Lying flat against the surface.
arching. Moderately curving.
areola (pl. ***areolae***). A small area between leaf veins.
areole. A small area formed by the interlocking of veins.
arillate. A small, spongy protuberance of some seeds.
aristate. Bearing an awn.
aristulate. Short-awned.
asymmetrical. Of different shape on the two sides.
attenuate. Gradually becoming narrowed.
auricle. An earlike process.
auriculate. Bearing an earlike process.
awn. A bristle usually terminating a structure.
axil. The angle between the base of a structure and the axis from which it arises.
axillary. Borne from an axil.
axis. The central support to which lateral parts are attached.

basal. Confined to the lowest part.
beak. A terminal projection.
berry. A type of fruit where the seeds are surrounded only by fleshy material.
bicostate. Having two veins.
bidentate. Having two teeth.
bifid. Two-cleft.
bilobed. Bearing two lobes.
bipinnate. Twice divided.
biseriate. In two rows or series.
bisexual. Referring to a flower which contains both stamens and pistils.
bivalved. With two valves, or coverings.
blade. The green, flat, expanded part of the leaf.
bract. An accessory structure at the base of many flowers, usually appearing leaflike.
bracteole. A secondary bract.
bristle. A stiff hair or hairlike growth; a seta.
bulb. An underground, vertical stem with both scaly and fleshy leaves.
bulblet. A small bulb.
bulbous. Bearing a swollen base.

calcareous. Referring to an alkaline condition.
callosity. Any hardened thickening.
callus. A hard swollen area at the outside base of a lemma or palea.
calyx. The outermost ring of structures of a flower, composed of sepals.
campanulate. Bell-shaped.
canescent. Grayish-hairy.
capillary. Threadlike.
capitate. Forming a head.
capsule. A dry, dehiscent fruit composed of more than one carpel.
carinate. Bearing a keel.
carpel. A simple pistil, or one member of a compound pistil.
cartilaginous. Firm but flexible.

carunculate. Referring to a spongy outgrowth of some seeds.
caudate. With a tail-like appendage.
caudex (pl. **caudices**). The woody base of a perennial plant.
caulescent. Having an aerial stem.
cauline. Belonging to a stem.
cavernous. Hollowed out.
cespitose. Growing in tufts.
chaffy. Covered with scales.
chartaceous. Papery.
cilia. Marginal hairs.
ciliate. Bearing cilia.
ciliolate. Bearing small cilia.
clasping. Referring to a leaf whose base encircles the stem.
claw. A narrow, basal stalk, particularly of a petal.
compressed. Flattened.
compound. Said of a structure which is divided into distinct units.
concave. Curved on the inner surface; opposed to convex.
conduplicate. Folded together lengthwise.
connate. Union of like parts.
connective. That portion of the stamen between the two anther halves.
connivent. Coming in contact; converging.
contiguous. Adjoining.
convex. Rounded on the outer surface; opposite of concave.
convolute. Rolled length wise.
cordate. Heart-shaped.
coriaceous. Leathery.
corm. An underground, vertical stem with scaly leaves, differing from a bulb by lacking fleshy leaves.
corolla. The ring of structures of a flower just within the calyx, composed of petals.
corymb. A type of inflorescence where the pedicellate flowers are arranged along an elongated axis but with the flowers all attaining about the same height.
costate. Veiny.
creeping. Spreading on the surface of the ground.
crenate. With round teeth.
crenulate. With small, round teeth.
crested. Bearing a ridge.
crispate. Referring to a leaf whose margins are wavy and crinkled.
crisped. Curled.
cross-striae. Markings perpendicular to the longitudinal axis.
cucullate. Hood-shaped.
culm. The stem which terminates in an inflorescence.
cuneate. Wedge-shaped or tapering at the base.
cuspidate. Terminating in a very short point.
cyme. A type of broad and flattened inflorescence in which the central flowers bloom first.
cymose. Bearing a cyme.

deciduous. Falling away.
decumbent. Lying flat, but with the tip ascending.
deflexed. Turned downward.
dehiscent. Splitting at maturity.
deltoid. Triangular.
dentate. With sharp teeth, the tips of which project outward.
denticulate. With small, sharp teeth, the tips of which project outward.
depauperate. Poorly developed.
dichotomous. Forked; two-branched.
diffuse. Loosely spreading.
digitate. With fingerlike projections.
dilated. Swollen; expanded.
dimorphic. Having two forms.
dioecious. With staminate flowers on one plant, pistillate flowers on another.
distended. Swollen.
distichous. Bearing two equal branches.
divaricate. Spreading.
divergent. Spreading apart.
dorsal. That surface turned away from the axis; abaxial.
drupaceous. Drupelike.
drupe. A type of fruit in which the seed is surrounded by a hard, dry covering which, in turn, is surrounded by fleshy material.

echinate. Having prickles.
eglandular. Without glands.
ellipsoid. Referring to a solid object which is broadest at the middle, gradually tapering to both ends.
elliptic. Broadest at middle, gradually tapering equally to both ends.
elliptic-ovate. Broadest midway between middle and base, tapering gradually to apex.
emarginate. Deeply notched at the tip.

emersed. Rising above the surface of the water.
endosperm. Food-storage tissue found outside the embryo.
ensiform. Sword-shaped.
entire. Without lobes or teeth, in reference to the margins of structures.
ephemeral. Lasting only a short time.
epicarp. The outermost layer of the fruit.
epunctate. Without dots.
erose. With an irregularly notched margin.
excentric. Off-center.

facial. Referring to the front surface.
falcate. Sickle-shaped.
fascicle. A cluster; a bundle.
fertile. Bearing reproductive parts.
fibrous. Referring to roots borne in tufts.
filament. That part of the stamen supporting the anther.
filiform. Threadlike.
flabellate. Fan-shaped.
flaccid. Weak; flabby.
flexible. Able to be bent readily.
flexuous. Zigzag.
foliaceous. Leaflike.
follicle. A type of dry, dehiscent fruit which splits along one side at maturity.
friable. Breaking easily into small particles.
frond. The vegetative structure in the Lemnaceae; the leaf of a fern.
funnelform. Shaped like a funnel.
fusiform. Spindle-shaped; tapering at both ends.

galea. A hooded portion of the perianth.
gemma. (pl. **gemmae**). An asexual bud.
geniculate. Bent.
glabrate. Becoming smooth.
glabrous. Without pubescence or hairs.
gland. An enlarged, spherical body functioning as a secretory organ.
glandular. Bearing glands.
glaucescent. Becoming covered with a whitish bloom which can be rubbed off.
glaucous. With a whitish covering which can be rubbed off.
glochidium. A process bearing barbs.
globose. Round; globular.
globular. Roundish.
glomerulate. Forming small heads.
glumaceous. Resembling a scale.
glume. The lowest sterile scales of a grass spikelet.
glutinous. Covered with a sticky secretion.
gynophore. A stalk beneath a pistillate flower.

head. A type of inflorescence in which several sessile flowers are clustered together at the tip of peduncle.
hemispherical. Half-spherical.
herbaceous. Not woody; dying back all the way to the ground in winter.
hirsute. With stiff hairs.
hirtellous. With minute stiff hairs.
hispid. With rigid hairs.
hispidulous. With minute rigid hairs.
hood. That part of an orchid flower which is strongly concave and arching.
hyaline. Transparent.

indehiscent. Not splitting open at maturity.
inferior. Referring to the position of the ovary when it is surrounded by the adnate portion of the floral tube or is embedded in the receptacle.
inflorescence. A cluster of flowers.
internode. The area between two adjacent nodes.
involucral. Referring to a circle of bracts which subtend a flower cluster.
involute. Rolled inward.
irregular. In reference to a flower having no symmetry at all.

keel. A central ridge.
keeled. Possessing a ridgelike process.

lanate. Woolly.
lanceolate. Lance-shaped; broadest near base, gradually tapering to the narrow apex.
lanceoloid. Referring to a solid object which is broadest near base, gradually tapering to the narrow apex.
lateral. From the side, in reference to the point of attachment of certain indusia.
lemma. A fertile scale in a grass spikelet.
ligule. A structure on the inner surface of the leaf of grasses and sedges at the junction of the blade and the sheath.
linear. Narrow and approximately the same width at either end and the middle.

linear-lanceolate. Somewhat broadened at base, very gradually tapering to the apex.
lip. The lowermost, often greatly modified petal, in the flower of an orchid.
lobe. A projection separated from each adjacent projection by a sinus.
lobed. Divided into rounded segments.
locular. Referring to the cells of a compound ovary.
locule. A cell or cavity of a compound ovary.
loculicidal. Said of a capsule which splits down the dorsal suture of each cell.
lodicule. A membranous scale found within some grass flowers, possibly representing the perianth.
lustrous. Shiny.

margin. Edge, referring to the blade.
marginate. With a definite margin.
mealy. Having a granular texture.
median. Pertaining to the middle.
megaspore. A spore that produces a female plant.
megasporangia. Spore cases that contain megaspores.
membranous. Like a membrane; thin.
microspore. A spore that produces a male plant.
microsporangia. Spore cases that contain microspores.
microsporocarp. A compound structure containing the microsporangia in *Marsilea* and *Azolla*.
moniliform. Constricted at regular intervals to resemble a string of beads.
monoecious. Bearing both sexes in separate flowers on the same plant.
mucro. A short abrupt tip.
mucronate. Said of a leaf with a short, terminal point.
mucronulate. Said of a leaf with a very short, terminal point.

nerve. Vein.
net-veined. Having veins forming closed meshes.
node. That place on the stem from which leaves and branchlets arise.
nodose. Knotty.
nutlet. A small nut.

oblanceolate. Reverse lance-shaped; broadest at apex, gradually tapering to narrow base.
oblique. One-sided; asymmetrical.
oblong. Broadest at the middle, and tapering to both ends, but broader than elliptic.
oblongoid. Referring to a solid object which, in side view, is nearly the same width throughout, but broader than linear.
obovate. Broadly rounded at apex, becoming narrowed below; broader than oblanceolate.
obovoid. Referring to a solid object which is broadly rounded at the apex, becoming narrowed below.
obpyramidal. Referring to an upside-down pyramid.
obsolete. Not apparent.
obtuse. Rounded at the apex.
once-pinnate. Divided once into distinct entire segments on either side of an axis.
opaque. Incapable of being seen through.
opposite. Referring to the condition of two like structures arising from the same point and across from each other.
orbicular. Round.
ovary. The lower swollen part of the pistil which produces the ovules.
ovate. Broadly rounded at base, becoming narrowed above; broader than lanceolate.
ovoid. Referring to a solid object which is broadly rounded at the base, becoming narrowed above.
ovule. The egg-producing structure found within the ovary.

palea. The scale opposite the lemma which encloses the flower.
palmately. Referring to a leaf whose segments radiate from a single point.
panduriform. Fiddle-shaped.
panicle. A type of inflorescence composed of several racemes.
papillate. Bearing small warts, or papillae.
papillose. Bearing pimplelike processes.
papule. A pimplelike projection.
parallel-veined. Having veins running in the same direction and not meeting.
pedicel. The stalk of a flower of an inflorescence.
pedicellate. Bearing a pedicel.
peduncle. The stalk of an inflorescence.
pedunculate. Bearing a peduncle.
pellucid. Being transparent, in reference to spots or dots.

peltate. Attached away from the margin, in reference to a leaf.
pendent. Suspended; overhanging.
pendulous. Hanging.
perennial. Living more than two years.
perfect. Bearing both stamens and pistils in the same flower.
perfoliate. Referring to a leaf which appears to have the stem pass through it.
perianth. Those parts of a flower including both the calyx and corolla.
pericarp. The ripened ovary wall.
persistent. Remaining attached.
petal. One segment of the corolla.
petaloid. Resembling a petal in texture and appearance.
petiolate. Bearing a petiole, or leafstalk.
petiole. The stalk of a leaf.
petiolule. A small stalk, referring to the stalk of a pinna.
phyllodia. Dilated petioles modified to resemble and function as leaves.
pilose. Bearing soft hairs.
pilosulous. Bearing short, soft hairs.
pinna. A primary division of a compound blade.
pinnate-pinnatifid. Divided once into distinct segments, with each segment further partially divided.
pinnatifid. Said of a simple leaf or leaf-part which is cleft or lobed only part way to its axis.
pinnule. The secondary segments of a compound blade.
pistil. The ovule-producing organ of a flower normally composed of an ovary, a style, and a stigma.
pistillate. Bearing pistils but not stamens.
plicate. Folded.
plumose. Feathery.
podogyne. A stalk in *Ruppia* on top at which the fruit is produced.
proliferous. Bearing asexual plants.
prophyll. A bracteole.
prophyllate. Bearing a bracteole at the base of a flower.
prostrate. Lying flat.
puberulent. With minute hairs.
pubescent. Bearing some kind of hairs.
pulviniform. Shaped like a small swelling.
punctate. Dotted.
punctation. A dot or dots.
pustular. Having raised surfaces like blisters.
quadrangular. Four-angled.
quadrate. Four-sided.

raceme. A type of inflorescence where pedicellate flowers are arranged along an elongated axis.
racemose. Bearing racemes.
rachilla. The axis bearing the flowers.
rachis. That portion of the leaf which is a continuation of the petiole.
ranked. Referring to the number of planes in which structures are borne.
receptacle. That part of the flower to which the perianth, stamens, and pistils are usually attached.
reflexed. Turned downward.
regular. Having radial symmetry (actinomorphic) or bilateral symmetry (zygomorphic).
reniform. Kidney-shaped.
resin. A usually sticky secretion found in various parts of certain plants.
resupinate. Upside down.
reticulate. Resembling a network.
retrorse. Pointing downward.
retuse. Shallowly notched at a rounded apex.
revolute. Rolled under at the margin.
rhizomatous. Bearing rhizomes.
rhizome. An underground horizontal stem, bearing nodes, buds, and roots.
rhombic. Becoming quadrangular.
ribbed. Nerved; veined.
root cap. A group of cells borne externally at the tip of a root.
rootlet. A small root.
rootstock. The underground stem, usually a rhizome.
rosette. A cluster of leaves in a circular arrangement at the base of a plant.
rotate. Flat and circular.
rudiment. A trace; a remnant.
rugose. Wrinkled.
rugulose. With small wrinkles.

saccate. Sac-shaped.
sagittate. Shaped like an arrowhead.
salverform. Referring to a tubular corolla which abruptly expands into a flat limb.
saprophyte. A type of plant living on dead or decaying organic matter, usually through the medium of mycorrhizae.
scaberulous. Minutely roughened; slightly rough to the touch.

scabrous. Rough to the touch.
scale. A minute epidermal outgrouth, sometimes becoming green and replacing the leaf in function.
scaly. Bearing scales, or minute epidermal outgrowths.
scape. A leafless stalk bearing a flower or inflorescence.
scapose. Possessing a leafless flowering stem.
scarious. Thin and membranous.
scurfy. Bearing scaly particles.
secund. Borne on one side.
sepal. One segment of the calyx.
sepaloid. Appearing like a sepal.
septa. (pl. **septae**). A cross wall.
septate. With cross walls.
septicidal. Said of a capsule which splits between the locules.
sericeous. Silky; bearing soft, appressed hairs.
serrate. With teeth which project forward.
serrulate. With very small teeth, the tips of which project forward.
sessile. Without a stalk.
seta. Bristle.
setaceous. Bearing bristles, or setae.
setose. Bearing bristles.
setulose. Bearing small setae.
sheath. A protective covering.
shoot. The developing stem with its leaves.
simple. Said of leaves which are not divided into distinct segments; as opposed to compound.
sinus. The cleft between two lobes or teeth.
sorus (pl. **sori**). An aggregation of sporangia.
spadix. A fleshy axis in which flowers are embedded.
spathe. A large sheathing bract subtending or usually enclosing an inflorescence.
spatulate. Oblong, but with the basal end elongated.
spicate. Bearing a spike.
spike. A type of inflorescence where sessile flowers are arranged along an elongated axis.
spikelet. The basic unit in a grass inflorescence.
spinescent. Becoming spiny.
spinule. A small spine.
spinulose. With small spines.
spiny-toothed. Bearing teeth with short, sharp-pointed projections.
sporangium. A structure bearing spores.
spore. That structure formed within a sporangium which gives rise to the gametophyte generation.
sporocarp. A compound structure containing the sporangia in *Marsilea* and *Azolla*.
sporophyll. A structure subtending the sporangia; in origin, it is a modified leaf.
stamen. The pollen-producing organ of a flower composed of a filament and an anther.
staminate. Bearing stamens but not pistils.
staminodium (pl. **staminodia**). A sterile stamen.
sterile. Not bearing any reproductive parts.
stigma. The apex of the pistil which receives the pollen.
stipe. The stalk or petiole of the leaf.
stipitate. Bearing a stipe or stalk.
stipule. A leaflike or scaly structure found at the point of attachment of a leaf to the stem.
stolon. A slender, horizontal stem on the surface of the ground.
stoloniferous. Bearing runners or slender horizontal stems on the surface of the ground.
stomate. An opening in the epidermis of the leaf.
stramineous. Straw-colored.
striate. Marked with grooves, such as the aerial stem of *Equisetum*.
strigose. With appressed, straight hairs.
style. That part of the pistil between the ovary and the stigma.
subacute. Nearly acute.
subcapitate. Nearly headlike.
subcoriaceous. Nearly leathery.
subglobose. Nearly round, in reference to solid objects.
subligneous. Nearly woody.
submersed. Covered with water.
suborbiculate. Nearly round.
subsessile. Nearly without a stalk.
subulate. Drawn to an abrupt sharp point.
subuloid. Referring to a solid object which is drawn to an abrupt short point.
suffused. Spread throughout; flushed.
superior. Referring to the position of the ovary when the free floral parts arise below the ovary.
supra-axillary. Borne above the axil.

taxa (sing. **taxon**). Plants or groups of plants of any taxonomic rank.
tendril. A spiraling coiling structure which enables a climbing plant to attach itself to a supporting body.
tenuous. Slender.

terete. Rounded in cross-section.
ternate. Divided into three principal parts.
thalloid. Possessing an undifferentiated plant body, that is, without roots, stems or leaves.
throat. The opening at the apex of a corolla tube where the limb arises.
translucent. Partly transparent.
trilocular. With three cavities.
trimorphic. Having three forms.
tripinnate. Divided three times into distinct segments.
truncate. Abruptly cut across.
tuber. An underground fleshy stem formed as a storage organ at the end of a rhizome.
tubercle. A small wartlike process.
tuberculate. Bearing small, rounded projections.
tubular. Shaped like a tube.
tunicated. Covered with concentric coats.
turbinate. Shaped like a top.
turgid. Tightly inflated.
turion. A specialized overwintering structure.

ultimate. The last of a series of divisions.
umbel. A type of inflorescence in which the flower stalks arise from the same level.
umbonate. With a stout projection at the center.
unarmed. Without prickles or spines.
undulate. Wavy.
unilocular. With one locule, or cavity.
unisexual. Bearing either stamens or pistils in one flower.
utricle. A small, one-seeded, indehiscent fruit with a thin covering.

valve. That part of a capsule which splits.
venation. The pattern of the veins.
ventral. That surface turned toward the axis; adaxial.
villous. With long soft erect hairs.
viscid. Sticky.

whorl. An arrangement of three or more structures at a point on the stem.
whorled. Referring to the condition of three or more like structures arising from the same point.
winged. Bearing a flat lateral outgrowth.

zygomorphic. Bilaterally symmetrical.

Illustration Credits

Illustrations 16, 25, and 93 were prepared by Phyllis K. Bick and are reprinted from *Steyermark's Flora of Missouri,* revised edition, volume 1, by George Yatskievych. Courtesy of the Flora of Missouri Project.

The follow illustrations were prepared by Miriam Meyer: 1, 6, 9, 10, 13, 14, 19, 20, 24, 28, 30, 32–34, 37, 38, 45, 47, 49, 50, 52–55, 57–59, 61–63, 65, 69–71, 75, 77, 78, 85–87, 95–98, 109, 110, 112, 116, 119, 122–24, 126, 127, 129, 131, 133–39, 141–44, 146–51, 153–55, 157, 159, 163, 164, 168, 169, 172, 174, 176, 179–81, 185–90, 192, 193, 195, 197–201, 203, 205, 206, 209–13, 215, 216, 220, and 221. Also prepared by Miriam Meyer: illustrations 3–5 and 7, originally published in *Ferns;* 27, 29, 36, 41, 51, 94, 178, 183, 184, 194, 208, 214, and 218, originally published in *Flowering Plants: Flowering Rush to Rushes;* 100–108, 111, 113–15, 117, 118, and 120, originally published in *Flowering Plants: Lilies to Orchids;* and 166, originally published in *Grasses: Panicum to Danthonia;* all from the Illustrated Flora of Illinois series, Southern Illinois University Press.

The following illustrations were prepared by Mark Mohlenbrock: 2, 17, 18, 21, 22, 26, 40, 43, 44, 46, 80, 81, 83, 84, 152, 158, 160, 161. Also prepared by Mark Mohlenbrock: illustration 12, originally published in *Ferns,* from the Illustrated Flora of Illinois series, Southern Illinois University Press.

The following illustrations were prepared by Paul Nelson: 23, 31, 35, 39, 42, 48, 60, 64, 67, 82, 88 a–d, 91 a–d, 170, 173, 175, 177, 182, 204, 217. Also prepared by Paul Nelson: illustrations 125 and 162, originally published in *Grasses: Bromus to Paspalum;* and 156, originally published in *Grasses: Panicum to Danthonia,* both from the Illustrated Flora of Illinois series, Southern Illinois University Press.

The following illustrations were prepared by K. L. Weik: 89 and 92. Also prepared by K. L. Weik: illustrations 90 and 91 e–o, originally published in *Flowering Plants: Flowering Rush to Rushes,* from the Illustrated Flora of Illinois series, Southern Illinois University Press.

Illustration 121 was prepared by Vera Ming Wong. Copyright 1991, 2005 by Vera Ming Wong. Originally published in *Orchids of Minnesota,* by Welby R. Smith, University of Minnesota Press, 1993.

Illustrations 8, 11, 15, 56, 68, 73, 74, 76, 79, 99, 128, 130, 132, 140, 145, 165, 167, 171, 191, 196, 202, 207, 219 are reprinted from *An Illustrated Flora of the Northern United States and Canada,* by Nathaniel Britton and Addison Brown. Charles Scribner's Sons, 1913; reprinted 1970 by Dover Publications.

Illustrations 66 and 72 are reprinted from *Aquatic and Wetland Plants of the Southeastern United States: Monocotyledons,* by Robert K. Godfrey and Jean W. Wooten. Copyright 1979 by the University of Georgia Press. Reprinted by permission of the University of Georgia Press.

Index to Genera and Species

Names in roman type are accepted names, while those in italics are synonyms and are not considered valid. Page numbers in bold refer to pages that have illustrations.

Index to Common Names

Robert H. Mohlenbrock taught botany at Southern Illinois University at Carbondale for thirty-four years, obtaining the title of Distinguished Professor. Since his retirement in 1990, he has served as senior scientist for Biotic Consultants Inc., teaching wetland identification classes around the country. Mohlenbrock has been named SIU Outstanding Scholar and has received the SIU Alumnus Teacher of the Year Award, the College of Science Outstanding Teacher Award, and the Meritorious Teacher of the Year Award from the Association of Southeastern Biologists. Since 1984, he has been a monthly columnist for *Natural History* magazine. He is the author of 53 books and more than 560 publications.